Lecture Notes in Mathematics

Edited by A. Dold and B. Eckmann

Subseries: Nankai Institute of Mathematics, Tianjin, P.R. China
vol. 3
Adviser: S. S. Chern

1306

S. S. Chern (Ed.)

Partial Differential Equations

Proceedings of a Symposium held in
Tianjin, June 23 – July 5, 1986

W9-ACV-003

Springer-Verlag

Berlin Heidelberg New York London Paris Tokyo

Editor

Shiing-shen Chern
Mathematical Sciences Research Institute
1000 Centennial Drive
Berkeley, CA 94720, USA

Mathematics Subject Classification (1980): 35C, 35H, 35J, 35K, 35L, 35S, 53C, 53E, 58G

ISBN 3-540-19097-X Springer-Verlag Berlin Heidelberg New York
ISBN 0-387-19097-X Springer-Verlag New York Berlin Heidelberg

This work is subject to copyright. All rights are reserved, whether the whole or part of the material is concerned, specifically the rights of translation, reprinting, re-use of illustrations, recitation, broadcasting, reproduction on microfilms or in other ways, and storage in data banks. Duplication of this publication or parts thereof is only permitted under the provisions of the German Copyright Law of September 9, 1965, in its version of June 24, 1985, and a copyright fee must always be paid. Violations fall under the prosecution act of the German Copyright Law.

© Springer-Verlag Berlin Heidelberg 1988
Printed in Germany

Printing and binding: Druckhaus Beltz, Hemsbach/Bergstr.
2146/3140-543210

CB-StK
SCIMON

FOREWORD

This volume contains a selection of papers presented at the 7th symposium on differential geometry and differential equations(=DD7), which took place at Nankai Institute of Mathematics, Tianjin, China, June 23 — July 5, 1986. The subject was partial differential equations. It was a culmination of a year-long activity in 1985-1986 at the institute. A list of the other papers presented at the symposium can be found at the end of this volume, some of which will be published elsewhere.

For the record I would like to give a list of the preceding DD-Symposia as follows:

	Subject	Date	Place	Publication
DD1	Differential geometry and differential equations	Aug. 18-Sept. 21, 1980	Beijing, China	Proceedings of the 1980 Beijing Symposium on Differential Geometry and Differential Equations, Science Press, Beijing, China, 1982
DD2	Differential geometry	Aug. 20-Sept. 13, 1981	Shanghai-Hefei, China	Proceedings of the 1981 Symposium on Differential Geometry and Differential Equations, Shanghai-Hefei, Science Press, Beijing, China, 1984
DD3	Partial differential equations	Aug. 23-Sept. 16, 1982	Changchun, China	Proceedings of the 1982 Changchun Symposium on Differential Geometry and Differential Equations, Science Press, Beijing, China, 1986
DD4	Ordinary differential equations	Aug. 29-Sept. 10, 1983	Beijing, China	Proceedings of the 1983.Beijing Symposium on Differential Geometry and Differential Equations, Science Press, Beijing, China, 1986
DD5	Computation on partial differential equations	Aug. 13-aug. 17, 1984	Beijing, China	Proceedings of the 1984 Beijing Symposium on Differential Geometry and Differential Equations, Science Press, Beijing, China, 1985
DD6	Differential geometry	Jun. 21-Jul. 6, 1985	Shanghai, China	Differential Geometry and Differential Equations, Proceedings, Shanghai 1985, Lecture Notes in Mathematics 1255, Springer-Verlag 1987

S.S. Chern

June 1987

Editorial board

S. Agmon	Institute of Math. and Computer Science, The Hebrew Univ. of Jerusalem, Givat Ram. 91904, Jerusalem, Israel
F. Almgren	Dept. of Math., Princeton Univ., Fine Hall-Box 37, Princeton, N.J. 08544, USA
R.W. Beals	Dept. of Math., Yale Univ., Box 2155, Yale Station, New Haven, Conn. 06520, USA
Chang Kung-ching	Dept. of Math., Beijing Univ., Beijing, China
Chi Min-you	Dept. of Math., Wuhan Univ., Wuhan, Hubei, China
Huang Yumin	Dept. of Math. Nankai Univ., Tianjin, China
Sun Hesheng	Institute of Applied Physics and Computational Math., Beijing, China
J.E. Taylor	Math. Dept., Rutgers Univ., New Brunswick, N.J. 08904, USA
A.J. Tromba	Dept. of Math., Univ. of California, Santa Cruz, CA. 95064, USA
Wang Rou-hwai	Dept. of Math., Jilin Univ., Changchun, Jilin, China
Zou Yulin	Institute of Applied Physics and Computational Math., Beijing, China

A TABLE OF CONTENTS OF DD7 SYMPOSIUM ON PARTIAL DIFFERENTIAL EQUATIONS – A SELECTION OF PAPERS

CO-AREA, LIQUID CRYSTALS, AND MINIMAL SURFACES[1]

F. Almgren, W. Browder, and E. H. Lieb
Department of Mathematics, Princeton University
Princeton, New Jersey 08544, USA

Abstract. Oriented n area minimizing surfaces (integral currents) in $\mathcal{M}^{m+n}$ can be approximated by level sets (slices) of nearly m-energy minimizing mappings $\mathcal{M}^{m+n} \to \mathbf{S}^m$ with essential but controlled discontinuities. This gives new perspective on multiplicity, regularity, and computation questions in least area surface theory.

In this paper we introduce a collection of ideas showing relations between co-area, liquid crystals, area minimizing surfaces, and energy minimizing mappings. We state various theorems and sketch several proofs. A full treatment of these ideas is deferred to another paper.

Problems inspired by liquid crystal geometries.[2] Suppose Ω is a region in 3 dimensional space $\mathbf{R}^3$ and f maps Ω to the unit 2 dimensional sphere $\mathbf{S}^2$ in $\mathbf{R}^3$. Such an f is a unit vectorfield in Ω to which we can associate an 'energy'

$$\mathcal{E}(f) = \left(\frac{1}{8\pi}\right) \int_\Omega |Df|^2 \, d\mathcal{L}^3;$$

here Df is the differential of f and $|Df|^2$ is the square of its Euclidean norm—in terms of coordinates,

$$|Df(x)| = \sum_{k=1}^{3} \sum_{i=1}^{3} \left(\frac{\partial f^k}{\partial x_i}(x) \right)^2$$

for each x. The factor $1/8\pi$ which equals 1 divided by twice the area of $\mathbf{S}^2$ is a useful normalizing constant. It is straightforward to show the existence of f's of least energy for given boundary values (in an appropriate function space).

Such boundary value problems have been associated with liquid crystals.[3] In this context, a "liquid crystal" in a container Ω is a fluid containing long rod like molecules whose directions are specified by a unit vectorfield. These molecules have a preferred alignment relative to each other—in the present case the preferred alignment is parallel. If we imagine the molecule orientations along

[1] This research was supported in part by grants from the National Science Foundation

[2] The research which led to the present paper began as an investigation of a possible equality between infimums of m-energy and the n area of area minimizing n dimensional area minimizing manifolds in $\mathbf{R}^{m+n}$ suggested in section VIII(C) of the paper, *Harmonic maps with defects* [**BCL**] by H. Brezis, J-M. Coron, and E. Lieb. Although the specific estimates suggested there do not hold (by virtue of counterexamples [**MF**][**W1**][**YL**]) their general thrust does manifest itself in the results of the present paper.

[3] See, for example, the discussion by R. Hardt, D. Kinderlehrer, and M. Luskin in [**HKL**].

$\partial\Omega$ to be fixed (perhaps by suitably etching container walls) then interior parallel alignment may not be possible. In one model the system is assumed to have 'free energy' given by our function $\mathcal{E}$ and the crystal geometry studied is that which minimizies this free energy.

If Ω is the unit ball and $f(x) = x$ for $|x| = 1$, then there is no continuous extension of these boundary values to the interior; indeed the unique least energy f is given by setting $f(x) = x/|x|$ for each x. It turns out that this singularity is representative, and the general theorem is that *least energy f's exist and are smooth except at isolated points p of discontinuity where 'tangential structure' is $\pm x/|x|$ (up to a rotation), e.g. f has local degree equal to ± 1* [SU] [BCL VII].

As a further step towards an understanding of the geometry of of energy minimizing f's one might seek estimates on the number of points of discontinuity which such an f can have—e.g. if the boundary values are not to wild must the number of points of discontinuity be not too big?[4] An alternative problem to this is to seek a lower bound on the energy when the points of discontinuity are prescribed together with the local degrees of the mapping being sought. This question has a surprisingly simple answer as follows.

THEOREM. *Suppose $p_1, \ldots, p_N$ are points in $\mathbf{R}^3$ and $d_1, \ldots, d_N \in \mathbf{Z}$ are the prescribed degrees with $\sum_{i=1}^N d_i = 0$. Let $\inf \mathcal{E}$ denote the infimum of the energies of (say, smooth) mappings from $\mathbf{R}^3 \sim \{p_1, \ldots, p_N\}$ to $\mathbf{S}^2$ which map to the 'south pole' outside some bounded region in $\mathbf{R}^3$ and which, for each i, map small spheres around p_i to $\mathbf{S}^2$ with degree d_i. Then $\inf \mathcal{E}$ equals the least mass $\mathbf{M}(T)$ of integral 1 currents T in $\mathbf{R}^3$ with*

$$\partial T = \sum_{i=1}^{N} d_i [\![p_i]\!].$$

This fact (stated in slightly different language) is one of the central results of [**BCL**]. We would like to sketch a proof in two parts: first by showing that $\inf \mathcal{E} \leq \inf \mathbf{M}$ (with the obvious meanings) and then by showing that $\inf \mathbf{M} \leq \inf \mathcal{E}$. The proof of the first part follows [**BCL**] while the second part is new. It is in this second part that the coarea formula makes its appearance.

Proof that $\inf \mathcal{E} \leq \inf \mathbf{M}$**.** The first inequality is proved by construction as illustrated in Figure 1. We there represent that case in which N equals 2 and p_1 and p_2 are distinct points with $d_1 = -1$ and $d_2 = +1$. We choose and fix a smooth curve C connecting these two points and orient C by a smoothly varying unit tangent vector field ς which points away from p_1 and towards p_2. The associated 1 dimensional integral current is $T = \mathbf{t}(C, 1, \varsigma)$ and its *mass* $\mathbf{M}(T)$ is the length of C since the density specified is everywhere equal to 1.[5] We now choose (somewhat arbitrarily)

[4] As it turns out, away from the boundary of Ω, the number of these points is bounded a *priori* independent of boundary values.

[5] Formally, a 1 current such as T is a linear functional on smooth differential 1 forms in $\mathbf{R}^3$. If φ is such a 1 form then

$$T(\varphi) = \int_{x \in C} \langle \varsigma(x), \varphi(x) \rangle \, d\mathcal{H}^1 x.$$

To each point p in $\mathbf{R}^3$ is associated the 0 dimensional current $[\![p]\!]$ which maps the smooth function ψ to the number $\psi(p)$. See Appendix A.4.

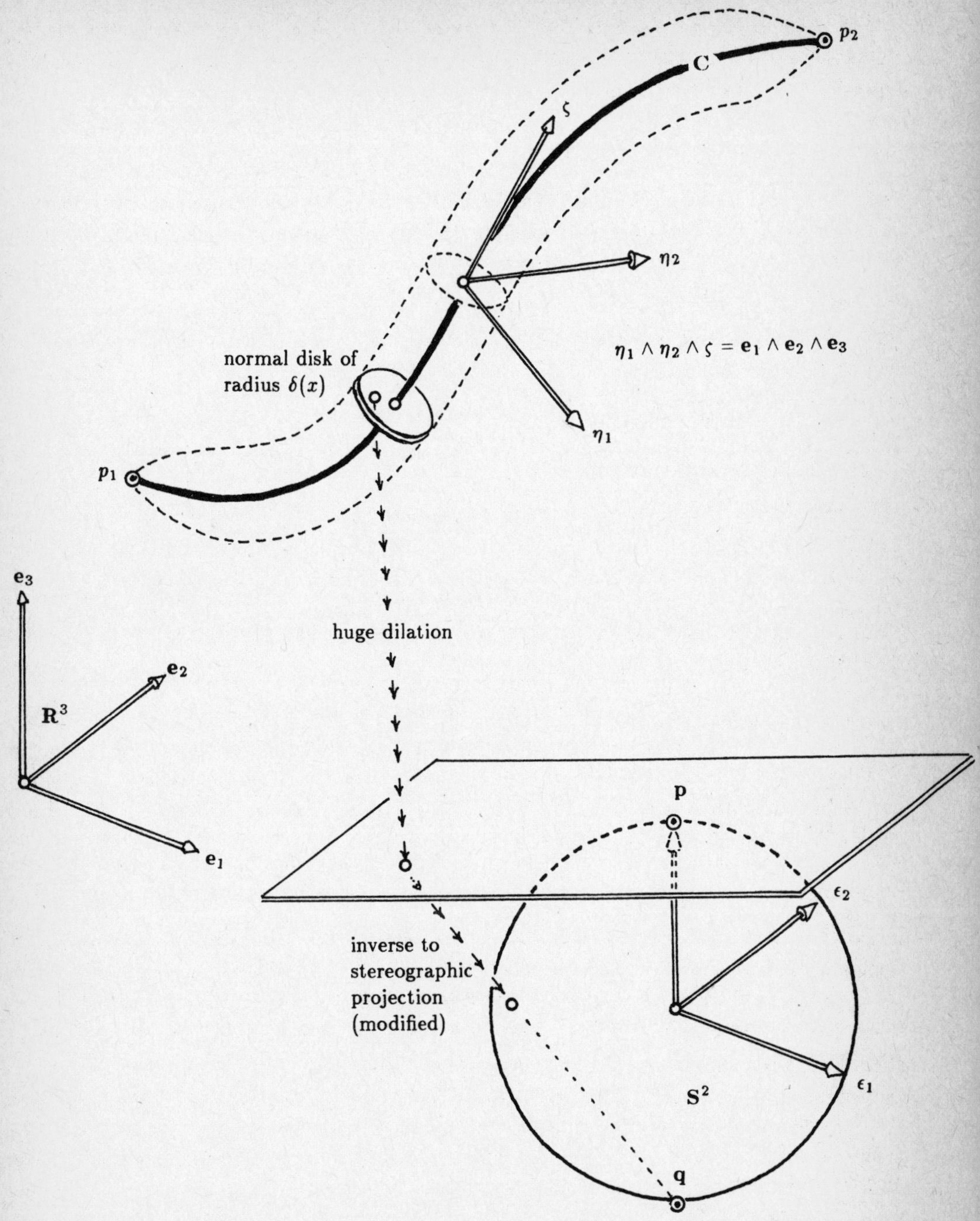

Figure 1. Construction of a mapping f (indicated by dashed arrows) from $\mathbf{R}^3$ to $\mathbf{S}^2$ having energy $\mathcal{E}(f)$ not much greater than the length of the curve C connecting the points p_1 and p_2. Small disks normal to C map by f to cover $\mathbf{S}^2$ once in a nearly conformal way. This implies that small spheres around p_1 map to $\mathbf{S}^2$ with degree -1 while small spheres around p_2 map with degree $+1$. The 1 current $\mathbf{t}(C, 1, \varsigma)$ is the slice $\langle \mathbf{E}^3, f, \mathbf{p}\rangle$ of the Euclidean 3 current $\mathbf{E}^3$ by the mapping f and the 'north pole' $\mathbf{p}$ of $\mathbf{S}^2$.

and fix two smoothly varying unit normal vector fields η_1 and η_2 along C which are perpendicular to each other and for which, at each point x of C, the 3-vector $\eta_1(x) \wedge \eta_2(x) \wedge \varsigma(x)$ equals the orienting 3-vector $\mathbf{e}_1 \wedge \mathbf{e}_2 \wedge \mathbf{e}_3$ for $\mathbf{R}^3$. These two vector fields are a 'framing' of the normal bundle of C.

We then construct a mapping γ of $\mathbf{R}^2$ onto the unit 2 sphere $\mathbf{S}^2$ which is a slight modification of the inverse to stereographic projection. To construct such γ we fix a huge radius R in $\mathbf{R}^2$ and require: (i) if $|y| \leq R$ then $\gamma(y)$ is that point in $\mathbf{S}^2$ which maps to y under stereographic projection $\mathbf{S}^2 \to \mathbf{R}^2$ from the south pole $\mathbf{q}$ of $\mathbf{S}^2$; (ii) if $|y| \geq 2R$ then $\gamma(y) = \mathbf{q}$; (iii) for $R < |y| < 2R$, $\gamma(y)$ is suitably interpolated. See Appendix A.2.

Next we choose some smoothly varying (and very small) radius function δ on C which vanishes only at the endpoints p_1 and p_2.

Finally, as our mapping f from $\mathbf{R}^3$ to $\mathbf{S}^2$ with which to estimate $\mathcal{E}(f)$ we specify the following. If p in $\mathbf{R}^3$ can be written $p = x + s\eta_1(x) + t\eta_2(x)$ for some x in C and some s and t with $s^2 + t^2 \leq \delta(x)^2$, then

$$f(p) = \gamma\left(\frac{2Rs}{\delta(x)}, \frac{2Rt}{\delta(x)}\right).$$

Otherwise, $f(p) = \mathbf{q}$. We leave it as an exercise to the reader to use the fact that γ is conformal for $|y| < R$ to check that $\mathcal{E}(f)$ very nearly equals $\mathbf{M}(T)$; see Appendix A.2. The remainder of the proof that $\inf \mathcal{E} \leq \inf \mathbf{M}$ is also left to the reader.

Proof that $\inf \mathbf{M} \leq \inf \mathcal{E}$. Suppose that f does map $\mathbf{R}^3$ to $\mathbf{S}^2$, has degree d_i at each p_i, and maps to the south pole outside some bounded region. From dimensional considerations one would expect that for most points w in $\mathbf{S}^2$ the inverse image $f^{-1}\{w\}$ would be a collection of curves connecting the various points $p_1, \ldots, p_N$. H. Federer's *coarea formula* is what enables one to quantify this idea; see Appendix A.5. This formula asserts

$$\int_{w \in \mathbf{S}^2} \mathcal{H}^1(f^{-1}\{w\})\, d\mathcal{H}^2 w = \int_{x \in \mathbf{R}^3} J_2 f(x)\, d\mathcal{L}^3 x;$$

here $\mathcal{H}^1$ and $\mathcal{H}^2$ are Hausdorff's 1 and 2 dimensional measures in $\mathbf{R}^3$ and $\mathcal{L}^3$ is Lebesgue's 3 dimensional measure for $\mathbf{R}^3$. Also $J_2 f(x)$ here denotes the 2 dimensional Jacobian of f at x and a key observation (as noted in [BCL]) is that $J_2 f(x)$ is always less than or equal to half of $|Df(x)|^2$ with equality only if the differential mapping $Df(x)\colon \mathbf{R}^3 \to \operatorname{Tan}(\mathbf{S}^2, f(x))$ is maximally conformal; see Appendix A.1.3. Also central to the present analysis is the manner in which the curves $f^{-1}\{w\}$ connect the various points $p_1, \ldots, p_N$ and how they relate to the prescribed degrees $d_1, \ldots, d_N$. This connectivity is naturally measured by the current structure of these $f^{-1}\{w\}$'s which comes from the slicing theory for currents; see Appendix A.5. To set this up we regard $\mathbf{R}^3$ as the Euclidean current $\mathbf{E}^3$ (oriented by the 3 vector $\mathbf{e}_1 \wedge \mathbf{e}_2 \wedge \mathbf{e}_3$). The *slice* of $\mathbf{E}^3$ by the map f at the point w in $\mathbf{S}^2$ is the current

$$\langle \mathbf{E}^3, f, w \rangle = \mathbf{t}(f^{-1}\{w\}, 1, \varsigma);$$

the meanings here are the same as for the current T discussed above. A check of orientations and degrees shows that

$$\partial\langle \mathbf{E}^3, f, w\rangle = \sum_{i=1}^{N} k_i [\![p_i]\!];$$

compare with our construction of η_1 and η_2 above. It follows immediately that

$$\begin{aligned} 4\pi \inf \mathbf{M}(T) &= \mathcal{H}^2(\mathbf{S}^2) \inf \mathbf{M}(T) \\ &\le \int_{w\in\mathbf{S}^2} \mathbf{M}(\langle \mathbf{E}^3, f, w\rangle)\, d\mathcal{H}^2 w \\ &= \int_{\mathbf{R}^3} J_2 f \, d\mathcal{L}^3 \\ &= \left(\frac{1}{2}\right) \int_{\mathbf{R}^3} |Df|^2 \, d\mathcal{L}^3. \end{aligned}$$

This finishes the proof that $\inf \mathbf{M} \le \inf \mathcal{E}$.

First Generalization. Since the methods used in the proofs of the two inequalities are quite general one might correctly suspect that considerable generalization is possible. Suppose, for example, we fix $B = \{p_1, \ldots, p_N\}$ as a general boundary set and let $\mathcal{F}_0$ be the family of those mappings f of $\mathbf{R}^3$ to $\mathbf{S}^2$ which are locally Lipschitzian except possibly on B, which map to the southpole outside some bounded region, and which have finite energy. Since deformations of mappings in $\mathcal{F}_0$ do not alter discrete combinatorial structures we are led to study properties of homotopy classes $\Pi(\mathcal{F}_0)$ of mappings in $\mathcal{F}_0$—it is most useful here if our homotopies $[0,1]\times\mathbf{R}^3 \to \mathbf{S}^2$ are permitted to have isolated point discontinuties; see Appendix A.3.

Our conditions about mapping degrees above generalize to requirements about degrees $\mathbf{d}(f, S)$ of f on general integral 2 dimensional cycles S in $\mathbf{R}^3 \sim B$. It turns out that such a degree $\mathbf{d}(f, S)$ depends only on the homotopy class of f and on the homology class of S.

It also turns out that the relative homology classes of the slices $\langle \mathbf{E}^3, f, w\rangle$ depend only on the homotopy class $[f]$ of f. We denote this homology class by $s[f]$.

The *Kronecker index* is a pairing between 2 dimensional cycles S in $\mathbf{R}^3 \sim B$ and 1 currents T having boundary in B. In general the Kronecker index $\mathbf{k}(S, T)$ is the sum over points of intersection of S and T of an index of relative orientations; see Appendix A.6

These various ideas are related in the following theorem.

THEOREM. *The diagram below is commutative. Furthermore, s is an isomorphism, and d and k are injections.*

$$\begin{array}{ccc} & & \mathbf{H}_1(\mathbf{R}^3, B; \mathbf{Z}) \\ & \nearrow s & \\ \Pi(\mathcal{F}_0) & & \downarrow k \\ & \searrow d & \\ & & \mathrm{Hom}(\mathbf{H}_2(\mathbf{R}^3 \sim B, \mathbf{Z}), \mathbf{Z}) \end{array}$$

Here

$s[f] = \text{"}[f^{-1}\{w\}]\text{"} = [\langle \mathbf{E}^3, f, w\rangle] =$ the integral homology class of the 1 current slice;

$d[f][S] = \mathbf{d}(f, S) =$ the degree of f on the 2 cycle S;

$k[T][S] = \mathbf{k}(S, T) =$ the Kronecker Index of the 2 cycle S and the 1 current T.

Our relations between energy minimization and area minimization become the following.

THEOREM. *Suppose that P is an integral 1 current in $\mathbf{R}^3$ with the support of ∂P in B. Suppose also that $T^{\mathbf{Z}}$ has least mass among all integral 1 currents which are homologous to P over the integers $\mathbf{Z}$ and that $T^{\mathbf{R}}$ has least mass among all integral 1 currents which are homologous to P over the real numbers $\mathbf{R}$. Then*

$$\mathbf{M}(T^{\mathbf{Z}}) = \inf\{\mathcal{E}(f) : s[f] = [P]\}$$

and

$$\mathbf{M}(T^{\mathbf{R}}) = \inf\{\mathcal{E}(f) : d[f] = k[P]\}.$$

Moreover, $\mathbf{M}(T^{\mathbf{Z}}) = \mathbf{M}(T^{\mathbf{R}})$ (because of our special situation).

Further generalizations. The essential ingredients of the analyses above remain, for example, if $\mathbf{R}^3$ is replaced by a general $m+n$ dimensional manifold $\mathcal{M}$ (without boundary) which is smooth, compact, and oriented (or $\mathcal{M} = \mathbf{R}^{m+n}$), and B is replaced by a sufficiently nice (possibly empty) compact subset of $\mathcal{M}$ of dimension $n-1$. To study n dimensional integral currents in $\mathcal{M}$ having boundary in B we consider mappings f of $\mathcal{M}$ to a sphere of the complementary dimension m. The spaces $\mathcal{F}$ and $\mathcal{F}_0$ of such mappings and the homotopy classes $\Pi(\mathcal{F})$ are specified in sections A.3.1 and A.3.2 of the Appendix. Some discontinuities are essential.[6] It seems worthwhile to consider three different energies $\mathcal{E}_1$, $\mathcal{E}_2$, and $\mathcal{E}_3$ for mappings in $\mathcal{F}_0$. $\mathcal{E}_1$ is a normalization of the usual 'n energy' of mappings, $\mathcal{E}_3$ is a normalized Jacobian integral associated with the coarea formula, and $\mathcal{E}_2$ is an intermediate energy; see Appendix A.3.2. As indicated above, mapping degrees and the Kronecker index have general meanings which are set forth in sections A.6 and A.7 of the Appendix. These various ideas are related as the following theorem shows.

THEOREM. *The diagram of mappings below is well defined and is commutative. In particular, the images of d and k and j in $\mathbf{Hom}(\mathbf{H}_m(\mathcal{M} \sim B, \mathbf{Z}), \mathbf{Z})$ are the same. Furthermore, s is an*

---

[6] Suppose $m = 2$ and $n = 5$ and $\mathcal{M} = \mathbf{R}^7$, and B is a smoothly embedded copy of 2 dimensional complex projective space $\mathbf{CP}(2)$. Then there are no continuous mappings f from the complement of B to $\mathbf{S}^2$ such that small 2 spheres S which link B once map to $\mathbf{S}^2$ with degree one. Any f satisfying such a linking condition for general position S's near B must have interior discontinuities of dimension at least 3.

isomorphism.

$$
\begin{array}{ccccc}
 & & \mathbf{H}_n(\mathcal{M},B;\mathbf{Z}) & \xrightarrow{c} & \mathbf{H}_n(\mathcal{M},B;\mathbf{R}) \\
 & \nearrow s & & \searrow c & \uparrow i \\
\Pi(\mathcal{F}) & & \downarrow k & & c[\mathbf{H}_n(\mathcal{M},B;\mathbf{Z})] \\
 & \searrow d & & \swarrow j & \\
 & & \mathrm{Hom}(\mathbf{H}_m(\mathcal{M}\sim B,\mathbf{Z}),\mathbf{Z}) & &
\end{array}
$$

Here

$s[f] = \text{“}[f^{-1}\{p\}]\text{”} = [\langle [\![\mathcal{M}]\!], f, p\rangle] =$ the integral homology class of the n current slice;

$d[f][S] = \mathbf{d}(f,S) =$ the degree of f on the m cycle S;

$k[T][S] = \mathbf{k}(S,T) =$ the Kronecker index of the m cycle S and the n current T;[7]

c is induced by the coefficient inclusion $\mathbf{Z} \to \mathbf{R}$;

i is the inclusion; and

j is defined by commutativity.

We defer proof of this theorem to our fuller treatment of this subject. The natural setting and generality of such relationships are still under investigation.

The relations between energy minimization and area minimization then become the following.

MAIN THEOREM. *Suppose P is an integral current in $\mathcal{M}$ with the support of ∂P contained in B so that the integral homology class $[P]$ of P belongs to $\mathbf{H}_n(\mathcal{M},B;\mathbf{Z})$. Let $T^{\mathbf{Z}}$ be an integral current of least mass among all integral currents belonging to the same integral homology class as P in $\mathbf{H}_n(\mathcal{M},B,\mathbf{Z})$, and let $T^{\mathbf{R}}$ be an integral current of least mass among all integral currents belonging to the same real homology class as P in $\mathbf{H}_n(\mathcal{M},B,\mathbf{R})$. Then*

$$\mathbf{M}(T^{\mathbf{Z}}) = \inf\{\mathcal{E}_1(f): s[f] = [P]\} = \inf\{\mathcal{E}_2(f): s[f] = [P]\} = \inf\{\mathcal{E}_3(f): s[f] = [P]\}$$

and

$$\mathbf{M}(T^{\mathbf{R}}) = \inf\{\mathcal{E}_1(f): d[f] = k[P]\} = \inf\{\mathcal{E}_2(f): d[f] = k[P]\} = \inf\{\mathcal{E}_3(f): d[f] = k[P]\}.$$

---

[7] Suppose $m = 2$ and $n = 1$ and $\mathcal{M}$ is a 3 dimensional real projective space $\mathbf{RP}(3)$ and $T = \mathbf{t}(\mathcal{N}, 1, \varsigma)$; here $\mathcal{N}$ is a 1 dimensional real projective space $\mathbf{RP}(1)$ sitting in $\mathbf{RP}(3)$ in the usual way and ς is some orientation function. Since T is not a boundary while $2T$ is, we conclude that the homology class

$$[T] \in \mathbf{H}_1(\mathcal{M}, \emptyset;\ \mathbf{Z}) = \mathbf{Z}_2$$

is not the 0 class although $\mathbf{k}(S,T) = 0$ for each 2 cycle S in $\mathcal{M}$. In particular, the mapping k is generally not an injection.

In general, of course, $\mathbf{M}(T^{\mathbf{R}}) < \mathbf{M}(T^{\mathbf{Z}})$. Although we again defer complete proofs to our fuller treatment of this subject, it does seem useful to sketch some of the main ideas.

Proof of the inequality "$\inf \mathcal{E} \leq \inf \mathbf{M}$". The proof here is again by construction. We will indicate the main ingredients in a special case. Suppose, say, $\mathcal{M} = \mathbf{R}^{m+n}$, B is polyhedral, and T is an integral n current which is mass minimizing subject to some appropriate constraints as in the Main Theorem above. We will construct a mapping $f: \mathbf{R}^{m+n} \to \mathbf{S}^m$ in the relevant homotopy class such that $\mathcal{E}_1(f), \mathcal{E}_2(f)$, and $\mathcal{E}_3(f)$ are nearly equal and are not much bigger that $\mathbf{M}(T)$. By virtue of the Strong Approximation Theorem for integral currents [FH1 4.2.20] we can modify T slightly to become simplicial with only a slight increase in mass.

Suppose then that we can express

$$T = \sum_{\alpha=1}^{M} \mathbf{t}(\Delta_\alpha^n, z_\alpha, \varsigma_\alpha)$$

as a 'simplicial' integral current (with the obvious interpretation). For each $k = 0, \ldots, n$ we denote by K_k the collection of closed k simplexes which occur as k dimensional faces of n simplexes among the Δ_α^n's. We then choose numbers $0 < \delta_n << \delta_{n-1} << \delta_{n-2} << \ldots << \delta_0 << 1$ and define sets $N_0, N_1, \ldots, N_n$ in $\mathbf{R}^{m+n}$ by setting

$$N_0 = \{x: \operatorname{dist}(x, \cup K_0) < \delta_0\}$$

and, for each $k = 1, \ldots, n$ set

$$N_k = \{x: \operatorname{dist}(x, \cup K_k) < \delta_k\} \sim (N_{k-1} \cup N_{k-2} \cup \ldots \cup N_0).$$

We assume that $\delta_0, \ldots, \delta_n$ have been chosen so that the distinct components of each N_k correspond to distinct k simplexes in K_k.

We now define mappings $f_{n+1}, f_n, \ldots, f_0 = f$ as follows.

First, the mapping $f_{n+1}: \mathbf{R}^{m+n} \sim (N_n \cup \ldots \cup N_0) \to \mathbf{S}^m$ is defined by setting $f_{n+1}(x) = \mathbf{q}$ for each x.

Second, the mapping $f_n: \mathbf{R}^{m+n} \sim (N_{n-1} \cup \ldots \cup N_0) \to \mathbf{S}^m$ is constructed geometrically in virtually the same manner as the mapping g in the example A.8 in the Appendix. Details are left to the reader.

Third, the mapping $f_{n-1}: \mathbf{R}^{m+n} \sim (N_{n-2} \cup \ldots \cup N_0) \to \mathbf{S}^m$ is constructed geometrically in a manner virtually identical with the construction of the mapping $f_{\delta,r}$ of example A.8 of the Appendix (with δ, r replaced by $\delta_n/2, \delta_{n-1}$ respectively there). The mapping f_{n-1} is Lipschitz across parts of $n-1$ simplexes which do not lie in B and is discontinuous on those $n-1$ simplexes which contain part of ∂T.

Assuming $f_{n+1}, f_n, \ldots, f_{k+1}$ have been constructed we define

$$f_k: \mathbf{R}^{m+n} \sim (N_{k-1} \cup \ldots \cup N_0) \to \mathbf{S}^m$$

as follows. Each point v in $N_k \sim (N_{k-1} \cup \ldots \cup N_0)$ can be written uniquely in the form $v = v_0 + (v - v_0)$ where v_0 is the unique closest point in $\cup K_k$ to v and $|v - v_0| < \delta_k$. If $v \neq v_0$ we note that

$$v_1 = v_0 + \delta_k \left(\frac{v - v_0}{|v - v_0|} \right) \in \operatorname{dmn}(f_{k+1})$$

and we set $f_k(v) = f_{k+1}(v_1)$. A direct extension of the estimates used for the example A.8 of the Appendix shows that the energies $\mathcal{E}_1(f), \mathcal{E}_2(f)$, and $\mathcal{E}_3(f)$ very nearly equal $\mathbf{M}(T)$.

Proof of the inequality "$\inf \mathbf{M} \leq \inf \mathcal{E}$". The argument here is a direct extension of the corresponding argument given above and is left to the reader.

Remarks.

(1) One of the main reasons for analyzing relations between the energy of mappings and the area of currents is that it provides a way to study n dimensional area minimizing *integral* currents (whose geometry is not specified ahead of time) by studying functions and integrals over the given ambient manifold. This seems the first such scheme which works in general codimensions. For *real* currents, however, differential forms play a role roughly analogous to that of our function spaces $\mathcal{F}_0$; in this regard see, for example, the paper of H. Federer, *Real flat chains, cochains, and variational problems* [**F2** 4.10(4), 4.11(2)]. Incidentally, in the language of [**F2** 5.12, page 400], examples show that the equation in question there is not always true under the alternative hypotheses of [**F2** 5.10].

(2) Suppose C consists of smooth simple closed curves in $\mathbf{R}^3$ oriented by ς. Suppose also for positive integers ν we have reasonable mappings f_ν from the complement of C in $\mathbf{R}^3$ to the circle $\mathbf{S}^1$ with the property that small circles which link C once are mapped to $\mathbf{S}^1$ by f_ν with degree ν. Because of the dimensions we have

$$\mathcal{E}_1(f_\nu) = \mathcal{E}_2(f_\nu) = \mathcal{E}_3(f_\nu) = \left(\frac{1}{2\pi}\right) \int |Df_\nu| \, d\mathcal{L}^3.$$

If f_ν is nearly $\mathcal{E}_1$ energy minimizing then for most w's in $\mathbf{S}^1$ the slice

$$T_\nu(w) = \langle \mathbf{E}^3 , f_\nu , w \rangle \in \mathbf{I}_2(\mathbf{R}^3)$$

will be defined with $\partial T_\nu(w) = \mathbf{t}(C , \nu , \varsigma)$ and will be nearly mass minimizing. H. Parks, in his memoir, *Explicit determination of area minimizing hypersurfaces, II* [**PH**], used a similar energy for mappings to the real numbers $\mathbf{R}$ (instead of to $\mathbf{S}^1$) and was able to exhibit an algorithm for finding area minimizing surfaces. The technique used by Parks requires that C be extreme, i.e. that it lie on the boundary of its convex hull. The analysis of our paper on the other hand applies to any collection of curves which, for example, may be knotted or linked in any way. One of our hopes is to develop a method of computation analogous to that of Parks.

(3) Suppose that C and the mappings f_ν have the same meaning as in (2) above. If θ denotes the usual (multiple-valued radian) angle function on $\mathbf{S}^1$ then $d\theta$ as a well defined closed 1-form

whose pullbacks $f_\nu^\sharp d\theta$ give closed 1 forms on the complement of C in $\mathbf{R}^3$ with $|f_\nu^\sharp d\theta| = |Df_\nu|$. For fixed x_0 in the complement of C we define functions g_ν mapping the complement of C to $\mathbf{S}^1$ by requiring that

$$\theta \circ g_\nu(x) = \theta \circ f_\nu(x_0) + \int_\gamma f_\nu^\sharp d\theta \pmod{2\pi}$$

for each x (with the obvious meanings); here $\gamma(x)$ denotes any oriented path in the complement of C starting at x_0 and ending at x. It is immediate to check that $g_\nu = f_\nu$ for each ν. If we write $\nu = \lambda \cdot \mu$ for some λ and μ and define $h_\lambda(x)$ in $\mathbf{S}^1$ by requiring

$$\theta \circ h_\lambda(x) = \int_\gamma \left(\frac{1}{\mu}\right) f_\nu^\sharp d\theta \pmod{2\pi}$$

for γ as above. The mapping h_λ maps small circles with the same degrees as does f_λ. Taking $\mu = \nu$ we readily conclude, for example, that

$$\inf\{\mathbf{M}(T): \partial T = \mathbf{t}(C, \nu, \varsigma)\} = \nu \cdot \inf\{\mathbf{M}(T): \partial T = \mathbf{t}(C, 1, \varsigma)\}$$

for each ν. This estimate implies that integral and real mass minimizing 2 currents having boundary $\mathbf{t}(C, 1, \varsigma)$ have the same masses [**F2** 5.8]; although this has been known for some time, the present proof by factoring mappings seems new and simpler. This fact (and our proof) extend to $n-1$ dimensional boundaries in general manifolds $\mathcal{M}$ of dimension $n+1$ with, for example, the property that each 1 cycle is a boundary. There are counterexamples to such equalities in higher codimensions given first by L. C. Young [**YL**] and later by F. Morgan [**MF**] and B. White [**W1**]. How *badly* such an equality can fail remains an important open question. It is not even known, for example, if the number

$$\inf\{\mathbf{M}(S)/\mathbf{M}(T): S, T \in \mathbf{I}_2(\mathbf{R}^4, \mathbf{R}^4) \text{ are mass minimizing with } 0 \neq \partial S = 2\partial T\}$$

is positive; note, however, the isoperimetric inequality [**A1** 2.6].

(4) Suppose $\mathcal{M}$ is a complex submanifold of some complex projective space $\mathbf{CP}(n)$ (or, more generally, $\mathcal{M}$ is a Kähler manifold). Then any complex analytic (meromorphic) function f from $\mathcal{M}$ to the Riemann Sphere $\mathbf{CP}(1) = \mathbf{S}^2$ has integral current slices which are absolutely mass minimizing in their integral homology classes [**F1** 5.4.19]. Such f's are thus necessarily maximally conformal and minimize each of the energies $\mathcal{E}_1$, $\mathcal{E}_2$, and $\mathcal{E}_3$ among functions in the same homotopy classes.

(5) In the context of this paper, if the mass minimizing current T being sought happens to be unique then most slices of nearly minimizing mappings will be close to that current. In a sense this describes the asymptotic behavior of a sequence $\{f_k\}_k$ of mappings in $\mathcal{F}_0$ converging towards energy minimization; in particular, the real currents

$$\left\{\left(\frac{1}{(m+1)\alpha(m+1)}\right) [\![\mathcal{M}]\!] \llcorner f_k^\sharp \sigma^*\right\}_k$$

must converge to T as $k \to \infty$. If $m = 2$ then the energy $\mathcal{E}_1$ is Dirichlet's integral which is widely studied in the general theory of harmonic mappings between manifolds pioneered by J. Eells and J. Sampson.

In any codimension m each n dimensional mass minimizing integral current is a *regular* minimal submanifold except possibly on a singular set of dimension not exceeding $n - 2$ as shown by F. Almgren in [**A2**]. It is not yet clear to what extent the present new setup will provide new tools for study of the regularity and singularity properties of mass minimizing integral currents. This could be one of its most important potential uses.

APPENDIX

When not otherwise specified we follow the general terminology of pages 669-671 of H. Federer's treatise, *Geometric Measure Theory* [**F1**] or the newer standardized terminology of the 1984 AMS Summer Research Institute in Geometric Measure Theory and the Calculus of Variations as summarized in pages 124-130 of F. Almgren's paper, *Deformations and multiple-valued functions* [**A1**].

A.1 Terminology.

A.1.1 We fix positive integers m and n and suppose that $\mathcal{M}$ is an $m + n$ dimensional submanifold (without boundary) of $\mathbf{R}^N$ (some N) which is smooth, compact, and oriented by the continuous unit $(m+n)$-vectorfield $\xi\colon \mathbf{M} \to \wedge_{m+n}\mathbf{R}^N$; alternatively $\mathcal{M} = \mathbf{R}^{m+n}$ with standard orthonormal basis vectors $\mathbf{e}_1, \ldots, \mathbf{e}_{m+n}$ and orienting $(m+n)$-vector $\mathbf{e}_1 \wedge \ldots \wedge \mathbf{e}_{m+n}$. We also suppose that B is a finite (possibly empty) union of various (curvilinear) $n - 1$ simplexes $\Delta_1, \Delta_2, \ldots, \Delta_J$ associated with some smooth triangulation of $\mathcal{M}$.

A.1.2 We denote by $\mathbf{S}^m$ the unit sphere in $\mathbf{R} \times \mathbf{R}^m = \mathbf{R}^{1+m}$ with its usual orientation given by the unit m-vectorfield $\sigma\colon \mathbf{S}^m \to \wedge_m \mathbf{R}^{1+m}$; in particular, for each $w \in \mathbf{S}^m \subset \mathbf{R}^{1+m} = \wedge_1 \mathbf{R}^{1+m}$, $\sigma(w) = *w$. It is convenient to let $z, y_1, \ldots, y_m$ denote the usual orthonormal coordinates for $\mathbf{R} \times \mathbf{R}^m$ and also let $\mathbf{p}, \varepsilon_1, \ldots, \varepsilon_m$ be the associated orthonormal basis vectors. In particular, $\sigma(\mathbf{p}) = *\mathbf{p} = \varepsilon_1 \wedge \ldots \wedge \varepsilon_m$. We regard $\mathbf{p}$ as the 'north pole' of $\mathbf{S}^m$. The 'south pole' is $\mathbf{q} = -\mathbf{p}$. We denote by σ^* the differential m form (the 'volume form') on $\mathbf{S}^m$ dual to σ.

A.1.3 If L is a linear mapping $\mathbf{R}^{m+n} \to \mathbf{R}^m$ then the polar decomposition theorem guarantees the existence of orthonormal coordinates for $\mathbf{R}^{m+n}$ and $\mathbf{R}^m$ with respect to which L has the matrix representation

$$L = \begin{pmatrix} \lambda_1 & 0 & \cdots & 0 & 0 & \cdots & 0 \\ 0 & \lambda_2 & \cdots & 0 & 0 & \cdots & 0 \\ \vdots & \vdots & \ddots & 0 & \vdots & \cdots & \vdots \\ 0 & 0 & \cdots & \lambda_m & 0 & \cdots & 0 \end{pmatrix}$$

with $\lambda_1 \geq \lambda_2 \geq \ldots \geq \lambda_m \geq 0$. In these coordinates we can express the Euclidean norm $|L|$ of L as

$$|L| = \left(\lambda_1^2 + \lambda_2^2 + \ldots + \lambda_m^2\right)^{\frac{1}{2}},$$

express the mapping norm $\|L\|$ of L as

$$\|L\| = \lambda_1,$$

and express the mapping norm $\| \wedge_m L\|$ of the linear mapping $\wedge_m L$ of m-vectors induced by L as

$$\| \wedge_m \| = \lambda_1 \cdot \lambda_2 \cdots \lambda_m.$$

Whenever $\lambda_1 \geq \lambda_2 \geq \cdots \geq \lambda_m \geq 0$ we have

$$\lambda_1 \cdot \lambda_2 \cdots \lambda_m \leq \frac{1}{m^{\frac{m}{2}}} (\lambda_1^2 + \lambda_2^2 + \ldots + \lambda_m^2)^{\frac{m}{2}} \leq \lambda_1^m \leq (\lambda_1^2 + \lambda_2^2 + \ldots + \lambda_m^2)^{\frac{m}{2}}.$$

The first two inequalities are equalities if and only if $\lambda_1 = \lambda_2 = \ldots = \lambda_m$. The right hand inequalitiy is an equality if and only if $\lambda_2 = \lambda_3 = \ldots = \lambda_m = 0$.

If f is a mapping and $L = Df(a)$ is the differential of f at a, then $|Df(a)|^2$ is of value of *Dirichlet's integrand* of f at a, and

$$J_m f(a) = \| \wedge_m Df(a)\|$$

is the m *dimensional Jacobian* of f at a.

A.2 Modified Stereographic Projection. Stereographic projection of $\mathbf{S}^m$ onto $\mathbf{R}^m$ from the south pole $\mathbf{q}$ maps $(z,y) \in \mathbf{S}^m \sim \{\mathbf{q}\}$ to $2y/(1+z) \in \mathbf{R}^m$ while the inverse mapping $\gamma_0 : \mathbf{R}^m \to \mathbf{S}^m$ sends $y \in \mathbf{R}^m$ to

$$\gamma_0(y) = \left(\frac{4 - |y|^2}{4 + |y|^2}, \quad \frac{4y}{4 + |y|^2} \right) \in \mathbf{S}^m \sim \{\mathbf{q}\}.$$

γ_0 is an orientation preserving conformal diffeomorphism between $\mathbf{R}^m$ and $\mathbf{S}^m \sim \{\mathbf{q}\}$ as is readily checked.

For convenience we let $\theta : \mathbf{S}^m \to [0, \pi]$ denote angular distance in radians (equivalently, geodesic distance in $\mathbf{S}^m$) to $\mathbf{p}$. General level sets of θ are thus $m-1$ spheres of constant latitude while $\theta(\mathbf{p}) = 0$ and $\theta(\mathbf{q}) = \pi$. Also for $(z,y) \in \mathbf{S}^m$ we have $z = \cos\theta(z,y)$ and $|y| = \sin\theta(z,y)$. Latitude lines on $\mathbf{S}^m$ are level sets of the function ω which maps $(z,y) \in \mathbf{S}^m \sim \{\mathbf{p},\mathbf{q}\}$ to

$$\omega(z,y) = \frac{y}{|y|} \in \mathbf{S}^{m-1} \subset \mathbf{R}^m.$$

Certain mappings derived from γ_0 are important in our constructions. If $0 < \delta << 1/2$ is a given very small number we fix $0 < r = r(\delta) << R < \infty$ by requiring that R be the radius of the sphere in $\mathbf{R}^m$ which γ_0 maps to the latitude sphere $\theta = \pi - \delta$ near $\mathbf{q}$ in $\mathbf{S}^m$ and that rR/δ be the radius of the sphere in $\mathbf{R}^m$ which γ_0 maps to the latitude sphere $\theta = \delta$ near $\mathbf{p}$.

We now modify γ_0 to obtain a mapping $\gamma = \gamma_\delta = \gamma_{\delta,1}$ which maps $\mathbf{R}^m$ onto all of $\mathbf{S}^m$ and which maps points y in $\mathbf{R}^m$ with norm less that r^2 to $\mathbf{p}$, maps points y in $\mathbf{R}^m$ with norm greater that 2δ to $\mathbf{q}$, maps points y in $\mathbf{R}^m$ with norm between r and δ to $\gamma_0(Ry/\delta)$ and suitable interpolates in the two remaining annular regions. More precisely, we set

$$\gamma(y) = \begin{cases} \mathbf{p} & \text{if } 0 \le |y| \le r^2 \\ \left(\cos\left(\delta\left(\frac{|y|-r^2}{r-r^2}\right)\right), \sin\left(\delta\left(\frac{|y|-r^2}{r-r^2}\right)\right)\frac{y}{|y|}\right) & \text{if } r^2 \le |y| \le r \\ \gamma_0(\frac{Ry}{\delta}) & \text{if } r \le |y| \le \delta \\ \left(\cos(\pi + |y| - 2\delta), \sin(\pi + |y| - 2\delta)\frac{y}{|y|}\right) & \text{if } \delta \le |y| \le 2\delta \\ \mathbf{q} & \text{if } 2\delta \le |y| < \infty. \end{cases}$$

In the region $0 \le |y| \le r$ we estimate that the Lipschitz constant of γ does not exceed $\delta/(r - r^2)$ which is less that $2\delta/r$ since $r < 1/2$. Hence

$$\int_{|y|\le r} |D\gamma|^m \, d\mathcal{L}^m \le m^{\frac{1}{2}} \left(\frac{2\delta}{r}\right)^m \alpha(m) r^m = 2^m m^{\frac{1}{2}} \alpha(m)\delta^m$$

which is small if δ is small.

Similarly, in the region $\delta \le |y| \le 2\delta$ we estimate that the local Lipschitz constants do not exceed 1. Hence

$$\int_{\delta\le|y|\le 2\delta} |D\gamma|^m \, d\mathcal{L}^m < m^{\frac{1}{2}} \alpha(m)(2\delta)^m = 2^m m^{\frac{1}{2}} \alpha(m)\delta^m$$

which is small is δ is small.

Finally we note that, in the region, $r < |y| < \delta$ the mapping γ is conformal so that

$$\int_{r\le|y|\le\delta} J_m\gamma \, d\mathcal{L}^m = \int_{r\le|y|\le\delta} \|D\gamma\|^m \, d\mathcal{L}^m = \frac{1}{m^{\frac{m}{2}}} \int_{r\le|y|\le\delta} |D\gamma|^m \, d\mathcal{L}^m = \mathcal{H}^m(\mathbf{S}^m \cap \theta^{-1}[\delta, \pi - \delta]).$$

Our mapping $\gamma_{\delta,1}$ from $\mathbf{R}^m$ to $\mathbf{S}^m$ preserves orientations and covers once. It is useful to have mappings $\gamma_{\delta,\nu}$ with similar conformal properties but covering ν times. To do this we fix a ratio $\rho = \left(r(\delta)^2/\delta\right)$ and let $\tau(z, y) = (-z, -y_1, y_2, \ldots, y_m)$ for $(z, y) \in \mathbf{S}^m$; the map τ thus interchanges the north and south poles of $\mathbf{S}^m$ while preserving orientation. We then define

$$\gamma_{\delta,\nu}(y) = \begin{cases} \gamma_\delta(y) & \text{if } \rho\delta \le |y| < \infty \\ \tau^k \circ \gamma_\delta(y/\rho^k) & \text{if } k \in \{1, \ldots, \nu - 2\} \text{ and } \rho^{k+1}\delta \le |y| \le \rho^k\delta \\ \tau^{\nu-1} \circ \gamma_\delta(y/\rho^{\nu-1}) & \text{if } 0 \le |y| \le \rho^{\nu-1}\delta. \end{cases}$$

A.3. Mappings and homotopies from $\mathcal{M}$ to $\mathbf{S}^m$ with contolled discontinuities.

A.3.1 Whenever $f: \mathcal{M} \to \mathbf{S}^m$ we denote by

$$C_f$$

the closure of the set of points of discontinuity of f. We then let

$$\mathcal{F}$$

be the collection of all functions $f : \mathcal{M} \to \mathbf{S}^m$ such that the closure of $C_f \sim B$ (recall A.1.1) has dimension not exceeding $n - 2$. In case m equals 1 we require that $C_f \subset B$ for functions f in $\mathcal{F}$. Also, if $\mathcal{M}$ is $\mathbf{R}^{m+n}$ we require that $f(x) = \mathbf{q}$ whenever $|x|$ is sufficiently large.

Similarly, whenever $h: [0,1] \times \mathcal{M} \to \mathbf{S}^m$ we denote by

$$C_h$$

the closure of the set of discontinuities of h. We then say that f and g in $\mathcal{F}$ are s-homotopic provided there is a function $h: [0,1] \times \mathcal{M} \to \mathbf{S}^m$ such that $h(0,\cdot) = f$ and $h(1,\cdot) = g$ and also

$$C_h \sim ((\{0\} \times C_f) \cup (\{1\} \times C_g) \cup ([0,1] \times B))$$

lies in $(0,1) \times \mathcal{M}$ and has dimension not exceeding $n-1$ (in case $\mathcal{M}$ is $\mathbf{R}^{m+n}$ we additionally require that $h(t,x) = \mathbf{q}$ for all t when $|x|$ is sufficiently large); such a function h is called an s-homotopy between f and g. We then denote by

$$\Pi(\mathcal{F})$$

the s-homotopy equivalence classes of $\mathcal{F}$.

A.3.2 We denote by

$$\mathcal{F}_0$$

those functions f in $\mathcal{F}$ for which $f|(\mathcal{M} \sim C_f)$ is locally Lipschitz and then associate to each such f three energies $\mathcal{E}_1(f)$, $\mathcal{E}_2(f)$, and $\mathcal{E}_3(f)$ given by setting

$$\mathcal{E}_1(f) = \frac{1}{m^{m/2}(m+1)\alpha(m+1)} \int_{\mathcal{M}} |Df|^m \, d\mathcal{H}^{m+n},$$

$$\mathcal{E}_2(f) = \frac{1}{(m+1)\alpha(m+1)} \int_{\mathcal{M}} \|Df\|^m \, d\mathcal{H}^{m+n},$$

$$\mathcal{E}_3(f) = \frac{1}{(m+1)\alpha(m+1)} \int_{\mathcal{M}} J_m f \, d\mathcal{H}^{m+n}.$$

For some analyses (beyond the scope of this present paper) it is important to recognize that

$$J_m f(x) = |\langle \sigma^*(f(x)), \wedge^m Df(x) \rangle|.$$

We also call the reader's attention to the paper *Homotopy classes in Sobolev spaces and the existence of energy minimizing mappings* [**W2**] by B. White in which p energy minimization is studied in homotopy classes of mappings which are not necessarily continuous.

A.3.3 A basic fact is the following

PROPOSITION.

(1) Each s-homotopy class in $\Pi(\mathcal{F})$ contains a representative f which belongs to $\mathcal{F}_0$ and for which each of the energies $\mathcal{E}_1(f)$, $\mathcal{E}_2(f)$, and $\mathcal{E}_3(f)$ is finite.

(2) Suppose f and g belong to $\mathcal{F}_0$ and are representatives of the same s-homotopy class in $\Pi(\mathcal{F})$. Suppose also that $\mathcal{E}_1(f)$ and $\mathcal{E}_1(g)$ are both finite. Then there is an s-homotopy h between f and g such that $h|\big([0,1] \times \mathcal{M} \sim C_h\big)$ is locally Lipschitz and

$$\int_{[0,1]\times\mathcal{M}} |Dh|^m \, d\mathcal{H}^{m+n+1} < \infty.$$

A.4 Currents. A general k (dimensional) current T is a continuous linear functional on an appropriate space of smooth differential k forms in $\mathbf{R}^N$. The boundary of a k current T is the $k-1$ current ∂T which maps a smooth differential $k-1$ form ω to the number $\partial T(\omega) = T(d\omega)$— Stokes's theorem becomes a definition. In this paper we are concerned with currents of the form $T = \mathbf{t}(\Sigma, \theta, \varsigma)$. In writing such an expression we mean that $\mathbf{set}(T) = \Sigma$ is a (bounded) $\mathcal{H}^k$ measurable and $(\mathcal{H}^k, k)$ rectifiable subset of $\mathcal{M}$, and that the *density function* $\theta\colon \Sigma \to \mathbf{R}^+$ is $\mathcal{H}^k \llcorner \Sigma$ summable, and that the *orientation* ς is an $\mathcal{H}^k \llcorner \Sigma$ measurable function whose simple unit k vector values are compatible with the tangent plane structure of Σ. Such a k current T maps a differential k form φ to the number

$$T(\varphi) = \int_{x\in\Sigma} \langle \varsigma(x), \varphi(x) \rangle \, \theta(x) \, d\mathcal{H}^k x.$$

Associated with $\mathcal{M}$ itself is the $m+n$ current

$$[\![\mathcal{M}]\!] = \mathbf{t}(\mathcal{M}, 1, \xi);$$

if $\mathcal{M} = \mathbf{R}^{m+n}$ a standard notation is

$$\mathbf{E}^{m+n} = \mathbf{t}(\mathbf{R}^{m+n}, 1, \xi)$$

with $\xi(x) = \mathbf{e}_1 \wedge \ldots \wedge \mathbf{e}_{m+n}$ for each x.

The area of a current $T = \mathbf{t}(\Sigma, \theta, \varsigma)$ weighted with its density gives its *mass*,

$$\mathbf{M}(T) = \int_\Sigma \theta \, d\mathcal{H}^k = \sup\{T(\varphi) \colon \|\varphi\| \le 1\}.$$

The theorems of this paper relate to minimization of this mass rather than, say, the k areas of the underlying set Σ (which is called the *size* of T and is denoted $\mathbf{S}(T)$). The measure $\|T\|$ associated with mass is thus $\mathcal{H}^k \llcorner \Sigma \wedge \theta$ so that $\mathbf{M}(T) = \|T\|(\mathcal{M}) = \|T\|(\mathbf{R}^N)$.

A general fact about such a current $T = \mathbf{t}(\Sigma, \theta, \varsigma)$ is that its general current boundary ignores closed sets of zero k-1 measure, e.g. if $U \subset \mathbf{R}^N$ is open and the support of ∂T inside U has zero $\mathcal{H}^{k-1}$ measure, then $\partial T(\omega) = 0$ for each ω supported in U [**F1** 4.1.20].

Suppose that $T = \mathbf{t}(\Sigma, \theta, \varsigma)$ is an n current such that the support of ∂T lies in B. Because of our special assumptions about B in **A.1.1** we can use [**F1** 4.1.31] together with our preceding remark to infer for each $k = 1, \ldots J$ the existence of nonnegative real numbers r_k and continuous orientation functions ς_k on Δ_k such that

$$\partial T = \sum_{k=1}^{J} \mathbf{t}(\Delta_k, r_k, \varsigma_k).$$

For general (possibly empty) subsets A and C of $\mathcal{M}$ with $C \subset A$ we denote by $\mathbf{R}_k(A, C)$ the vector space of those k currents $T = \mathbf{t}(\Sigma, \theta, \varsigma)$ with the closure of Σ contained in A such that $\partial T = \mathbf{t}(\Sigma', \theta', \varsigma')$ for some $\Sigma', \theta', \varsigma'$ with the closure of Σ' contained in C. We further let $\mathbf{I}_k(A, C)$ denote the subgroup of those currents $T = \mathbf{t}(\Sigma, \theta, \varsigma)$ in $\mathbf{R}_k(A, C)$ such that θ assumes only positive integer values. It follows from [**F1** 4.2.16(2)] that $\partial T \in \mathbf{I}_{k-1}(C, \emptyset)$ whenever $T \in \mathbf{I}_k(A, C)$.

When convenient we will denote by $\operatorname{spt} T$ the support of a current T.

A.5 The coarea formula and slices of currents. A key ingredient of the present paper is slicing the current $[\![\mathcal{M}]\!]$ by mappings $f \colon \mathcal{M} \to \mathbf{S}^m$ belonging to $\mathcal{F}_0$ and use of the *coarea formula* to estimate the masses of these slices in terms of the energy $\mathcal{E}_3(f)$. As a consequence of [**F1** 3.2.22, 4.3.8, 4.3.11] we infer that for $\mathcal{H}^m$ almost every $w \in \mathbf{S}^m$ the slice

$$\langle [\![\mathcal{M}]\!]\,,\, f\,,\, w \rangle = \mathbf{t}(f^{-1}\{w\}\,,\, 1\,,\, \varsigma)$$

is well defined as an n dimensional current. Here, for $\mathcal{H}^n$ almost every $x \in f^{-1}\{w\}$, if $\eta(x)$ is that simple unit m vector associated with the m plane $\ker Df(x)^{\perp}$ in $\operatorname{Tan}(\mathcal{M}, x)$ for which

$$\langle \eta(x)\,,\, \wedge_m Df(x) \rangle \bullet \sigma(w) > 0$$

then we specify $\varsigma(x)$ to be that simple unit n vector associated with $\ker Df(x)$ in $\operatorname{Tan}(\mathcal{M}, x)$ for which $\xi(x) = \eta(x) \wedge \varsigma(x)$; we have used the symbol $\bullet$ to denote the inner product in $\wedge_m \mathbf{R}^{m+1}$.

We further infer from the coarea formula [**F1** 3.2.22] that

$$(m+1)\alpha(m+1)\mathcal{E}_3(f) = \int_{w \in \mathbf{S}^m} \mathbf{M}(\langle [\![\mathcal{M}]\!]\,,\, f\,,\, w \rangle)\, d\mathcal{H}^m w.$$

Since $\partial [\![\mathcal{M}]\!] = 0$ we readily infer from [**F1** 4.3.1] together with **A.3.2** and **A.4** above that for $\mathcal{H}^m$ almost every $w \in \mathbf{S}^m$, $\partial \langle [\![\mathcal{M}]\!]\,,\, f\,,\, w \rangle$ belongs to $\mathbf{I}_{m-1}(B, \emptyset)$.

A.6 Kronecker indices of integral currents. Whenever $S \in \mathbf{I}_m(\mathcal{M}, \mathcal{M})$ and $T \in \mathbf{I}_n(\mathcal{M}, \mathcal{M})$ with

$$\emptyset = \operatorname{spt}\partial S \cap \operatorname{spt} T = \operatorname{spt} S \cap \operatorname{spt}\partial T,$$

there is naturally defined the *Kronecker index* of S and T in $\mathcal{M}$, denoted

$$\mathbf{k}(S,T) = \mathbf{k}(S,T;\mathcal{M}) \in \mathbf{Z}.$$

which is a direct extension of the definitions in [**F1** 4.3.20]. For 'sufficiently regular' such currents

$$S = \mathbf{t}(\Sigma_1, \theta_1, \varsigma_1) \qquad \text{and} \qquad T = \mathbf{t}(\Sigma_2, \theta_2, \varsigma_2)$$

in 'general position', we can write

$$\mathbf{k}(S,T) = \sum_{x \in \Sigma_1 \cap \Sigma_2} \theta_1(x) \cdot \theta_2(x) \cdot \operatorname{sign}(\varsigma_1(x) \wedge \varsigma_2(x) \bullet \xi(x)).$$

Among the important facts about the Kronecker index is its ability to characterize real homology classes. We have the following.

PROPOSITION. *Suppose* $T_1, T_2 \in \mathbf{I}_n(\mathcal{M}, B)$ *with* $\partial T_1 = \partial T_2$ *and*

$$\mathbf{k}(S, T_1) = \mathbf{k}(S, T_2)$$

for each $S \in \mathbf{I}_m(\mathcal{M}, \emptyset)$ *for which both Kronecker indices are defined. Then there is* $Q \in \mathbf{R}_{n+1}(\mathcal{M}, \mathcal{M})$ *such that* $\partial Q = T_1 - T_2$.

Proof. In view of [**F1** 4.4.1] it is sufficient to verify the assertion in the context of Lipschitz singular chains of algebraic topology. Moreover it is sufficient to check than an n cycle T in $\mathcal{M}$ is a boundary in case its general position intersections with m cycles S in $\mathcal{M}$ all have Kronecker index zero. This is well known.

A.7 Degrees of mappings of currents. Suppose $f \in \mathcal{F}_0$ and

$$S = \mathbf{t}(\Sigma\,,\,\theta\,,\,\varsigma) \in \mathbf{I}_m(\mathcal{M} \sim C_f, \emptyset).$$

Then the m current $f_\sharp S$ in $\mathbf{S}^m$ is naturally defined in accordance with [**F1** 4.1.14, 4.1.15] with $\partial f_\sharp S = 0$ since $\partial S = 0$. We then infer from [**F1** 4.1.31] the existence of an integer $\mathbf{d}(f,S)$ such that

$$f_\sharp S = \mathbf{t}(\mathbf{S}^m\,,\,\mathbf{d}(f,S)\,,\,\sigma).$$

We call $\mathbf{d}(f,S)$ the *degree* of f on S. If f and S are 'sufficiently regular' then, for $\mathcal{H}^m$ almost every $w \in \mathbf{S}^m$,

$$\mathbf{d}(f,S) = \sum_{x \in \Sigma \cap f^{-1}\{w\}} \theta(x) \operatorname{sign}\left(\langle \varsigma(x), \wedge_m Df(x)\rangle \bullet \sigma(w)\right).$$

Basic properties of degrees are the following.

PROPOSITION.

(1) The degree $\mathbf{d}(f,S)$ *depends only on the real homology class of* S *in* $\mathcal{M} \sim B$. *More precisely, if* $f \in \mathcal{F}_0$, *and* $S_1, S_2 \in \mathbf{I}_m(\mathcal{M} \sim C_f, \emptyset)$, *and* $Q \in \mathbf{R}_{m+1}(\mathcal{M} \sim B, \mathcal{M} \sim B)$ *with* $\partial Q = S_1 - S_2$, *then* $\mathbf{d}(f, S_1) = \mathbf{d}(f, S_2)$.

(2) The degree $\mathbf{d}(f,S)$ depends only on the s-homotopy class of f. More precisely, if $f,g \in \mathcal{F}_0$ are s-homotopic and $S \in \mathbf{I}_m(\mathcal{M} \sim (C_f \cup C_g \cup B), \emptyset)$, then $\mathbf{d}(f,S) = \mathbf{d}(g,S)$.

A.8 An example showing relations between integral current slices and boundaries, Kronecker indices, and mapping degrees. Suppose, as illustrated in Figure 2, the following.

(a) $\mathcal{M} = \mathbf{R}^{m+n}$ with its usual orthonormal basis, and

$$U = \mathbf{U}^{m+1}(0,1) \times \mathbf{U}^{n-1}(0,1),$$

is an open set, and

$$\Delta = \{0\} \times \mathbf{U}^{n-1}(0,1)$$

is an $n-1$ disk with orientation function

$$\beta: \Delta \to \{\mathbf{e}_{m+2} \wedge \ldots \wedge \mathbf{e}_{m+n}\}.$$

(b) K and $z_1, \ldots, z_K$ are positive integers and $\epsilon_1, \ldots, \epsilon_K \in \{-1, +1\}$.

(c) For each k the vectors

$$p(k), \eta_1(k), \ldots, \eta_m(k) \in \mathbf{S}^m \times \{0\} \subset \mathbf{R}^{m+1} \times \mathbf{R}^{n-1}$$

are an orthonormal family such that

$$\eta_1(k) \wedge \ldots \wedge \eta_m(k) \wedge p(k) = \mathbf{e}_1 \wedge \ldots \wedge \mathbf{e}_{m+1}$$

and also $p(1), \ldots, p(K)$ are distinct.

(d) For each k we let Π_k denote the n plane spanned by $p(k)$ and $\{0\} \times \mathbf{R}^{n-1}$ and define the n half disk

$$\Delta_k = \Pi_k \cap U \cap \{x: x \bullet p(k) \le 0\}$$

with orientation function

$$\varsigma: \Delta_k \to \{\epsilon_k\, p(k) \wedge \mathbf{e}_{m+2} \wedge \ldots \wedge \mathbf{e}_{m+n}\}.$$

(e) $0 < \delta << r \le s << 1$ are very small numbers and

$$N = U \cap \{x: \operatorname{dist}(x, \Delta) < r\} \quad \text{and} \quad N_k = (U \sim N) \cap \{x: \operatorname{dist}(x, \Delta_k < 2\delta\}$$

for each k; we assume that δ is small enough so that the sets $N_1, \ldots, N_k$ are positive distances apart.

(e) We denote by Σ the small m sphere

$$\Sigma = \partial \mathbf{B}^{m+1}(0,s) \times \{0\}$$

with the standard continuous orientation function $\tau: \Sigma \to \wedge_m \mathbf{R}^{m+n}$ determined by requiring

$$x \wedge \tau(x) = s \cdot \mathbf{e}_1 \wedge \ldots \wedge \mathbf{e}_{m+1}$$

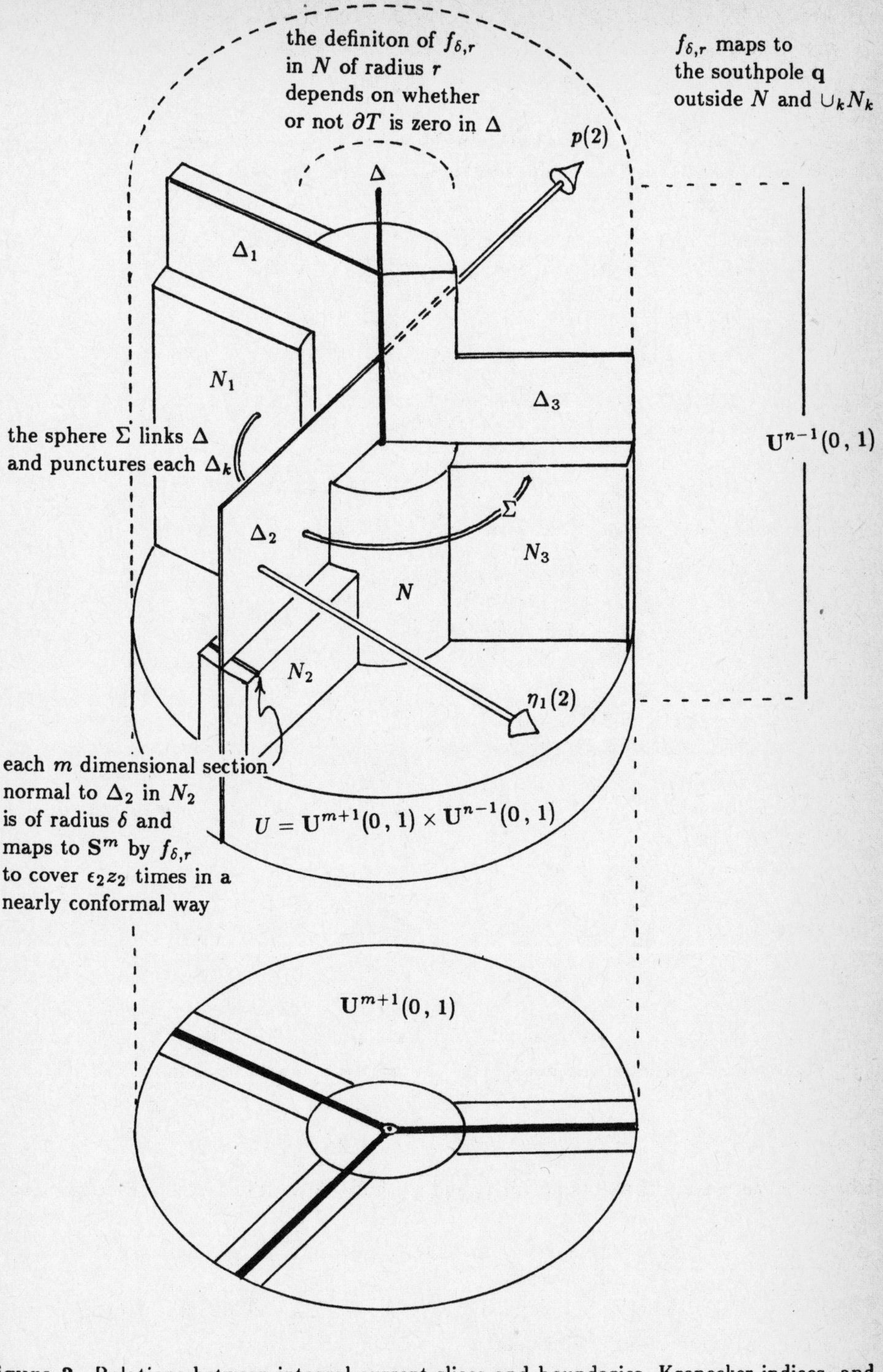
the definiton of $f_{\delta,r}$
in N of radius r
depends on whether
or not ∂T is zero in Δ
$f_{\delta,r}$ maps to
the southpole $\mathbf{q}$
outside N and $\cup_k N_k$
$p(2)$
Δ
Δ_1
N_1
Δ_3
the sphere Σ links Δ
and punctures each Δ_k
$\mathbf{U}^{n-1}(0, 1)$
Σ
Δ_2
N_3
N
N_2
$\eta_1(2)$
each m dimensional section
normal to Δ_2 in N_2
is of radius δ and
maps to $\mathbf{S}^m$ by $f_{\delta,r}$
to cover $\epsilon_2 z_2$ times in a
nearly conformal way
$U = \mathbf{U}^{m+1}(0, 1) \times \mathbf{U}^{n-1}(0, 1)$
$\mathbf{U}^{m+1}(0, 1)$

Figure 2. Relations between integral current slices and boundaries, Kronecker indices, and mapping degrees are illustrated by example in Appendix A.8.

for each x in Σ; it follows that

$$\tau(-s \cdot p(k)) = (-1)^{m+1}\eta_1(k) \wedge \ldots \wedge \eta_m(k)$$

for each k. Here $\cdot$ denotes scalar multiplication of a vector. The m sphere Σ 'links' the $n-1$ disk Δ in U while 'puncturing' each Δ_k at the point $-s \cdot p(k)$.

We then set

$$T = \sum_{k=1}^{K} \mathbf{t}(\Delta_k\,,\, z_k\,,\, \epsilon_k \cdot \varsigma_k) \qquad \text{and} \qquad S = \mathbf{t}(\Sigma\,,1\,,\tau)$$

and estimate

(1) The boundary of T inside U is given by

$$\partial T \llcorner U = \sum_{k=1}^{K} \mathbf{t}(\Delta\,,\, z_k\,,\, \epsilon_k \cdot \beta)$$

[**F1** 4.1.8] so that $\partial T \llcorner U = 0$ if and only if $\sum_{k=1}^{K} \epsilon_k z_k = 0$.

(2) The Kronecker index of S and T is given by

$$\begin{aligned}\mathbf{k}(S,T) &= \sum_{k=1}^{K} z_k \cdot \tau(-s \cdot p(k)) \wedge \varsigma(-s \cdot p(k)) \bullet \mathbf{e}_1 \wedge \ldots \wedge \mathbf{e}_{m+n} \\ &= \sum_{k=1}^{K} z_k(-1)^{m+1}\eta_1(k) \wedge \ldots \eta_m(k) \wedge \epsilon_k \cdot p(k) \wedge \mathbf{e}_{m+1} \wedge \ldots \wedge \mathbf{e}_{m+n} \bullet \mathbf{e}_1 \wedge \ldots \wedge \mathbf{e}_{m+n} \\ &= (-1)^{m+1} \sum_{k=1}^{K} \epsilon_k z_k\end{aligned}$$

so that $\mathbf{k}(S,T) = 0$ if and only if $\partial T \llcorner U = 0$.

We now assume $r = s$ and will construct a mapping $g\colon U \sim N \to \mathbf{S}^m$. We first set $g(x) = \mathbf{q}$ (the southpole) if x lies outside both N and all the N_k's. Each point in each N_k can be written uniquely in the form

$$x + y_1\eta_1(k) + \ldots + y_m\eta_m(k)$$

where x is the unique closest point in Δ_k and $y \in \mathbf{B}^{m+1}(0, 2\delta)$; for each such point we set

$$g(x + y_1\eta_1(k) + \ldots y_m\eta_m(k)) = \gamma_{\delta, z_k}(\epsilon_k \cdot y_1, y_2, \ldots, y_m).$$

Since $r \le s < 1$ our function g is defined on Σ and there is a well defined mapping degree $\mathbf{d}(g, S)$ (with the obvious meaning). Since each γ_{δ, z_k} is orientation preserving (and δ is very small) the orientation of g on Σ near $p(k)$ is determined by ϵ_k and by the inner product

$$\eta_1(k) \wedge \ldots \wedge \eta_m(k) \bullet \tau(-s \cdot p(k)),$$

and we compute

(3) The degree of g on S is given by

$$\mathbf{d}(g,S) = \sum_{k=1}^{K} z_k \epsilon_k \cdot \eta_1(k) \wedge \ldots \wedge \eta_m(k) \bullet \tau(-s \cdot p(k))$$
$$= (-1)^{m+1} \sum_{k=1}^{K} \epsilon_k z_k$$

so that $\mathbf{d}(g,S) = 0$ if and only if $\partial T \llcorner U = 0$.

The extension of g to a mapping $f = f_{\delta,r}$ on all of U depends on which of two cases occurs.

Case 1. If $\mathbf{d}(g,S) = 0$ we infer from Hurewicz's theorem the existence of a Lipschitz mapping $h: \mathbf{B}^{m+1}(0,r) \to \mathbf{S}^m$ such that

$$h(w) = \begin{cases} g(w,0) & \text{if } |w| = r \\ \mathbf{q} & \text{if } |w| \leq r/2. \end{cases}$$

We then define our mapping $f: U \to \mathbf{S}^m$ by setting

$$f(x) = \begin{cases} g(x) & \text{if } x \notin N \\ h(x_1, \ldots, x_{m+1}) & \text{if } x \in N. \end{cases}$$

Case 2. If $\mathbf{d}(g,S) \neq 0$ we define a discontinuous mapping $h: \mathbf{B}^{m+1} \to \mathbf{S}^m$ by setting

$$h(w) = g\left(\frac{rw}{|w|}, 0\right)$$

for each w and, as above, define $f: U \to \mathbf{S}^m$ by setting

$$f(x) = \begin{cases} g(x) & \text{if } x \notin N \\ h(x_1, \ldots, x_{m+1}) & \text{if } x \in N. \end{cases}$$

With the obvious interpretation of $\mathcal{E}_1, \mathcal{E}_2$, and $\mathcal{E}_3$ for function on U, each of these energies of mappings $f_{\delta,r}$ nearly equals the mass of T when δ and r are small (and reasonable choices are made for h in Case 1). More precisely, we have.

$$\lim_{\delta,r\downarrow 0} \mathcal{E}_1(f_{\delta,r}) = \lim_{\delta,r\downarrow 0} \mathcal{E}_2(f_{\delta,r}) = \lim_{\delta,r\downarrow 0} \mathcal{E}_3(f_{\delta,r}) = \mathbf{M}(T) = \sum_{k=1}^{K} z_k \mathcal{H}^n(\Delta_k).$$

It is also straight forward to check that for $\mathcal{H}^m$ almost every $w \in \mathbf{S}^m$ the slice

$$T_w = \langle \mathbf{E}^{m+n} \llcorner U \,,\, f_{\delta,r} \,,\, w \rangle$$

exists with

$$\partial T_w \llcorner U = \partial T \llcorner U,$$

and also if a sequence of δ's and r's converging to 0 is fixed then, for $\mathcal{H}^m$ almost every w in $\mathbf{S}^m$,

$$\lim_{\delta, r \downarrow 0} \langle \mathbf{E}^{m+n} \llcorner U, f_{\delta,r}, w \rangle = T.$$

REFERENCES

[A1] F. Almgren, *Deformations and multiple-valued functions*, Geometric Measure Theory and the Calculus of Variations, Proc. Symposia in Pure Math. **44** (1986), 29-130.

[A2] ________, Q *valued functions minimizing Dirichlet's integral and the regularity of area minimizing rectifiable currents up to codimension two*, preprint.

[BCL] H. Brezis, J-M Coron, E. Lieb, *Harmonic maps with defects*, Comm. Math. Physics, 1987; see also C. R. Acad. Sc. Paris **303** (1986), 207-210.

[F1] H. Federer, *Geometric Measure Theory*, Springer-Verlag, 1969, XIV + 676 pp.

[F2] ________, *Real flat chains, cochains and variational problems*, Indiana U. Math. J. **24** (1974), 351–407.

[HKL] R. Hardt, D. Kinderlehrer, and M. Luskin, *Remarks about the mathematical theory of liquid crystals*, Institute for Mathematics and its Applications, preprint, 1986.

[MF] F. Morgan, *Area-minimizing currents bounded by higher multiples of curves*, Rend. Circ. Matem. Palermo, (II) **33** (1984), 37–46.

[PH] H. Parks, *Explicit determination of area minimizing hypersurfaces, II*, Mem. Amer. Math. Soc. **60**, March 1986, iv + 90 pp.

[SU] R. Schoen and K. Uhlenbeck, *A regularity theory for harmonic maps*, J. Diff. Geom. **60** (1982), 307–335.

[W1] B. White, *The least area bounded by multiples of a curve*, Proc. Amer. Math. Soc. **90** (1984), 230-232.

[W2] ________, *Homotopy classes in Sobolev spaces and the existence of energy minimizing maps*, preprint.

[YL] L. C. Young, *Some extremal questions for simplicial complexes V. The relative area of a Klein bottle*, Rend. Circ. Matem. Palermo, (II), **12** (1963), 257–274.

NONELLIPTIC PROBLEMS AND COMPLEX ANALYSIS

Richard Beals*
Yale University

Introduction

The purpose of these lectures is to describe some natural hypoelliptic but non-elliptic differential equations, and to show how elliptic methods may be modified in order to attack them. The hypoelliptic problems considered here are quite special as a subset of the set of all hypoelliptic problems; it is an important open question to obtain a satisfactory understanding of other cases. Here "satisfactory" should be understood with respect to the elliptic case as the standard. As a point of comparison and of departure, the first lecture outlines a treatment of some classical elliptic problem *via* the pseudodifferential calculus.

1. Some classical problems.

The most classical elliptic operator is the Laplacian in $\mathbb{R}^n$,

$$\Delta = \sum_{j=1}^{n} \left(\frac{\partial}{\partial x_j} \right)^2 , \tag{1.1}$$

and the most classical associated problem is the generalized Dirichlet problem for a bounded region U:

$$\Delta u = f \text{ in } U, \quad u = g \text{ on } \mathrm{bdy}(U). \tag{1.2}$$

There are associated local and global regularity problems: Given some regularity of Δu in U, how regular is u? Given some regularity of f and g, how regular is the solution to (1)?

On a Riemannian manifold M there is a natural generalization, the Laplace-Beltrami operator

$$\Delta = \sum_{j,k=1}^{n} g^{-1/2} \frac{\partial}{\partial x_j} \left(g^{1/2} g^{jk} \frac{\partial}{\partial x_k} \right). \tag{1.3}$$

If M is compact, Δ is a negative self-adjoint operator with compact resolvent, so there is an orthonormal basis of eigenfunctions $(\varphi_\nu)_{\nu=1}^{\infty}$ with eigenvalues $(-\lambda_\nu)_{\nu=1}^{\infty}$, $\lambda_\nu \to +\infty$. The λ_ν are (global) geometric invariants for M.

---

*Research partially supported by NSF grant DMS-8402637

One way to study Δ is to examine the semigroup $(e^{t\Delta})_{t>0}$ or, equivalently, the heat operator $\frac{\partial}{\partial t} - \Delta$. The kernel of $e^{t\Delta}$ is

$$G_t(x,y) = \Sigma\, e^{-\lambda_\nu t} \varphi_\nu(x)\, \overline{\varphi_\nu(y)}$$

so its trace is

$$\mathrm{tr}(e^{t\Delta}) = \Sigma\, e^{-\lambda_\nu t} . \tag{1.4}$$

Asymptotic behavior of λ_ν as $\nu \to \infty$ is related to behavior of this trace as $t \to 0+$. Therefore, rather vaguely, one wants to study the trace (1.4) as $t \to 0+$ and to relate its behavior to geometric invariants.

Similar global questions arise for the deRham complex on a compact Riemannian manifold M:

$$0 \to \mathcal{E}^0 \overset{d_0}{\to} \mathcal{E}^1 \overset{d_1}{\to} \mathcal{E}^2 \to \dots \to \mathcal{E}^n \to 0 \tag{1.5}$$

where $\mathcal{E}^j$ is the space of smooth j-forms on M and d_j is the exterior derivative. The deRham cohomology

$$\mathcal{H}^j = \ker d_j / \mathrm{ran}\, d_{j-1} \tag{1.6}$$

is the kernel of

$$\Delta_j = d_j^* d_j + d_{j-1} d_{j-1}^* \tag{1.7}$$

which is a system whose second order part is scalar and coincides with the second order part of $-\Delta$ with Δ as in (3).

<u>The classical pseudodifferential calculus</u>. If U is a domain $\mathbb{R}^n$, let $S_m(U)$, $m \in \mathbb{Z}$, denote the space of functions $q \in C^\infty(U \times [\mathbb{R}^n \backslash 0])$ with the homogeneity property

$$q(x,\lambda\xi) = \lambda^m q(x,\xi), \quad x \in U, \quad \xi \in \mathbb{R}^n \backslash 0, \quad \lambda > 0. \tag{1.8}$$

Let $S^m(U)$ denote the space of functions $q \in C^\infty(U \times \mathbb{R}^n)$ which have an asymptotic expansion

$$q \sim \sum_{j=0}^{\infty} q_{m-j}, \quad q_{m-j} \in S_{m-j}(U) \tag{1.9}$$

where (1.9) means that the functions

$$\sup_{|\xi| \ge 1} (|\xi|^{N+|\alpha|-m} D_\xi^\alpha D_x^\beta [q - \sum_{j<N} q_{m-j}])$$

are locally bounded on U. Associated to the <u>symbol</u> $q \in S^m(U)$ is an operator

$Q = \mathrm{Op}(q)$ from $\mathcal{D}(U)$ to $C^\infty(U)$:

$$Qu(x) = \int_{\mathbb{R}^n} e^{ix\cdot\xi} q(x,\xi)\hat{u}(\xi)đ\xi, \qquad đ\xi = (2\pi)^{-n}\, d\xi \,. \tag{1.10}$$

Modulo a small technical consideration, the composition of $\mathrm{Op}(q_1)$ and $\mathrm{Op}(q_2)$ with symbols $q_j \in S^{m_j}(U)$ has symbol

$$\begin{cases} q_1 \circ q_2 \in S^{m_1+m_2}(U), \\ q_1 \circ q_2 \sim \Sigma \frac{1}{\alpha!} \left(\frac{\partial}{\partial\xi}\right)^\alpha q_1 D_x^\alpha q_2 \\ \qquad = q_1 q_2 + \text{(lower order terms)}. \end{cases} \tag{1.11}$$

As examples: The identity operator has symbol 1 (by the Fourier inversion formula). If $q \in S^m(U)$ is a <u>polynomial</u> in ξ, then $Q = \mathrm{Op}(q)$ is a differential operator of order m, and conversely. If Q_1 is a differential operator the formula (1.11) is exact and terminates at $|\alpha| = m_1$.

A pseudodifferential operator $Q = \mathrm{Op}\, q$ has a (distribution) kernel K:

$$\begin{cases} Qu(x) = \int K(x,x-y)u(y)dy, \\ K(x,z) = \int e^{iz\cdot\xi}\, q(x,\xi)đ\xi \,. \end{cases} \tag{1.12}$$

This kernel has an asymptotic expansion $K \sim \Sigma K_{m-j}$ where K_{m-j} corresponds to q_{m-j} and the expansion means

$$K - \sum_{j<M} K_{m-j} \in C^N(U \times U) \quad \text{if } N \geq M = M(N). \tag{1.13}$$

An operator (for example a differential operator) $P = \mathrm{Op}(p)$, $p \in S^m(U)$ is <u>elliptic</u> of order m if the principal symbol p_m is nowhere zero on $U \times [\mathbb{R}^m\backslash 0]$. Such an operator has a <u>left parametrix</u> $Q = \mathrm{Op}(q)$, $q \in S^{-m}(U)$. This means that Q inverts P modulo a smoothing operator:

$$QP = I - R, \quad R\colon \mathcal{D}'(U) \to C^\infty(U). \tag{1.14}$$

The idea is first to find the asymptotic expansion Σq_{-m-j}. From (1.11) we need $p_m q_{-m} = 1$ so $q_{-m} = 1/p_m$ and the remaining terms are calculated recursively from (1.11). A technical trick then provides a symbol $q \in S^{-m}(U)$ which has this asymptotic expansion.

The class of operators $\mathrm{Op}(S^m(U))$ is invariant under diffeomorphisms of U, so the construction can be carried over to a manifold M.

<u>A parabolic pseudodifferential calculus.</u> Just as operators like Δ may suggest the previous calculus with its characteristic homogeneities (1.8),

consideration of $\frac{\partial}{\partial t} - \Delta$ leads to parabolic homogeneities

$$q(x,\lambda\xi,\lambda^2\tau) = \lambda^{-m}q(x,\xi,\tau), \quad x \in U, \quad (\xi,\tau) \in \mathbb{R}^{n+1}\backslash 0 . \tag{1.15}$$

Let $S_{m,p}(U \times \mathbb{R})$ denote the space of smooth functions satisfying (1.15) for all $\lambda \in \mathbb{R}\backslash 0$ and which extend to be holomorphic in τ, $\operatorname{Im} \tau < 0$. Let $S^m_p(U \times \mathbb{R})$ denote the smooth functions q with asymptotic expansion Σq_{m-j}, $q_{m-j} \in S_{m-j,p}(U \times \mathbb{R})$. The associated operator on $\mathcal{D}(U \times \mathbb{R})$ is

$$Qu(x,t) = \int_{\mathbb{R}^{n+1}} e^{ix\cdot\xi+it\tau} q(x,\xi,\tau)\hat{u}(\xi,\tau)đ\xi \, d\tau .$$

Such operators compose exactly as in (1.11). The corresponding kernel $K(x,z,t)$ has an expansion $K \sim \Sigma K_{m-j}$ where

$$\begin{cases} K_{m-j}(x,z,t) = 0, & t < 0 \\ K_{m-j}(x,\lambda z,\lambda^2 t) = \lambda^{-m+j-n-2} K_{m-j}(x,z,t), & \lambda \in \mathbb{R}\backslash 0. \end{cases} \tag{1.16}$$

<u>Applications to the classical problems</u>. The Laplacian and the Laplace-Beltrami operator are elliptic of order 2, so Δ has a left parametrix $Q = \operatorname{Op}(q)$, $q \in S^{-2}$. The local regularity question (up to C^∞) becomes the question: Given some regularity of f, how regular is Qf? The expansion of the kernel of K gives all the information necessary to answer this question. The key fact is that the principal term K_{-2} is in $C^\infty(U \times \mathbb{R}^n\backslash 0)$ and satisfies (for $n > 2$)

$$K_{-2}(x,\lambda z) = \lambda^{2-n} K_{-2}(x,z), \quad \lambda > 0.$$

For example, one obtains for the Sobolev spaces and classical Lipschitz spaces:

$$\begin{cases} \Delta u \in L^p_{k,loc}(U) \Rightarrow u \in L^p_{k+2,loc}(U) \\ \Delta u \quad \Lambda_{\alpha,loc}(U) \Rightarrow u \in \Lambda_{\alpha+2,loc}(U). \end{cases} \tag{1.17}$$

If M is a compact Riemannian manifold and $P = \frac{\partial}{\partial t} - \Delta$ with Δ the Laplace-Beltrami operator, then the parabolic calculus allows us to find an actual inverse Q and to calculate its asymptotic expansion. On the kernel side, the kernel $K(x,z,t)$ of Q is for fixed $t > 0$ the kernel of $\operatorname{ext}(t\Delta)$. Therefore one can deduce from (1.16) the asymptotic expansion

$$\operatorname{tr}(e^{t\Delta}) \sim t^{-n/2} \sum_{j=0}^{\infty} c_j t^j \tag{1.18}$$

where

$$c_j = \int_M K_{m-2j}(x,0,1)dx. \tag{1.19}$$

The coefficients c_j in (1.18) are global geometric invariants, while the integrands (1.19) can be shown to be local geometric invariants.

Similar considerations apply to the deRham complex and the operators $\frac{\partial}{\partial t} + \Delta_j$. One can deduce various geometric facts from these considerations: the vanishing of the Euler characteristic for odd-dimensional manifolds, the Chern-Gauss-Bonnet formula, and the index theorem.

Remarks and references. All the material in this section is well-known, and it is far beyond the scope of these notes to give a complete history, discussion, or references. The following can be consulted for classical elliptic theory, pseudo-differential operators, and the heat equation and geometry.

L. Bers, F. John, and M. Schechter, Partial Differential Equations, Interscience, New York, 1964.

M. E. Taylor, Pseudodifferential Operators, Princeton Univ. Press, Princeton, 1981.

F. Treves, Introduction to Pseudodifferential and Fourier Integral Operators, vol. I, Plenum, New York, 1980.

P. Gilkey, Invariance Theory, the Heat Equation, and the Atiyah-Singer Index Theorem, Publish or Perish, Wilmington, 1984.

2. Some nonelliptic problems.

We begin with some examples of nonelliptic operators. First, in $\mathbb{R}^3 = \{(x,y,t)\}$ let

$$P_1 = \frac{\partial}{\partial t} - \left(\frac{\partial}{\partial x}\right)^2 - \left(\frac{\partial}{\partial y}\right)^2,$$

$$P_2 = -\left(\frac{\partial}{\partial x}\right)^2 - \left(\frac{\partial}{\partial y}\right)^2,$$

$$P_3 = -\left[\frac{\partial}{\partial x} - y\frac{\partial}{\partial t}\right]^2 - \left[\frac{\partial}{\partial y} + x\frac{\partial}{\partial t}\right]^2.$$

The first is just the heat operator $\frac{\partial}{\partial t} - \Delta$ in $\mathbb{R}^2 \times \mathbb{R}$; it has an inverse Q_1 which can be constructed explicitly, and it is hypoelliptic:

$$u \in \mathcal{D}'(U), \quad P_1 u \in C^\infty(U) \Rightarrow u \in C^\infty(U).$$

The second is certainly not hypoelliptic, since $P_2 u = 0$ whenever u depends only on t. The third, like the second, has the form $-X_1^2 - X_2^2$ where the X_j are vector fields. Moreover, on the line $x = y = 0$, P_3 coincides with P_2. Nevertheless P_3 is hypoelliptic and has much in common with P_1.

One source of operators and systems of this type is several complex variables. Suppose that M is the (smooth) boundary of a domain $U \subset \mathbb{C}^{n+1}$. The complex tangent bundle $T_{\mathbb{C}}M$ contains a subbundle $T_{1,0}$ consisting of vectors of type $(1,0)$,

$$v = \sum_{j=1}^{n+1} a_j \frac{\partial}{\partial z_j} ,$$

which are tangent to M. This gives a CR-structure: $T_{1,0}$ has complex dimension n and

$$\begin{cases} T_{1,0} \cap T_{0,1} = (0) \quad \text{where} \quad T_{0,1} = \overline{T}_{1,0}; \\ \text{if } Z_1, Z_2 \text{ are sections of } T_{0,1}, \text{ so is } [Z_1,Z_2]. \end{cases} \tag{2.1}$$

Projecting the complexified deRham complex for M onto its quotient by the annihilator of $T_{0,1}$, one obtains the $\bar{\partial}_b$-complex of Kohn and Rossi:

$$0 \to E^{0,0} \xrightarrow{\bar{\partial}_b} E^{0,1} \xrightarrow{\bar{\partial}_b} \cdots \to E^{0,n} \to 0 \quad . \tag{2.2}$$

Given a hermitian structure on $T_{\mathbb{C}}M$ such that $T_{0,1} \perp T_{1,0}$, one obtains the corresponding Kohn laplacians

$$\square_{b,q} = \bar{\partial}^*_{b,q} \bar{\partial}_{b,q} + \bar{\partial}_{b,q-1} \bar{\partial}^*_{b,q-1} \quad . \tag{2.3}$$

Depending on U, or M, these operators are hypoelliptic for some values of q but not for other values. They are systems whose "top order" parts look like the examples P_1, P_3 in various respects.

Based on these examples we consider here the following local model. Suppose U is a domain in $\mathbb{R}^{n+1}$ and $V \subset TU$ is a subbundle whose fibers have codimension 1. Consider a second order operator having the form

$$P = -\sum_{j=1}^{n} X_j^2 + \lambda(x)X_0 \tag{2.4}$$

where $X_1,\ldots,X_n$ is a frame for V and $X_0,X_1,\ldots,X_n$ is a frame for TU. Under certain conditions on the coefficient λ, P will be <u>hypoelliptic with loss of one derivative</u> (with respect to the elliptic case), i.e.

$$u \in \mathcal{D}'(U), \; Pu \in L^p_{k,loc}(U) \Rightarrow u \in L^p_{k+1,loc}(U). \tag{2.5}$$

Our goal is to construct a parametrix for such P within an appropriate class of pseudodifferential operators.

<u>An adapted pseudodifferential calculus</u>. With V and $(X_j)_{j=0}^n$ as above, let $\sigma_j(x,\xi)$ be the symbol of the operator $-\sqrt{-1}\,X_j$. For each $x \in U$, $\{\sigma_j(x,\cdot)\}$ is a linearly independent set of linear functionals on $\mathbb{R}^{n+1}$, so any function

$q:U \times \mathbb{R}^{n+1} \to \mathbb{C}$ can be written as

$$q(x,\xi) = f(x,\sigma(x,\xi)). \tag{2.6}$$

Given $m \in \mathbb{Z}$ let $S_{V,m}(U)$ denotethe space of functions q of this form, where

$$\begin{aligned} &f \in C^\infty(U \times [\mathbb{R}^{n+1}\backslash 0]), \\ &f(x,\lambda^2\sigma_0,\lambda\sigma_1,\ldots,\lambda\sigma_n) = \lambda^m f(x,\sigma), \quad \lambda > 0. \end{aligned} \tag{2.7}$$

Let $S_V^m(U)$ denote the space of symbols $q \in C^\infty(U \times \mathbb{R}^{n+1})$ which have an asymptotic expansion

$$q \sim \sum_{j=0}^{\infty} q_{m-j}, \quad q_{m-j} \in S_V^{m-j}(U), \tag{2.8}$$

where (2.8) has a meaning analogous to (1.9). Corresponding pseudodifferential operators are defined as before. Note that P of (2.4) has symbol $p \in S_V^2(U)$.

It turns out that the class of operators $\mathrm{Op}\, S_V^m(U)$ so defined depends only on the bundle V, not on the choice of a frame (X_j) nor on the choice of coordinates in U. Therefore analogous classes may be defined on manifolds. It also turns out that $\cup S_V^m(U)$ is an algebra. There is a composition result which is analogous to - but more complicated than - the classical formula (1.11): if $p \in S_V^m$ and $q \in S_V^r$ then

$$\begin{aligned} q \circ p &= q_r \# p_m + (\text{terms of lower order}) \\ &= q_r \# p_m \mod S_V^{m+r-1}. \end{aligned} \tag{2.9}$$

Here q_r and p_m are the principal symbols, and the composition $\#$ is discussed next.

Pointwise approximation and the composition of principal symbols. According to the classical pseudodifferential calculus, the principal symbol of a composition is the pointwise product of the principal symbols of the factors. When suitably interpreted, the same idea explains the composition $\#$ between principal symbols in (2.9). In the classical case: at a point x, approximate $Q = \mathrm{Op}(q)$ by the operator Q^x with symbol $q^x(y,\xi) = q(x,\xi)$. This is a euclidean convolution operator and the symbol of the composition Q^xP^x is the product Q^xP^x. Now letting x vary, we obtain the principal symbol of QP. (We are overlooking some technical points here and in the discussion which follows.) The idea in the present case is the same: approximate $Q = \mathrm{Op}(q)$ at x by a (simpler) operator Q^x and let $q \# p$ at x be the symbol (at x) of Q^xP^x. We do this by letting Q^x be the operator whose symbol (when q has the form (2.5)) is

$$q^x(y,\xi) = f(x, \sigma^x(y,\xi)) \tag{2.10}$$

where $\sigma^x = (\sigma_0^x, \sigma_1^x, \ldots, \sigma_n^x)$ is a pointwise approximation to $(\sigma_0, \ldots, \sigma_n)$. An appropriate pointwise approximation is obtained by taking

$$\begin{cases} \sigma_0^x(y,\xi) \equiv \sigma_0(x,\xi) \\ \sigma_j^x(y,\xi) \equiv \sigma_j(x,\xi) + \ell_x(y-x)\sigma_0(x,\xi), \quad j > 0, \end{cases} \tag{2.11}$$

where $\ell_x(\cdot)$ is a linear functional chosen so that

$$\begin{aligned} |\sigma_j^x(y,\xi) - \sigma_j(y,\xi)| &= 0(|x-y|^2 \, |\sigma_0(x,\xi)| \\ &\quad + \sum_{k=1}^{n} |x-y| \, |\sigma_j(x,\xi)|). \end{aligned} \tag{2.12}$$

As an example, if $n = 2$ and

$$X_0 = \frac{\partial}{\partial x_0}, \quad X_1 = \frac{\partial}{\partial x_1} + x_1 \frac{\partial}{\partial x_2} - \sin x_2 \frac{\partial}{\partial x_0},$$
$$X_2 = (1 + x_2)\frac{\partial}{\partial x_2} + (x_1^2 + x_1)\frac{\partial}{\partial x_0},$$

then the approximating vector fields at $x = 0$ are

$$X_0^0 = \frac{\partial}{\partial x_0}, \quad X_1^0 = \frac{\partial}{\partial x_1} - x_2 \frac{\partial}{\partial x_0}, \quad X_2^0 = \frac{\partial}{\partial x_2} + x_1 \frac{\partial}{\partial x_0}.$$

One may note that these are left-invariant vector fields for a certain Heisenberg group structure on $\mathbb{R}^3$. It is always the case that the approximating vector fields (X_j^x) are left-invariant with respect to an abelian or two-step nilpotent Lie group structure on the affine space $\mathbb{R}^{n+1}$, say with x as identity element. Thus Q^x is a convolution operator with respect to this group structure. (It is important to note that the isomorphism class of the group may vary from point to point.)

Parametrices and local theory. Granted the pseudodifferential calculus just described, we try to construct a parametrix Q for the operator P of (2.4), with $Q = \mathrm{Op}(q)$, $q \in S_V^{-2}(U)$. As in the classical case we can do this recursively if and only if we can solve the principal symbol equation

$$q_{-2} \# p_2 = 1. \tag{2.13}$$

Considering the meaning of the composition $\#$, this means solving

$$Q^x P^x = I \tag{2.14}$$

for each $x \in U$. Unlike the classical case this problem is not necessarily trivial (because the associated group is not necessarily abelian) but it is tractable (because the associated group is not too complicated). It turns out that (2.14) is solvable at $x \in U$ if and only if the coefficient $\lambda(x)$ avoids a certain (possibly discrete) subset of $i\mathbb{R}$.

When the parametrix exists it can be used to obtain the local regularity theory, because once again the asymptotic expansion of the symbol of Q implies an asymptotic expansion of the kernel which gives a complete description of the singularity of the kernel. For any given $x \in U$ there is a choice of coordinates in U such that at x,

$$Qu(x) = \int K(x,x-y)u(y)dy$$

where

$$K(x,z) = K_{-2}(x,z) + (\text{less singular terms}),$$
$$K_{-2}(x,\lambda^2 z_0, \lambda z_1,\ldots, \lambda z_n) = \lambda^{-n} K_{-2}(x,z).$$

For example, for $1 < p < \infty$ one obtains (when the parametrix Q exists) the regularity result, which is clearly optimal:

$$Pu \in L^p_{loc}(U) \Rightarrow u, X_j u \in L^p_{loc}(U), \quad \text{all } j;$$
$$X_j X_k u \in L^p_{loc}(U), \quad \text{all } j,k > 0.$$

Global theory: the heat equation. The operator $\frac{\partial}{\partial t} + P$ can be treated in analogy with the classical case,by making a parabolic enlargement of the class of symbols. The building blocks are symbols

$$q(x, \xi,\tau) = f(x,\sigma(x,\xi),\tau)$$

where $f \in C^\infty(U \times [R^{n+2} \ 0])$ and

$$f(x, \lambda^2\sigma_0,\lambda\sigma_1,\ldots,\lambda\sigma_n,\lambda^2\tau)$$
$$= \lambda^m f(x,\sigma,\tau), \quad \lambda \in R\ 0.$$

Moreover, f is assumed to extend holomorphically in τ, $\text{Im}\ \tau < 0$. This leads to results for the $\bar\partial_b$-complex analogous to some of the Riemannian results.

Remarks and references. As noted, the prototype of operators like (2.4) is $\Box_b$, introduced by Kohn, who obtained basic L^2 estimates. Operators like (2.4) with λ real, but considerably more general, were studied by Hormander (Acta Math. 1967) and further important results are due to Folland-Stein (Comm. Pure Appl. Math. 1974), Rothschild-Stein (Acta Math. 1976), Fefferman-Phong (Comm. Pure Appl. Math. 1981),Bolley-Camus-Helffer-Nourrigat, Comm. P.D.E. 1982), and, in the analytic category, to Tartakoff (Acta Math. 1980) and Treves (Comm. P.D.E. 1978). These results also spawned an interest in left-invariant operators on nilpotent Lie groups: see various papers of Rothschild, Helffer-Nourrigat, Corwin, Lipsman, and others. Regularity theory (L^2 or even L^p estimates) is well understood in some generality; full symbolic and kernel calculi are still open questions in most cases. The calculus described here is due to Beals-Greiner and the heat equation extension to Beals-Greiner-Stanton (J. Diff. Geom.) For a much more complete discussion and references see:

R. Beals, and P. Greiner, Calculus on Heisenberg Manifolds, Princeton Univ. Press Ann. Math. Studies, Princeton, 1987.

M. Taylor, Noncommutative microlocal analysis, Part I, Amer. Math. Soc. Memoirs no. 313, Providence 1984.

3. Boundary value problems.

In the first lecture we noted the classical Dirichlet problem

$$\Delta u = f \quad \text{in} \quad U, \quad u = g \quad \text{on} \quad bU \tag{3.1}$$

where U is a bounded domain in $\mathbb{R}^{n+1}$ with smooth boundary bU. The classical Neumann problem is

$$\Delta u = f \quad \text{in} \quad U, \quad \frac{\partial u}{\partial n} = h \quad \text{on} \quad bU. \tag{3.2}$$

Problem (3.2) arises, for example, from the deRham complex (1.5) if we take $E^j = E^j(\overline{U})$, forms smooth up to bU. Let n_0 denote the unit inward normal vector on bU. The domain of the formal adjoint d_j^* (when intersected with E^j) is

$$\text{dom } d_j^* = \{u \in E^j : n_0 \lrcorner\, u = 0 \text{ on } bU\}.$$

Similarly

$$\text{dom } \Delta_j = \{u \in E^j : n_0 \lrcorner\, u = 0,\ n_0 \lrcorner\, d_j u = 0\}.$$

In particular the functions in $\text{dom } \Delta_0$ are those with $\partial u/\partial n = 0$ on bU.

There is an analogous problem in several complex variables. Let U be a bounded domain in $\mathbb{C}^{n+1}$ with smooth boundary bU and let $E^{0,q} = E^{0,q}(\overline{U})$ denote the smooth $(0,q)$ forms

$$u = \sum_{|J|=q} u_J (d\bar{z})^J, \quad u_J \in C^\infty(\overline{U}).$$

The associated Dolbeault complex is

$$0 \to E^{0,0} \xrightarrow{\bar{\partial}_0} E^{0,1} \xrightarrow{\bar{\partial}_1} E^{0,2} \to \dots \to E^{0,n+1} \to 0. \tag{3.3}$$

Again let n_0 be the unit inward normal to bU and let n_1 be its projection to $T^{0,1}$ in the complexified tangent space:

$$n_1 = \frac{1}{2}(\bar{n}_0, i\bar{n}_0) \in T^{0,1} \subset \mathbb{C}^{2n+1}.$$

Then

$$\begin{cases} \operatorname{dom} \bar{\partial}_q^* = \{u \in E^{0,q} : n_1 \lrcorner u = 0 \text{ on } bU\}, \\ \operatorname{dom} \square_q = \{u \in E^{0,q} : n_1 \lrcorner u = 0 \text{ and } n_1 \lrcorner \bar{\partial}_q u = 0\}. \end{cases} \tag{3.4}$$

Kohn's solution to the $\bar{\partial}$ problem on $\bar{U}$,

$$\bar{\partial}_q u = v \in E^{0,q+1}, \tag{3.5}$$

assuming $\bar{\partial}_{q+1} v = 0$ and $n_1 \lrcorner v = 0$, is the solution orthogonal to $\ker(\bar{\partial}_q)$, which is the solution of

$$\square_q u = \bar{\partial}_q^* v. \tag{3.6}$$

This last is the $\bar{\partial}$-Neumann problem, and it implicitly contains the $\bar{\partial}$-Neumann boundary conditions

$$n_1 \lrcorner u = 0 \text{ on } bU, \quad n_1 \lrcorner \bar{\partial}_q u = 0 \text{ on } bU. \tag{3.7}$$

The pseudodifferential approach to boundary problems. Problems (3.1) and (3.2) are easily reduced to homogeneous versions

$$\Delta u = 0 \text{ in } U, \quad u = g \text{ on } bU; \tag{3.8}$$

$$\Delta u = 0 \text{ in } U, \quad \frac{\partial u}{\partial n} = h \text{ on } bU. \tag{3.9}$$

Suppose $u = Jg$ denotes the solution of (3.8). Then (3.9) can be reduced to

$$u = Jg, \quad Tg = h, \tag{3.10}$$

where T is a classical pseudodifferential operator (of order 1) on bU. In fact every differential boundary value problem for Δ on U is equivalent to a problem (3.10) for some choice of classical pseudodifferential T. We can say that the problem is elliptic if the associated T is elliptic. Thus (3.8), where $T = I$, and (3.9) are elliptic, as are the generalizations associated to the deRham complex. However the $\bar{\partial}$-Neumann problem is not elliptic and is, therefore, more delicate.

Here is a sketch of the construction of an approximation to J and the reduction of (3.9) to (3.10). Working locally and allowing Δ to have variable coefficients we may take $U = \mathbb{R}_+^{n+1} = \mathbb{R}^n \times \mathbb{R}_+ = \{(x,r) : r > 0\}$. Assume that the coordinates are chosen so that on bU, $-\Delta$ has principal symbol $a(x,\xi) + \rho^2$ with $a(x,\cdot)$ a positive definite quadratic form on $\mathbb{R}^n$. Let $Q = Op(q)$ be a parametrix for Δ on $\mathbb{R}^{n+1}$.

Given $f \in \mathcal{D}(\mathbb{R}^n)$, consider it as a density on bU, i.e. consider $f \otimes \delta(r) \in \mathcal{D}'(\mathbb{R}^{n+1})$. Then $u = Q(f \otimes \delta)$ is an approximate solution of $\Delta u = 0$ on $\mathbb{R}^{n+1}$ bU. On $U = \mathbb{R}^{n+1}_+$,

$$\begin{aligned} u(x,r) &= \int_{\mathbb{R}^n} e^{ix\xi} j_0'(x,r,\xi)\hat{f}(\xi)d\xi \\ &= J_{0,r}' f(x) \end{aligned} \tag{3.12}$$

where

$$j_{0,r}'(x,r,\xi) = \frac{1}{2\pi} \int_{\mathbb{R}} e^{ir\rho} q(x,r,\xi,\rho)d\rho .$$

As $r \to 0+$ the pseudodifferential operator $J_{0,r}'$ converges to a pseudodifferential operator J_0' whose principal symbol is $\frac{1}{2}a(x,\xi)^{-1/2}$. Thus to a first approximation we obtain a solution to (3.8) by $u = J_1 g$ with

$$J_1 g(x,r) = \int_{\mathbb{R}^n} e^{-ix\xi} 2j_0'(x,r,\xi) \sqrt{a(x,\xi)}\ \hat{g}(\xi)d\xi. \tag{3.13}$$

Corrections can be made recursively, leading to a full Poisson operator J,

$$Jg(x,r) = \int_{\mathbb{R}^n} e^{-ix\xi} j(x,r,\xi)\hat{g}(\xi)d\xi \tag{3.13}$$

where

$$\begin{cases} j(x,r,\xi) \sim \sum_{k=0}^{\infty} j_k(x,r,\xi), \\ j_0(x,r,\xi) = e^{-r\sqrt{a(x,\xi)}} \end{cases} \tag{3.14}$$

and j_k is a finite sum of terms of the form

$$r^m b(x,\xi)e^{-r\sqrt{a(x,\xi)}}, \quad b \in S_{m-k}(\mathbb{R}^n).$$

Therefore

$$\begin{aligned} \frac{\partial}{\partial r} Jg(x,0) &= \int_{\mathbb{R}^n} e^{-ix\cdot\xi} \frac{\partial}{\partial r} j(x,0,\xi)\hat{g}(\xi)d\xi \\ &\equiv Tg(x) \end{aligned}$$

where T is a pseudodifferential operator with principal symbol $-\sqrt{a(x,\xi)}$.

<u>Application to the $\bar{\partial}$-Neumann problem</u>. The operator $\square_q$ in $\mathbb{C}^{n+1}$ is scalar; in fact it is $-\frac{1}{4}\Delta$ where Δ is the Laplacian in $\mathbb{R}^{2n+2} = \mathbb{C}^{n+1}$. Modulo solving the (form-valued) Dirichlet problem (3.1), the $\bar{\partial}$-Neumann problem (3.6) reduces easily to a problem of the form

$$\Delta u = 0 \text{ in } U, \quad B_{\bar\partial} u = h \text{ on } bU \tag{3.15}$$

where $B_{\bar\partial} u = (n_1 \lrcorner \bar\partial_q u)|_{bU}$. As above this becomes

$$u = Jg, \quad \Box_+ g = h \tag{3.16}$$

where $\Box_+$ is the classical pseudodifferential operator of order 1,

$$\Box_+ = B_{\bar\partial} J. \tag{3.17}$$

Now $\Box_+$ is not elliptic, so it is not easily inverted. It can be shown that there is a second classical pseudodifferential operator $\Box_-$ of order 1 such that

$$\Box_- \Box_+ = \Box_b + \text{(smaller terms)} \tag{3.18}$$

where $\Box_b = \Box_{b,q}$ is the operator associated with the $\bar\partial_b$-complex on bU. Therefore when $\Box_b$ has a parametrix, we can also expect $\Box_- \Box_+$ to have a parametrix Q. If so, then $Q\Box_-$ is a parametrix for $\Box_+$ and

$$u = JQ\Box_- h \tag{3.19}$$

gives an approximate solution of (3.15).

Actually one can find a parametrix for $\Box_+$ under more general conditions than those which allow a parametrix for $\Box_b$. The appropriate condition is known as Condition $Z(q)$. Let $r\colon \mathbb{C}^{n+1} \to \mathbb{R}$ be smooth with $dr \neq 0$ on bU and $U = \{r < 0\}$. Condition $Z(q)$ says that at each point p of bU the restriction to $\{z \in \mathbb{C}^{n+1} : z \cdot dr(p) = 0\}$ of the hermitian form with matrix

$$\left(\frac{\partial^2 r}{\partial z_j \partial \bar z_k} \right)_{j,k=1}^{n+1}$$

has at least $n+1-q$ positive eigenvalues or at least $q+1$ negative eigenvalues. (U is said to be strictly pseudoconvex if this form is positive definite on the subspace above.) When Condition $Z(q)$ holds, $\Box_+$ has a left parametrix which belongs to the algebra of pseudodifferential operators generated by the classical operators and those discussed in the previous lecture.

Once again, the existence of approximate solution operators in well defined classes of operators allows one to prove local and global regularity results, as well as asymptotic and geometric results connected with the heat operator $\frac{\partial}{\partial t} + \Box_q$. For example one has

$$u \in \operatorname{dom} \Box_q \Rightarrow \|u\|_{L_1^p(U)} \le C_p \|\Box_q u\|_{L^p(U)} \tag{3.20}$$

for $1 < p < \infty$ and $1 \leq q \leq n$ when U is strictly pseudoconvex. Thus the solution operator N_q for $\square_q$ extends to map $L^p(U)$ to the Sobolev space $L_1^p(U) = W^{1,p}(U)$. Similar estimates hold in Lipschitz spaces.

As obtained by the procedure described here, the solution operator N_q is a sum of compositions of operators, some of whose kernels have an isotropic homogeneity like $K(x,\lambda z) = \lambda^{-m}K(x,z)$ and some have non-isotropic homogeneities like $K(x,\lambda^2 z_0,\lambda z') = \lambda^{-m}K(x,z_0,z')$. A close analysis shows that the kernels of the composed operators can themselves be written as sums of products having asymptotically such homogeneities. This fact gives another approach to proving estimates.

Finally, we note that Condition $Z(q)$ implies the existence of an asymptotic expansion as $t \to 0+$,

$$\operatorname{tr}(e^{-t\square_q}) \sim t^{-n-1}\left[c_0 + \sum_{j=1}^{\infty} (c_j + c_j' \log t)t^{\frac{1}{2}j}\right] \tag{3.21}$$

where the c_j are the integrals of local geometric invariants. As in the classical case, (3.21) and the Karamata Tauberian theorem give an estimate for $N(\lambda)$, the number of eigenvalues of $\square_q$ which are $\leq \lambda$:

$$N(\lambda) \sim \tilde{c}_0 \lambda^{n+1} \quad \text{as } \lambda \to +\infty. \tag{3.22}$$

Unlike the classical case the constants c_0 and $\tilde{c}_0$ involve the boundary bU.

Remarks and references. The $\bar{\partial}$-Neumann problem was formulated by Spencer. The basic L^2-estimates were proved by Kohn (ann. Math. 1963, 1964) in the strictly pseudoconvex case by Hörmander (Acta Math. 1965) under Condition $Z(q)$. The reduction of boundary problems to pseudodifferential problems on the boundary is due Calderón; its use in the study of the $\bar{\partial}$-Neumann problem is due to Greiner-Stein, who obtained sharp regularity results for $(0,1)$ forms on strictly pseudo-convex domains. For $(0,q)$ forms under the weaker Condition $Z(q)$ these results have been proved by Beals-Greiner-Stanton (preprint). Phong (Proc. Nat. Acad. Sci. USA 1979) obtained a representation of the solution operator N_q as a sum of products for the special case of the Siegel domain. The asymptotic expansion (3.21) and its geometric interpretation are due to Beals-Stanton (preprints). The eigenvalue estimate (3.22) was first obtained by Metivier (Duke Math. J. 1981) for domains in $\mathbb{C}^{n+1}$.

For the $\bar{\partial}$-Neumann problem and the Greiner-Stein results, and other questions, see the survey article M. Beals-Fefferman-Grossman (Bull. Amer. Math. Soc. 1983) and:

G. B. Folland and J. J. Kohn, The $\bar{\partial}$-Neumann Problem for the Cauchy-Riemann Complex, Princeton Univ. Press Ann. Math. Studies, Princeton, 1972

P. C. Greiner and E. M. Stein, Estimates and Existence Theorems for the $\bar{\partial}$-Neumann Problem, Princeton Univ. Press Mathematical Notes, Princeton, 1977.

The techniques above also give regularity results for the $\bar{\partial}$-problem (3.5); see Greiner-Stein and Beals-Greiner-Stanton. For a different approach to the $\bar{\partial}$ and $\bar{\partial}$-Neumann problem and estimates based on integral representations such as those of Henkin and Ramirez, see Lieb-Range (Annals Math. 1986) and the references there.

SMOOTHNESS OF SHOCK FRONT SOLUTIONS FOR SYSTEMS OF CONSERVATION LAWS

Chen Shuxing
Dept. of Math., Fudan Univ.,
Shanghai, China.

§1. Introduction.

There have been many works on analyzing singularities of solutions of nonlinear partial differential equations recently, and the technique of paradifferential operators,developed by J.M. Bony, has been borne out as a general and powerful tool to deal with nonlinear problems (see[1]). For a nonlinear equation of m-th order in n dimensional space, if u is an H^s solution with $s>\frac{n}{2}+m+2$, then the singularities of u with strength stronger than $(2s-\frac{n}{2}-m-1)$–th order propagate according to the rule of propagation of singularities in the linear case. However, for weaker singularities, the situation is more complicated because of the appearance of interaction. In the latter case, one describes singularities of solutions of nonlinear equations not only by wave front sets, but also by conormal distributions. Since the surface,depending on which singularities of conormal type are described, is just a weakly discontinuous surface, so the corresponding rule on propagation of singularities clearly shows how arise and propagate the weakly discontinuous surfaces of a solution as progressing waves, see e.q.[2-5]. Recently, S.Alinhac introduced paracomposition in [6] to prove a theorem on the evolution of a simple progressing wave for general nonlinear equations. Certainly, this progressing wave is also a weakly discontinuous surface.

In this paper we are going to use paradifferential operators to study propagation of a strongly discontinuous surface of solutions for a nonlinear system. For a nonlinear system of conservation laws, if there exists a solution with shock front, with it and the solution on both sides having some given smoothness, and if the shock front and the initial data on both sides are in C^∞, then in the determinate region of the initial data, the shock front and the solution on both sides are still in C^∞. Besides paradifferential operators, in our proof we mainly use the technique of the energy integral introduced by A.Majda, which is a developement of Kreiss's energy method on general boundary value problems for hyperbolic equations in [10].

The main result in the paper is described in §2, Paraproducts and paradifferential operators with parameter η are studied in §3 and §4; we believe that they are also useful in studying other problems for nonlinear hyperbolic equations. In §5 the system of conservation laws and the boundary condition on the shock front are paralinearized. Finally, the energy estimates are established in §6 and the proof of the main theorem is completed there.

§2. Main result.

Let N, M be integers, $N_1=N+1$, Ω be a domain in Euclidean space R^N with variables $(t,x_1,\dots,x_N)$, $\Omega \subseteq \{t>-T_o\}$, $T_o>0$. In Ω a system of conservation laws

$$\frac{\partial u}{\partial t} + \sum_{j=1}^{N} \frac{\partial}{\partial x_j} F_j(u) = 0 \tag{2.1}$$

is given, where $u={}^t(u_1,\dots,u_M)$, $F_j(u)$ for each j is a C^∞ function of u, $\frac{\partial F_j}{\partial u}=A_j(u)$ is an M×M matrix. A surface S in Ω is given with equation $x_1=\phi(t,x')$, where $x'=(x_2,\dots,x_N)$. We take $\Omega^+=\{(t,x)\in\Omega,\ x_1\geqq\phi(t,x')\}$, $\Omega^-=\{(t,x)\in\Omega,\ x_1\leqq\phi(t,x)\}$, and denote the restriction of u on $\Omega^\pm$ by $u^\pm$.

A function u is called a solution of system (2.1) with S as its strongly discontinuous surface, if:

1) $u^\pm \in H^s(\Omega^\pm)$, $\phi\in H^{s+1}$ with $s>\frac{N_1}{2}$.

2) $u^\pm$ satisfy system (2.1) on $\Omega^\pm$ respectively.

3) $$n_t(u^+-u^-)+\sum_{j=1}^{N} n_j(F_j(u^+)-F_j(u^-))=0, \text{ on } S \tag{2.2}$$

where $n=(n_t,n_1,\dots,n_N)$ is the normal direction to S.

4) The uniform stable condition for shock fronts is satisfied.

The detailed explanation of the uniform stable condition for shock front can be found in [8]; which is satisfied for each system appearing in gas dynamics. Here we write down the explicit form as follows.

By a coordinate transformation χ: $(\bar t,\bar x_1,\bar x')\to(t,x_1,x')$:

$$\begin{aligned}\bar x_1 &= x_1-\phi(t,x'),\\ \bar x' &= x',\quad \bar t = t,\end{aligned} \tag{2.3}$$

(2.1), (2.2) can be written as

$$\frac{\partial u}{\partial \bar t} + \sum_{j=2}^{N} A_j(u)\frac{\partial u}{\partial \bar x_j} + (A_1(u)-\phi_t I-\sum_{j=2}^{N}\phi_{x_j}A_j(u))\frac{\partial u}{\partial \bar x_1}=0, \tag{2.4}$$

$$\phi_t(u^+-u^-)-(F_1(u^+)-F_1(u^-))+\sum_{j=2}^{N}\phi_{x_j}(F_j(u^+)-F_j(u^-))=0, \tag{2.5}$$

where (2.4) can be divided into two systems for u^+ and u^- in Ω^+ and Ω^- respectivily. For the second system, after changing variable $\bar x_1$ to $-\bar x_1$ and freezing coefficients we may derive a system for $v={}^t(u^+,u^-)$ with constant coefficients:

$$\frac{\partial\tilde{v}}{\partial t}+\sum_{j=2}^{N}\begin{pmatrix}A_j(u^+) & \\ & A_j(u^-)\end{pmatrix}\frac{\partial\tilde{v}}{\partial\bar{x}_j}$$

$$+\begin{pmatrix}A_1(u^+)-\phi_t I-\sum_{j=2}^{N}\phi_{x_j}A_j(u^+) & \\ & -A_1(u^-)+\phi_t I+\sum_{j=2}^{N}\phi_{x_j}A_j(u^-)\end{pmatrix}\frac{\partial\tilde{v}}{\partial\bar{x}_1}=0. \qquad (2.6)$$

solutions of (2.6) have the form

$$\tilde{v}=\sum_{\ell} e^{i\xi'\bar{x}'+\kappa_\ell\bar{x}_1+\tau\bar{t}}\, p_\ell(\bar{x}_1)V_\ell, \qquad (2.7)$$

where p_ℓ are polynomials of variables $\bar{x}_1,\xi'\in R^{N-1}$, $\tau\in\mathbb{C}$, κ_ℓ are the roots of the characteristic equation of (2.6). The vectors in (2.7) corresponding to Re $\kappa_\ell<0$ form a subspace $\tilde{E}^+(\tau,\xi')$; then the uniform stable condition for a shock front can be written as: there is $\gamma>0$, such that

$$\inf_{\substack{\mathrm{Re}\,\tau\geq 0\\ \tau^2+|\xi'|^2=1}} |(b_\circ\tau+\sum_{j=2}^{N} b_j(i\xi'\ \lambda+M\tilde{v}\ |^2\geq\gamma^2(|\tilde{v}|^2+\lambda^2)\ , \qquad (2.8)$$

for any $\tilde{v}\in\tilde{E}^+(\tau,\xi')$, $\lambda\in R$, where b_0,b_j and M are determined by the coefficients of (2.4) (2.5), i.e.,

$$\begin{aligned} b_0 &= u^+-u^-,\\ b_j &= F_j(u^+)-F_j(u^-),\ j=2,\dots,N, \qquad (2.9)\\ M(v^+,v^-) &= -(A_1(u^+)-\phi_t I-\sum_{j=2}^{N}\phi_{x_j}A_j(u^+))v^+ +(A_1(u^-)\\ &\quad -\phi_t I-\sum_{j=2}^{N}\phi_{x_j}A_j(u^-))v^- \end{aligned}$$

Let $\Sigma=\bar{\Omega}\cap\{t=-T_0\}$, Σ_1 be a space-like surface in Ω, $\bar{\Sigma}_1\cap\partial\Omega=\bar{\Sigma}_1\cap\Sigma$, $\tilde{\Omega}$ be the domain in between Σ_1 and Σ, $\tilde{\Omega}^{\pm}=\Omega^{\pm}\cap\tilde{\Omega}$. The result in this paper is

Theorem: Suppose that u is a solution of system (2.1) with S as its strongly discontinuous surface with $s>\frac{N_1}{2}+3$, and that S is a C^∞ surface in $\tilde{\Omega}\cap\{t<0\}$, $u^{\pm}\in C^\infty(\tilde{\Omega}^{\pm}\cap\{t<0\})$. Then in $\tilde{\Omega}\cap\{t>0\}$ S is still of type C^∞, and $u^{\pm}$ are C^∞ smooth up to the boundary S.

§3. Paraproducts with parameter η.

In this section we study a class of paraproducts with parameter, which exhibit homogeneity with respect to the dual variables ξ of space R^n_x and the parameter called η briefly.

Taking function $\phi(\xi,\eta) \in C_o^\infty(R^n\times R)$, satisfying supp $\phi \subset \{k^{-1} < |(\xi,\eta)| < 2k\}$, $\phi_j(\xi,\eta) = \phi(2^{-j}\xi, 2^{-j}\eta)$, we have supp $\phi_j \subset \{2^j k^{-1} < |(\xi,\eta)| < 2^{j+1}k\}$. Without loss of generality we may assume $\Sigma\phi_j \equiv 1$. Denoting $\psi(\xi,\eta) = \sum_{-\infty}^{-1} \phi_j(\xi,\eta)$, we introduce for each $u \in \mathcal{S}'(R^n)$

$$\begin{aligned} u_{j,\eta} &= F^{-1}_{\xi\to x}(\phi_j(\xi,\eta)\hat{u}(\xi)), \qquad j \geq 0 \\ u_{-1,\eta} &= F^{-1}_{\xi\to x}(\psi(\xi,\eta)\hat{u}(\xi)). \end{aligned} \tag{3.1}$$

Here $F^{-1}_{\xi\to x}$ is the inverse Fourier transform operator.

For $s \geq 0$, $\eta > 0$, denoting H^s norm by $\| \ \|_s$, and

$$\|u\|^2_{s(\eta)} = \eta^{2s}\|u\|^2_o + \|u\|^2_s, \tag{3.2}$$

We have

Lemma 3.1. If $u \in H^s$, then

$$\|u_{j,\eta}\|^2_o \leq 2^{-2js}(C^2_{j\eta} + \eta^{2s}C'^2_{j\eta}), \tag{3.3}$$

where

$$\Sigma C^2_{j\eta} \leq \|u\|^2_s, \qquad \Sigma C'^2_{j\eta} \leq \|u\|^2_o. \tag{3.4}$$

Conversely, if $u = \Sigma u_{j,\eta}$, supp $\hat{u}_{j,\eta} \subset \{2^j k^{-1} < |(\xi,\eta)| < 2^{j+1}k\}$, $u_{j,\eta}$ satisfy (3.3) with $\{c_{j\eta}\}$, $\{c'_{j\eta}\} \in \ell_2$, then $u \in H^s$, and

$$\|u\|^2_{s(\eta)} \leq C(\|\{c_{j\eta}\}\|^2_{\ell_2} + \eta^{2s}\|\{c'_{j\eta}\}\|^2_{\ell_2}). \tag{3.5}$$

Proof. If $u \in H^s$, then

$$\begin{aligned} \|u_{j,\eta}\|^2_o &= \|\phi_j(\xi,\eta)\hat{u}(\xi)\|^2_o \\ &\leq C2^{-2js}\| |(\xi,\eta)|^s \phi_j(\xi,\eta)\hat{u}(\xi)\|^2_o \\ &\leq C2^{-2js}(\eta^{2s}\|\phi_j\hat{u}\|^2_o + \|\phi_j|\xi|^s\hat{u}(\xi)\|^2_o) \\ &\leq C2^{-2js}(\eta^{2s}(c'_{j\eta})^2 + c^2_{j\eta}). \end{aligned}$$

Since $\phi_j = 1$, we have (3.4).

Conversely, if u has decomposition $\Sigma u_{j,\eta}$, then according to the property of supp $\hat{u}_{j,\eta}$, we have

$$\|u_{j,\eta}\|^2_{s(\eta)} \leq C2^{2js}\|u_{j,\eta}\|^2_o \leq C(c^2_{j\eta} + \eta^{2s}(c'_{j\eta})^2).$$

Hence, $\{c_{j\eta}\}$, $\{c'_{j\eta}\}$ and $\|u\|^2_{s(\eta)} \leq C \sum_j \|u_{j,\eta}\|^2_{s(\eta)}$ imply (3.5).

Lemma 3.2 For any given function $a \in H^r (r > \frac{n}{2})$ with compact support, we can define an operator T'_a :

$$T'_a u = \sum_{-1 \leq p \leq q-N_0} a_p u_{q,\eta} \tag{3.6}$$

which is a map from H^s to H^s, keeping $\|\cdot\|_{s(\eta)}$ bounded. Using $\|\cdot\|_{s(\eta)}$ as the norm in H^s space, the norm of the operator T'_a is dominated by $C\|a\|_r$, where C is independent of η. Moreover, for different choice of N_0, the difference for T'_a is

$$\|\Delta T'_a u\|_{(s+r-\frac{n}{2}-\varepsilon)(\eta)} \leq C\|a\|_r \|u\|_{s(\eta)} . \tag{3.7}$$

where $\varepsilon > 0$ is arbitrarily small.

Proof:

$$T'_a u = \sum_q \Big(\sum_{-1 \leq p \leq q-N_0} a_p \Big) u_{q,\eta}$$

$$= \sum_q f_{q,\eta},$$

$$\|f_{q,\eta}\|_0 \leq C \sum_p |a_p| \|u_{q,\eta}\|_0$$

$$\leq C \sum_p \|a_p\|_r \|u_{q,\eta}\|_0$$

$$C \|a\|_r \|u_{q,\eta}\|_0 .$$

In view of lemma 3.1, $\|u_{q,\eta}\|_0 \leq 2^{-qs}(C_{q\eta} + \eta^s C'_{q\eta})$ with $\{c_{q\eta}\}$, $\{c'_{q\eta}\}$ in ℓ^2, therefore, there are such estimates for $\|f_{q,\eta}\|_0$. Besides, the support of $\hat{f}_{q,\eta}$ satisfies the requirement in Lemma 3.1, if N_0 is large enough. This yields

$$\|T'_a u\|_{s(\eta)} \leq C\|a\|_r \|u\|_{s(\eta)}. \tag{3.8}$$

Furthermore, when N_0 is replaced by N_0+1, we have the estimate for the difference $\Delta f_{q,}$,

$$\|\Delta f_{q,\eta}\| \leq C|a_{q-N_0+1}| \|u_{q,\eta}\|_0$$

$$\leq C\|a_{q-N_0+1}\|_{\frac{n}{2}+\varepsilon} \|u_{q,\eta}\|_0$$

$$\leq C \cdot 2^{-(q-N_0+1)(r-\frac{n}{2}-\varepsilon)} \|a_{q-N_0+1}\|_r \|u_{q,\eta}\|_0$$

$$\leq C \cdot 2^{-q(r-\frac{n}{2}-\varepsilon)} \|a\|_r \|u_{q,\eta}\|_0$$

Combining assumption on the support, we obtain (3.7).

Assume $\chi(\theta,\zeta,\eta)$ is a C^∞ function on $(R^n \times R^n \times R) \setminus 0$, homogeneous of degree 0, and for $0<\varepsilon_1<\varepsilon_2$

$$\chi(\theta,\zeta,\eta)=1\ , \qquad \text{if} \quad |\theta| \leq \varepsilon_1|(\zeta,\eta)|\ ,$$

$$\chi(\theta,\zeta,\eta)=0\ , \qquad \text{if} \quad |\theta| > \varepsilon_2|(\zeta,\eta)\ .$$

where $|(\zeta,\eta)|=(|\zeta|^2+\eta^2)^{\frac{1}{2}}$, then for $a\in H^r(r>\frac{n}{2})$, we can define T_a as

$$(T_a u)\hat{}(\xi)=\int\chi(\xi-\zeta,\zeta,\eta)\hat{a}(\xi-\zeta)\hat{u}(\zeta)d\zeta, \tag{3.9}$$

which has the following properties.

<u>Lemma 3.3</u>. For any $\varepsilon>0$, the difference $T_a-T'_a$ is an $(r-\frac{n}{2}-\varepsilon)$-regular operator; that means, for any $s\geqq 0$, and $u\varepsilon H^s$,

$$\|(T_a-T'_a)u\|_{s+r-\frac{n}{2}-\varepsilon} \leqq C\|u\|_{s(\eta)}. \tag{3.10}$$

Moreover, a different choice of χ in (3.9) only causes a difference of an $(r-\frac{n}{2}-\varepsilon)$-regular operator.

<u>Proof</u>. Obviously,

$$(T_a u)\hat{}(\xi) = \sum_{p,q}\int\chi(\xi-\zeta,\zeta,\eta)\hat{a}_p(\xi-\zeta)\hat{u}_{q,\eta}(\zeta)d\zeta.$$

Since $2^q k^{-1} \leqq |(\zeta,\eta)|\leqq k2^{q+1}$ on supp $\hat{u}_{q,\eta}$, and $2^p k^{-1}\leqq|\xi-\zeta|\leqq k2^{p+1}$ on supp $\hat{a}_p$, then there are integers $0<N_1<N_2$, such that $\chi=1$ if $p\leqq q-N_2$, and $\chi=0$ if $p>q-N_1$. Therefore,

$$(T_a-T'_a)u(x)=\sum_{q-N_2<p<q-N_1}\int e^{ix(\theta+\zeta)}\chi(\theta,\zeta,\eta)\hat{a}_p(\theta)\hat{u}_{q,\eta}(\zeta)d\zeta d\theta.$$

Choose $x_1(\theta,\zeta,\eta)$ as a C^∞ function, equal to χ on $2^{-N_2}k^{-1}\leqq|(\theta,\zeta,\eta)|\leqq 4k$ and supported on $|(\theta,\zeta,\eta)|\leqq 8k$. Using the homogeneity of $\chi(\theta,\zeta,\eta)$, we have

$$(T_a-T'_a)u(x)=\sum_{q-N_2<p<q-N_1}\int e^{ix(\theta+\zeta)}\chi(2^{-q}\theta,2^{-q}\zeta,2^{-q}\eta)\hat{a}_p(\theta)\hat{u}_{q,\eta}(\zeta)d\zeta d\theta.$$

Noting that on the support of $\hat{a}_p(\theta)\hat{u}_{q,\eta}(\zeta)$,

$$2^q k^{-1}\leqq|(\zeta,\eta)|\leqq 2^{q+1}k,$$

$$2^{-N_2}2^q k^{-1}\leqq k^{-1}2^p\leqq|\theta|\leqq 2k\cdot 2^p\leqq 2k\cdot 2^q\ ,$$

which imply

$$2^{-N_2}k^{-1}\leqq|2^{-q}(\theta,\zeta,\eta)|\leqq 4k,$$

we can assert $(T_a-T'_a)u$ unchanged if χ is replaced by χ_1. Obviously, the inverse Fourier transform $h(s,t,\eta)$ of $\chi_1(\theta,\zeta,\eta)$ does not vanish, if only η is in a fixed compact set. Now we write (3.11) as follows:

$$(T_a-T'_a)u(x)=\sum_{q-N_2<p<q-N_1}\int h(s,t,2^{-q}\eta)a_p(x-2^{-q}s)u_{q,\eta}(x-2^{-q}t)dsdt = \sum_q f_q,$$

$$\|f_q\|_o \leq C \sum_{q-N_2<p<q-N_1} |a_p| \|u_{q,\eta}\|_o \int |h(s,t,2^{-q}\eta)| dsdt.$$

Since $2^{-q}\eta$ is in a compact set independent of q, then the integral is dominated by a constant independent of η. Therefore, in the same way as that in Lemma 2, we can prove that $T_a - T'_a$ is an $(r-\frac{n}{2}-\varepsilon)$-regular operator with $\varepsilon>0$, and

$$\|(T_a - T'_a)u\|_{s+r-\frac{n}{2}-\varepsilon} \leq C\|a\|_r \cdot \|u\|_{s(\eta)}.$$

The second part of the lemma can be easily derived from the first part.

As in [1], for paraproducts with parameter η , the following propositions hold.

Lemma 3.4 Let a and b be H^r functions with compact support, $\frac{n}{2}+1 \geq r > \frac{n}{2}$, then for any $\varepsilon>0$, $T_{ab} - T_a T_b$ is an $(r-\frac{n}{2}-\varepsilon)$-regular operator, and the norm of the difference operator is dominated by $C\|a\|_r \|b\|_r$ Moreover, if $r>\frac{n}{2}+1$, then the difference operator is a 1 - regular operator.

Lemma 3.5 Let a be a function as in lemma 3.4, $\frac{n}{2}+1 \geq r > \frac{n}{2}$, then the adjoint operator T^*_a is a map from H^s to H^s, $T^*_a - T_a$ is an $(r-\frac{n}{2}-\varepsilon)$-regular operator for any $\varepsilon>0$, with norm dominated by $C\|a\|_r$. Moreover, if $r>\frac{n}{2}+1$, then the difference operator is a 1-regular operator.

Lemma 3.6 Assume $a \in H^r (r>\frac{n}{2})$, $u \in H^s$, with compact support, then for $s \geq 0$ and $\varepsilon>0$,

$$a u = T_a u + R \tag{3.12}$$

with

$$\|R\|_{(s+r-\frac{n}{2}-\varepsilon)(\eta)} \leq C\|a\|_r \|u\|_{s(\eta)}, \tag{3.13}$$

The proofs of lemma 3.4 to 3.6 are similar to those of the corresponding propositions in [1]; we omit them here.

Lemma 3.7 Assume $a \in H^r$, $r>\frac{n}{2}+k$ has compact support, $u \in H^s$, $j(\xi,\eta) \in C^\infty$. If J_ε is the operator with $j(\varepsilon\xi,\varepsilon\eta)$ as Fourier multiplier, then

$$[T_a, J_\varepsilon]u = \sum_{0<|\alpha|<k} \frac{\varepsilon^{(\alpha)}}{\alpha!} T_{D_x^\alpha a} J_\varepsilon^\alpha u + R_k(\varepsilon)u, \tag{3.14}$$

where $R_k(\varepsilon)$ satisfies

$$\|R_k(\varepsilon)u\|_{s(\eta)} \leq C\varepsilon^k \|u\|_{s(\eta)}. \tag{3.15}$$

Proof. Denoting the Littlewood's decompositions of a, u by $\{a_p\}, \{u_{q,\eta}\}$, according to (3.9) we have

$$(T_a u)\hat{}(\xi) = \sum_{p,q} \int \chi(\xi-\zeta,\zeta,\eta) \hat{a}_p(\xi-\zeta) \hat{u}_{q,\eta}(\zeta) d\zeta,$$

$$[J_\varepsilon, T_a]u(\xi) = \sum_{p,q} \int (j(\varepsilon\xi,\varepsilon\eta) - j(\varepsilon\zeta,\varepsilon\eta))\chi(\xi-\zeta,\zeta,\eta)\hat{a}_p(\xi-\zeta)\hat{u}_{q,\eta}(\zeta)d\zeta$$

$$= \sum_{p,q}\int \sum_{0<|\alpha|<k} (\xi-\zeta)^\alpha \frac{\varepsilon^{(\alpha)}}{\alpha!} \partial_\xi^\alpha j(\varepsilon\zeta,\varepsilon\eta)\chi(\xi-\zeta,\zeta,\eta)\hat{a}_p(\xi-\zeta)\hat{u}_{q,\eta}(\zeta)d\zeta$$

$$+ \sum_{p,q}\int r_{p,q,k}(\xi,\zeta,\varepsilon)\ (\xi-\zeta,\zeta,\eta)\hat{a}_p(\xi-\zeta)\hat{u}_{q,\eta}(\zeta)d\zeta, \tag{3.16}$$

where

$$r_{p,q,k}(\xi,\zeta,\varepsilon) = k\int_0^1 \sum_{|\alpha|=k} \frac{(\xi-\zeta)^\alpha \varepsilon^k}{\alpha!} \partial_\xi^\alpha j(\nu\varepsilon\xi+(1-\nu)\varepsilon\xi,\varepsilon\eta)(1-\nu)^{k-1}d\nu. \tag{3.17}$$

Obviously, taking the inverse Fourier transform of the first term on the right-hand side of (3.14). Next we consider the remainder in (3.16), and write

$$R_{p,q,k}(\varepsilon)u = \iint \chi(\xi-\zeta,\zeta,\eta)r_{p,q,k}(\xi,\zeta,\varepsilon)\hat{a}_p(\xi-\zeta)\hat{u}_{q,\eta}(\tau)e^{i\xi x}d\zeta d\xi,$$

$$R_k(\varepsilon)u = \sum_q R_{q,k}(\varepsilon)u = \sum_q(\sum_p R_{p,q,k}(\varepsilon)u).$$

In the expression of $R_{p,q,k}(\varepsilon)u$, the integrand vanishes, when $|(\zeta,\eta)|$ is not in $[\frac{2^q}{3}, 3\cdot 2^{q+1}]$; then by the property of χ, the support of $\widehat{R_{q,k}(\varepsilon)u}$ is in

$$\frac{2^q}{4} \leqq |(\xi,\eta)| \leqq 4\cdot 2^{q+1}\ .$$

Therefore, in order to obtain (3.15), we only need to estimate $\|R_{q,k}(\varepsilon)u\|_0$. By (3.17)

$$\|\widehat{R_{p,q,k}(\varepsilon)u}\|_0 = k\varepsilon^k(\int|\int \sum_{|\alpha|=k} \frac{1}{\alpha!}\int_0^1 j^{(\alpha)}(\varepsilon\zeta+t\varepsilon(\xi-\zeta),\varepsilon\eta)$$

$$(1-t)^{k-1}dt\cdot\chi(\xi-\zeta,\zeta,\eta)\widehat{D^\alpha a_p}(\xi-\zeta)\hat{u}_{q,\eta}(\zeta)d\zeta|^2 d\xi)^{\frac{1}{2}}$$

$$\leqq C\varepsilon^k \sum_{|\alpha|=k} \|\widehat{D^\alpha a_p}\|_{L^1} \|\hat{u}_{q,\eta}\|_0$$

$$\leqq C\varepsilon^k \|a_p\|_{k+\frac{n}{2}+\delta} \|\hat{u}_{q,\eta}\|_0\ ,$$

where $\delta>0$ is arbitrarily small, so it can be chosen, such that $k+\frac{n}{2}+2\delta<r$. Hence

$$\|a_p\|_{k+\frac{n}{2}+\delta} \leqq C_p\|a\|_r\cdot 2^{-p\delta}\ ,$$

$$\|u_{q,\eta}\|_0 \leqq C'_q\|u\|_{s(\eta)} 2^{-qs}\ ,$$

where Σc_p^2, $\Sigma c'^2_q$ are independent of a and u. Thus

$$\|R_{p,q,k}(\varepsilon)u\|_0 = \|\widehat{R_{p,q,k}(\varepsilon)u}\|_0$$

$$\leqq C\ c_p\ c'_q\|a\|_r\|u\|_{s(\eta)} 2^{-p\delta-qs}\varepsilon^k\ .$$

Summing up with respect to p, we have

$$\|R_{q,k}(\varepsilon)u\|_0 \leqq C\, c'_q \varepsilon^{k_2-qs}\|a\|_r\|u\|_{s(\eta)} ,$$

whence (3.15).

In what follows we denote

$$\|u\|^2_{s(\eta),\delta} = \int |\hat{u}(\xi)|^2(1+|\xi|^2+\eta^2)^{s+1}(1+\delta^2|\xi|^2+\delta^2\eta^2)^{-1}d\xi, \qquad (3.18)$$

and assume that the C^∞ function $j(\xi,\eta)$ satisfies:

for some $\nu>0$, $j=0(|(\xi,\eta)|^\nu)$, when $(\xi,\eta)\to 0$, (3.19)

$j(t\varepsilon,t\eta)=0$ for any real t implies $(\xi,\eta)=0$.

$j(t\xi,t\eta)$ is analytic with respect to t. (3.20)

Then we have

Lemma 3.8 If (3.19), (3.20) hold for $\nu>s-s_1+1$, then there exist constants C_1, C_2, independent of u,δ, such that

$$C_1\|u\|^2_{s(\eta),\delta} \leqq \int_o^1 \|J_\varepsilon u\|^2_{s_1(\eta)}\varepsilon^{-2(s-s_1+1)}(1+\frac{\delta^2}{\varepsilon^2})^{-1}\frac{d\varepsilon}{\varepsilon} + \|u\|^2_{s(\eta)}$$

$$\leqq C_2\|u\|^2_{s(\eta),\delta}. \qquad (3.21)$$

The proof is similar to that of Theorem 2.4.1 of [14], so we omit it here.

Lemma 3.9 Assume a, J_ε are given as in Lemma 3.7, and the corresponding $j(\varepsilon,\eta)$ satisfies (3.19), (3.20) with $\nu>s-s_1+1$ $(s_1\leqq s-1)$. Then there exists a constant C_3 such that

$$\int_0^1 \|[T_a\, J_\varepsilon]u\|^2_{s_1(\eta)}\ \varepsilon^{-2(s-s_1+1)}(1+\frac{\delta^2}{\varepsilon^2})^{-1}\ \frac{d\varepsilon}{\varepsilon}$$

$$\leqq C_3\|u\|^2_{(s-1)(\eta),\delta}. \qquad (3.22)$$

Proof. By lemma 3.7 we have the expression (3.14) of $[T_a,J_\varepsilon]u$. In view of (3.15) and (3.21),

$$\int_0^1 \|R_k(\varepsilon)u\|^2_{s_1(\eta)}\ \varepsilon^{-2(s-s_1+1)}(1+\frac{\delta^2}{\varepsilon^2})^{-1}\ \frac{d\varepsilon}{\varepsilon}$$

$$\leqq C\int_0^1 \|u\|^2_{s_1(\eta)}\ \varepsilon^{2k}\varepsilon^{-2(s-s_1+1)}(1+\frac{\delta^2}{\varepsilon^2})^{-1}\ \frac{d\varepsilon}{\varepsilon}$$

$$\leqq C\ \|u\|^2_{(s-1)(\eta),\delta},$$

when $k\geqq s-s_1+1$. Besides,

$$\int_0^1 \|\ \frac{\varepsilon^{|\alpha|}}{\alpha!}\ T_{D_x^\alpha a}\ J_\varepsilon^\alpha\ u\ \|^2_{s_1(\eta)}\varepsilon^{-2(s-s_1+1)}(1+\frac{\delta^2}{\varepsilon^2})^{-1}\ \frac{d\varepsilon}{\varepsilon}\ \leqq$$

$$\leq C\int_0^1 \|J_\varepsilon^\alpha u\|^2_{s_1(\eta)}\, \varepsilon^{2|\alpha|-2(s-s_1+1)}(1+\frac{\delta^2}{\varepsilon^2})^{-1}\frac{d\varepsilon}{\varepsilon}$$

Since $j^{(\alpha)}(\xi,\eta)$ is $O(|(\xi,\eta)|^{\nu-|\alpha|})$, when $(\xi,\eta)\to 0$, and $\nu>s-s_1+1$ implies $\nu-|\alpha|>s-|\alpha|-s_1+1$, then by Lemma 3.8 we know

$$\int_0^1 \|J_\varepsilon^\alpha u\|^2_{s_1(\eta)}\, \varepsilon^{2|\alpha|-2(s-s_1+1)}(1+\frac{\delta^2}{\varepsilon^2})^{-1}\frac{d\varepsilon}{\varepsilon} \leq C\|u\|^2_{(s-1)(\eta),\delta}\,.$$

whence (3.22).

§4. Paradifferential operators with parameter

Assume that $\ell(x,\xi,\eta)$ is in $H^s(s>\frac{n}{2})$ with respect to $(x_1,\dots,x_n)$, with compact support, it is in $C^\infty(R^{n+1}\setminus 0)$ with respect to $(\xi_1,\dots,\xi_n,\eta)$ and positively homogeneous of degree m. Then we can make a spherical harmonic factorization on S^n

$$\ell = \Sigma a_\nu(x)h_\nu(\xi,\eta), \tag{4.1}$$

where $a_\nu\in H^s$ with uniform bounded norm $\|a_\nu\|_s$, $h_\nu(\xi,\eta)$ is a $C^\infty(R^{n+1}\setminus 0)$ function, positively homogeneous of degree m, and for any M, $\|h_\nu\|_{C^M(S^n)}$ rapidly decreases with respect to ν. By using this factorization, we can define a paradifferential operator with parameter η as follows:

$$T_\ell u = \sum_\nu T_{a_\nu}(h_\nu(D,\eta)S(D,\eta))u \tag{4.2}$$

where $S(D,\eta)$ is pseudodifferential operator with symbol $S(\xi,\eta)$, which vanishes in $|\xi|^2+\eta^2<\frac{1}{2}$ and is equal to in $|\xi|^2+\eta^2>\frac{1}{2}$ identically.

Denote by $\Sigma^{m,\eta}_{(s)}$ the set of above-mentioned symbols $\ell(x,\xi,\eta)$ and by $OP(\Sigma^{m,\eta}_{(s)})$ the class of corresponding paradifferential operators. Next we are going to derive various estimates on $OP(\Sigma^{m,\eta}_{(s)})$ operators. An in §3, if there is no special explanation, all constants in these estimates are independint of η.

<u>Lemma 4.1</u> Assume $\ell(x,\xi,\eta)\in\Sigma^{m,\eta}_{(s)}$, $s>\frac{n}{2}$, then

$$\|T_\ell u\|_{\mu(\eta)} \leq C\|u\|_{(m+\mu)(\eta)}. \tag{4.3}$$

<u>Proof</u>. If $h(\varepsilon,\eta)$ is a positively homogeneous function of degree m, and is C^∞ except at 0, then

$$\begin{aligned}\|h(D,\eta)u\|_{\mu(\eta)} &= \|h(\xi,\eta)\hat{u}(\xi)<(\xi,\eta)>^\mu\|_{L^2(R^n_\xi)}\\ &= C\,\|\hat{u}(\xi)<(\xi,\eta)>^{\mu+m}\|_{L^2(R^n_\xi)}\\ &= C\,\|u\|_{(m+\mu)(\eta)}.\end{aligned}$$

Using lemma 3.2 we obtain (4.3)

It is possible to obtain the following lemma from Lemmas 3.4 and 3.5; the details of the proof are omitted.

Lemma 4.2 If $h(\xi,\eta)$ is a positively homogeneous function of degree m, and is C^∞ except at 0, and if $a(x)\in H^s$ with $s>\frac{n}{2}$ has compact support, then for large M,

$$\|[h(D,\eta),T_a]v\|_{\mu(\eta)} \leq C\|a\|_s \|h\|_{C^M(S^n)} \|v\|_{(m+\mu-1)(\eta)} \tag{4.4}$$

Lemma 4.3 Assume $a(x,\varepsilon,\eta)\in\Sigma^{1,\eta}_{(s)}$, $s>\frac{n}{2}+1$, then

$$\|(T^*_a - T_{\bar{a}})v\|_o \quad C\|v\|_o \tag{4.5}$$

Proof. Expressing $T_a v$, by (4.2) and using (3.5), we have

$$T^*_a v = \sum_\nu \bar{h}(D,\eta)(T_{\bar{a}_\nu} v + R_\nu v),$$

where R_ν is a 1-regular operator. Moreover, Lemma 4.2 shows that

$$\|[\bar{h}_\nu(D,\eta),\ T_{\bar{a}}]v\|_o \leq C\|v\|_o \|h_\nu\|_{C^M(S^n)} \|a_\nu\|_s .$$

According to the properties of spherical harmonic expansion, $\|h_\nu\|_{C^M(S^n)}$ temperately increases with respect to ν, and $\|a_\nu\|_s$ rapidly decreases with respect to ν, hence

$$\|\sum_\nu [\bar{h}_\nu(D,\eta),\ T_{\bar{a}_\nu}]v\|_o \leq C\|v\|_o .$$

Using the estimate on R_ν, we obtain (4.5).

Lemma 4.4 Assume $a(x,\xi,\eta)\in\Sigma^{m_1,\eta}_{(s)}$, $b(x,\xi,\eta)\in\Sigma^{m_2,\eta}_{(s)}$, $s>\frac{n}{2}+1$, then

$$T_a T_b = T_{ab} + R, \tag{4.6}$$

where R is a $-(m_1+m_2-1)$-regular operator.

Proof. By (4.2)

$$\begin{aligned} T_a T_b &= \sum_{\nu\mu} T_{a_\nu} h_\nu(D,\eta) T_{b_\mu} h_\mu(D,\eta) \\ &= \sum_{\nu\mu} T_{a_\nu} T_{b_\mu} h_\nu(D,\eta)\ h_\mu(D,\eta) + R_1 \\ &= \sum_{\nu\mu} T_{a_\nu b_\mu} h_\nu(D,\eta) h_\mu(D,\eta) + R_1 + R_2. \end{aligned}$$

The last summation after the new combination is just the expression of T_{ab}, Lemma 4.2 shows that R_1 is a $-(m_1+m_2-1)$-regular operator, and Lemma 3.4 shows that R_2 is a $-(m_1+m_2-1)$-regular operator.

Lemma 4.5 Assume $a(x,\xi,\eta)\in\Sigma^{o,\eta}_{(s)}$, $s>\frac{n}{2}+1$ and $a(x,\xi,\eta)\geq\delta>0$, then for any compact set F, there are C, $C_1>0$, such that for $\eta>C_1$,

$$\mathrm{Re}(T_a\, u,u) \geq \frac{\delta}{2}(u,u), \qquad \forall\ u\in C_0^\infty(F), \tag{4.7}$$

Moreover, if $a(x,\xi,\eta)$ is a matrix of $\Sigma_{(s)}^{o,\eta}$ symbols, and $a(x,\xi,\eta)\geq\delta I$, then (4.7) still holds.

Proof. By the conditions $a(x,\xi,\eta)-\frac{3\delta}{4} \geq \frac{\delta}{4} > 0$, setting $b(x,\xi,\eta)=(a(x,\xi,\eta)-\frac{3\delta}{4})^{\frac{1}{2}}$, we have $b\in\Sigma_{(s)}^{o,\eta}$. In view of Lemma 4.3 and Lemma 4.4 we have

$$0 \leq (T_b^* T_b\, u,u) = (T_a\, u,u)+(Ru,u) - \frac{3\delta}{4}(u,u),$$

where R is a 1-regular operator. Therefore,

$$\eta\|Ru\|_o \leq C\|u\|_o ,$$

and for $\eta>\frac{4c}{\delta}$,

$$\mathrm{Re}(T_a u,u) \geq \frac{3\delta}{4}(u,u)-|(Ru,u)| \geq \frac{\delta}{2}(u,u).$$

Lemma 4.6 Assume $a(x,\xi,\eta)\in\Sigma_{(s)}^{1,\eta}$, $s>\frac{n}{2}+2$, $a(x,\xi,\eta)\geq0$, then for any compact set F, there are C, $C_1>0$, such that for $n_1>C_1$

$$\mathrm{Re}(T_a u,u) \geq -C(u,u), \qquad \forall\ u\in C_0^\infty(F), \tag{4.8}$$

Proof. (4.8) is called the sharp Gårding inequality for paradifferential operators with parameter η. We are going to prove it by Cordoba-Fefferman's method (see [11]). First, write T_a in the form (4.2): $T_a=\Sigma T_{a_\nu} h_\nu(D,\eta)$, where $a_\nu\in C^2$. By Lemma 3.6, $T_{a_\nu}-a_\nu$ is a 1-regular operator, and by the properties of spherical harmonic expanions,

$$\|(T_{a_\nu} - a_\nu)u\|_{1(\eta)} \leq C\|u\|_o,$$

therefore

$$\|(T_a - a(x,D,\eta))u\|_o \leq C\|u\|_o. \tag{4.9}$$

Denote by C_n the constant satisfying $C_n^2(2\pi)^n\int e^{-2|w|^2}dw=1$, and define w_η and w_η^* as follows,

$$(w_\eta u)(x,\xi,\eta)=C_n|(\xi,\eta)|^{\frac{n}{4}}\int e^{i(x-y)\xi}e^{-|(\xi,\eta)||x-y|^2}u(y)dy,$$

$$(w_\eta^* F)(x)=C_n\int e^{i(x-z)\xi}e^{-|(\xi,\eta)||x-z|^2}|(\xi,\eta)|^{\frac{n}{4}}F(z,\xi,\eta)dz d\xi.$$

We have

$$w_\eta^* a(z,\xi,\eta)w_\eta u=(2\pi)^{-n}\int e^{-i(x-y)\xi}t(x,y,\xi,\eta)u(y)dyd\xi,$$

where

$$t(x,y,\xi,\eta) = (2\pi)^n C_n^2|(\xi,\eta)|^{\frac{n}{4}}\int e^{-|(\xi,\eta)|(|x-z|^2+|y-z|^2)}a(z,\xi,\eta)dz,$$

satisfying

$$|D_x^{\alpha} D_y^{\beta} D_{\xi,\eta}^{\gamma} t(x,y,\xi,\eta)| \leq C_{\alpha,\beta,\gamma} |(\xi,\eta)|^{1-\gamma-\frac{\alpha+\beta}{2}} \tag{4.10}$$

Therefore, $\tilde{A}=W_\eta^* a W_\eta$ is a pseudodifferential operator in $OPS^1_{1,\frac{1}{2}}$ with parameter η, and satisfies (4.8). Besides, $\tilde{A}$ can be written as

$$\tilde{A} = t(x,x,D,\eta) + R(x,D,\eta),$$

where $R(x,\xi,\eta)$ can be determined by the asymptotic expansion of $t(x,y,\xi,\eta)$ with the leading term $\sum_{j=1}^{n} \frac{1}{i} \frac{\partial^2 t}{\partial \xi_j \partial y_j}\Big|_{y=x}$ Noting that

$$\frac{\partial t}{\partial y_j} = (2\pi)^n C_n^2 |(\xi,\eta)|^{\frac{n}{2}} \int e^{-|(\xi,\eta)|(|x-\zeta|^2 + y-\zeta|^2)} \cdot$$

$$\cdot(-2|(\xi,\eta)|(y_j-\zeta_j))a(\zeta,\xi,\eta)d$$

$$= (2\pi)^n C_n^2 |(\xi,\eta)|^{\frac{n}{2}} \int e^{-|(\xi,\eta)||y-\zeta|^2} \frac{\partial}{\partial \zeta_j}(e^{-|(\xi,\eta)||x-\zeta|^2} a(\zeta,\xi,\eta))d$$

$$= (2\pi)^n C_n^2 |(\xi,\eta)|^{\frac{n}{2}} (\int e^{-|(\xi,\eta)|(|x-\zeta|^2+|y-\zeta|^2)} \cdot$$

$$\cdot(-2|(\xi,\eta)|(\zeta_j-x_j))a(\zeta,\xi,\eta)d + \int e^{-|(\xi,\eta)|(|x-\zeta|^2+|y-\zeta|^2)} a_{\zeta_j}(\zeta,\xi,\eta)d\zeta)$$

and substituting y by x, we have

$$\frac{\partial t}{\partial y_j}\Big|_{y=x} = \frac{1}{2}(2\pi)^n C_n^2 |(\xi,\eta)|^{\frac{n}{2}} \int e^{-2|(\xi,\eta)||x-\zeta|^2} a_{\zeta_j}(\zeta,\xi,\eta)d\zeta,$$

which yields $\frac{\partial^2 t}{\partial y_j \partial \xi_j}\Big|_{y=x} \in S^o_{1,\frac{1}{2}}$. As for the terms of lower order in $R(x,D,\eta)$, they belong to $S^o_{1,\frac{1}{2}}$ according to the calculus of symbols $S^m_{\rho,\delta}$. Therefore, $R(x,D,\eta)$ is a bounded operator from L^2 to L^2 with bound independent of η.

Let $r(x,D,\eta)=t(x,x,D,\eta)-a(x,D,\eta)$; then by the choice of C_n, we have

$$\gamma(x,\xi,\eta)=(2\pi)^n C_n^2 |(\xi,\eta)|^{\frac{n}{2}} \int e^{-2|(\xi,\eta)|w^2} [a(x-w,\xi,\eta)-a(x,\xi,\eta)]dw$$

$$=(2\pi)^n C_n^2 |(\xi,\eta)|^{\frac{n}{2}} \int e^{-2|(\xi,\eta)|w^2} (-\sum_{j=1}^{n} w_j a_{x_j}(x,\xi,\eta) +$$

$$+ \frac{1}{2} \sum_{k,j=1}^{n} \int_o^1 a_{x_j x_k}(x-tw,\xi,\eta)(1-t)dt(w_j w_k)dw.$$

Since the first integral is 0, then

$$|D^{\alpha}_{\xi,\eta} r(x,\xi,\eta)| \leq C'_{\alpha} |(\xi,\eta)|^{-\alpha}$$

$$|D^{\alpha}_{\xi,\eta}(r(x_1,\xi,\eta)-r(x_2,\xi,\eta))| \leq C''_{\alpha} |x_1-x_2|^{\delta} |(\xi,\eta)|^{-\alpha}, \quad \delta<s-\frac{n}{2}-2 \tag{4.11}$$

Using Coifman-Meyer's theorem in [12], we know $r(x,D,\eta)$ is a bounded operator from L^2 to L^2, and it is easy to see that the norm is independent of η.

Now we have

$$\begin{aligned}(T_a u,u) &= ((T_a - a(x,D,\eta))u,u) + (a(x,D,\eta)u,u)\\ &= ((T_a - a(x,D,\eta))u,u) + (w_\eta^* a w_\eta u,u) -\\ &\quad -(R(x,D,\eta)u,u) - (r(x,D,\eta)u,u),\end{aligned}$$

and noting the positive definiteness, we get

$$\operatorname{Re}(T_a u,u) \geqq -C\|u\|^2$$

finally.

Corollary 4.7 Assume $a(x,\xi,\eta)\in\Sigma^{1,\eta}_{(s)}$, $s>\frac{n}{2}+2$, $a(x,\xi,\eta)\geqq\delta\eta I$, then for any compact set F, there is $C_1>0$, such that for $\eta>C_1$.

$$\operatorname{Re}(T_a u,u) \geqq \frac{\delta}{2}\eta(u,u), \quad \forall\, u\in C_0^\infty(F), \tag{4.12}$$

Proof. Taking $a_1(x,\xi,\eta)=a(x,\xi,\eta)-\delta\eta I$, and applying Lemma 4.6, we have

$$\operatorname{Re}(T_a u,u) \geqq -C(u,u),$$

if $\eta>C'$. Setting $C_1=\max(C',\frac{2c}{\delta})$, (4.12) is obtained.

§5. Paralinearization

Now we are going to paralinearize (2.4), (2.5). Let χ be the transformation (2.3), $G^\pm$ and G^0 be the inverse image of $\Omega^\pm$ and $\Omega\cap S$. $u\circ\chi$ satisfies (2.4) in $G^\pm$, $u\circ\chi|_{\bar{x}_1=0}$ and ϕ satisfies (2.5). In the sequal we still denote $\bar{t}$, $\bar{x}_1$, $\bar{x}'$ by t, x_1, x' and denote $u\circ\chi$ by u for the notational convenience. Besides, we take the following natations as well:

$$\tilde{A}_j = \begin{pmatrix} A_j(u^+) & \\ & A_j(u^-)\end{pmatrix} \qquad j=2,\ \dots\ ,N$$

$$\tilde{A}_1 = \begin{pmatrix} \phi_t I+\sum_{j=2}^N\phi_{x_j}A_j(u^+)-A_1(u^+) & \\ & -\phi_t I-\sum_{j=2}^N\phi_{x_j}A_j(u^-)+A_1(u^-)\end{pmatrix}$$

$$\tilde{P}_o = \begin{pmatrix} P_o^+ & \\ & -P_o^-\end{pmatrix} = \tilde{A}_1$$

$$\begin{aligned}\tilde{P} &= \begin{pmatrix} P^+ & \\ & P^-\end{pmatrix}\\ &= i\xi_o I+\sum_{j=1}^N i\xi_j\tilde{A}_j+\begin{pmatrix}\sum_{j=2}^N A_j'(u^+)\frac{\partial u^+}{\partial x_j}+(A_1'(u^+)-\sum_{j=2}^N\phi_{x_j}A_j'(u^+))\frac{\partial u^+}{\partial x_1} & \\ & \sum_{j=2}^N A_j'(u^-)\frac{\partial u^-}{\partial x_j}-(A_1'(u^-)-\sum_{j=2}^N\phi_{x_j}A_j'(u^-))\frac{\partial u^-}{\partial x_1}\end{pmatrix}\end{aligned}$$

$$\tilde{q} = \begin{pmatrix} q^+\\ q^-\end{pmatrix} = i\begin{pmatrix} -\frac{\partial u^+}{\partial x_1}\xi_o - \sum_{j=2}^N A_j(u^+)\frac{\partial u^+}{\partial x_1}\xi_j\\ \frac{\partial u^-}{\partial x_1}\xi_o + \sum_{j=2}^N A_j(u^-)\frac{\partial u^-}{\partial x_1}\xi_j\end{pmatrix}$$

$$h = b_\circ \xi_\circ + \sum_{j=2}^{N} b_j \xi_j = i(u^+ - u^-)\xi_\circ + i \sum_{j=2}^{N} (F_j(u^+) - F_j(u^-))\xi_j .$$

As in [1], when a symbol is not compactly supported, we can also define a corresponding properly supported paradifferential operator with ℓ as its symbol by using a cut-off function. In what follows we still denote it by T_ℓ . With s_1 as parameter, we paralinearize (2.4), (2.5) in space (t,x'). Since $u^\pm \in H^s$, $\nabla\phi \in H^s$, then

$$D_{x_1}^k u^\pm \in L^\infty(x_1, H^{s'-k}), \quad (k \leqq s'); \ \phi \in H^{s'+1} \tag{5.1}$$

with $s' \leqq s - \frac{1}{2}$. Hence, if $s > \frac{N}{2} + \frac{3}{2}$, then (2.4) can be paralinearized as

$$T_{p_0^\pm} \frac{\partial u^\pm}{\partial x_1} + T_{p^\pm} u^\pm + T_{q^\pm}\phi = r_1^\pm \tag{5.2}$$

where $r_1^\pm \in H_{loc}^{2s'-\frac{N}{2}-1-\rho}$ holds for any $\rho > 0$. In particular, $s' > \frac{N}{2} + 2$ implies $r_1^\pm \in H_{loc}^{s'+1}$.

Paralinearizing (2.5) gives us

$$T_h \phi + T_M \tilde{u} = r_2 , \tag{5.3}$$

where $u = {}^t(u^+, u^-)$, and $M = (a^+, a^-)$ is a matrix given in (2.9), $r_2 \in H_{loc}^{2s'-\frac{N}{2}-\rho}$ holds for any $\rho > 0$. In particular, $s' > \frac{N}{2} + 2$ implies $r_2 \in H_{loc}^{s'+2}$

In order to write down the estimates for $r_1^\pm$ and r_2, we need more notations. Denote by π_1, π the inverse image of Σ_1, Σ under the transformation χ, by $\pi_{1\Delta}$ the surface obtained by moving π_1 for distance Δ in the opposite direction of t-axis, by G_Δ the domain surrounded by $\pi_{1\Delta}$ and $t=0$. Let $G_\Delta^\pm = G_\Delta \cap \{ {x_1 > 0 \atop x_1 < 0} \}$, $G_{\Delta,a} = G_\Delta \cap \{x_1 = a\}$, we have

$$\|r_1^\pm\|_{s'+1, G_{\Delta,x_1}} \leqq C(\|u^\pm\|_{s', G_{x_1}} + \|\frac{\partial u^+}{\partial x_1}\|_{s'-1, G_{x_1}} + \|\phi\|_{s'+1, G_\circ})$$

In view of the assumption that the boundary $x_1 = 0$ is not characteristic, so $\det|p_\circ^\pm| \neq 0$ hence by (2.4) we obtain

$$\|r_1^\pm\|_{s'+1, G_{\Delta,x_1}} \leqq C(\|u^\pm\|_{s', G_{x_1}} + \|\phi\|_{s'+1, G_0}), \tag{5.4}$$

Similarly,

$$\|r_2\|_{s'+2, G_{\Delta,0}} \leqq C(\|u^\pm\|_{s', G_{x_1}} + \|\phi\|_{s'+1, G_0}). \tag{5.5}$$

In order to obtain estimates of solutions of (2.4), (2.5) by estimates on the initial data, we need to localize (5.2), (5.3). First, make a set of $C_0^\infty(G)$ functions $\{\zeta_k\}$, such that $\{\text{supp } \zeta_k\}$ form a covering of $G_{\frac{\Delta}{2}}$ and $\Sigma\zeta_k \equiv 1$ on $G_{\frac{\Delta}{2}}$. Moreover, we may assume that each supp ζ_k is small enough, such that on each subdomain there is a symmetrizer as mentioned in the next section.

Setting $\phi_k = \phi\zeta_k$, $u_k^\pm = u^\pm\zeta_k$, $r_{1k}^\pm = r_1^\pm\zeta_k$, $r_{2k} = r_2\zeta_k$, and expanding $T_{p^\pm}(\zeta_k u^\pm)$, we obtain

$$T_{p_o^\pm}\frac{\partial u_k^\pm}{\partial x_1} + T_{p^\pm} u_k^\pm + T_{q^\pm}\phi_k = r_{1k}^\pm + r'^{\pm}_{1k} \tag{5.6}$$

$$T_h\phi_k + T_{a^+}u_k^+ - T_{a^-}u_k^- = r_{2k} + r'_{2k} \tag{5.7}$$

where $r_{1k}^\pm$, r_{2k} satisfy estimates similar to (5.3), (5.4). Since ζ_k is given C^∞ function, then by the calculus of paradifferential operators.

$$\| T_{p^\pm}(\zeta_k u^\pm) - \zeta_k T_{p^\pm} u^\pm \|_{\mu, G_{\Delta, x_1}}$$

$$\leqq C \| p^\pm \|_{\frac{N}{2}+1+\varepsilon, G_{x_1}} \| u^\pm \|_{\mu, G_{\Delta, x_1}}$$

$$\leqq C(\| u^\pm \|_{s', G_{x_1}} + \| \phi \|_{s', G_o}) \| u^\pm \|_{\mu, G_{\Delta, x_1}} ,$$

Then regarding $\| u^\pm \|_{s', G_{x_1}}$ and $\| \phi \|_{s', G_o}$ as known, we have

$$\begin{aligned} \| r'^{\pm}_{1k} \|_{\mu, G_{\Delta, x_1}} &\leqq C(\| u^\pm \|_{\mu, G_{\Delta, x_1}} + \| \phi \|_{\mu, G_{\Delta, o}}), \\ \| r'_{2k} \|_{\mu, G_{\Delta, o}} &\leqq C(\| u^\pm \|_{\mu-1, G_{\Delta, x_1}} + \| \phi \|_{\mu, G_{\Delta, o}}, \end{aligned} \tag{5.8}$$

and (5.6), (5.7) can be also written as

$$T_{\tilde{p}_o}\frac{\partial \tilde{u}_k}{\partial x_1} + T_{\tilde{p}}\tilde{u}_k + T_{\tilde{q}}\phi_k = \tilde{f}_k , \tag{5.9}$$

$$T_h\phi_k + T_M\tilde{u}_k = g_k , \tag{5.10}$$

where $\tilde{u}_k = {}^t(u_k^+, u_k^-)$, and the expressions of $\tilde{f}_k$, g_k are easily obtained.

Using the transformation of unknown variables $\tilde{w} = e^{-\eta t}\tilde{u}_k$, $\psi = e^{-\eta t}\phi_k$ (we omit the subscript k, if there is no confusion), we have

$$T_{\tilde{p}_o}\frac{\partial \tilde{w}}{\partial x_1} + T_{\tilde{p}+\eta}\tilde{w} + T_{\tilde{q}+q_o\eta}\psi = e^{-\eta t}\tilde{f} , \tag{5.11}$$

$$T_{h+b_o\eta}\psi + T_M\tilde{w} = e^{-\eta t}g . \tag{5.12}$$

In the next section we begin with them to derive the desired energy estimates.

§6. Energy estimates and the proof of smoothness theorem.

We are going to estimate solutions by means of $\| \cdot \|_{\mu(\eta)}$, Set

$$\langle g \rangle_{\mu, \eta} = \| e^{-\eta t} g \|_{\mu(\eta)} ,$$

$$|f|^2_{\mu(\eta)} = \sum_{k=o}^{\mu} \int_o^\infty \| D^k_{x_1} f \|^2_{(\mu-k)(\eta)} dx_1 ,$$

$$|f|^2_{\mu,\eta} = \sum_{k=o}^{\mu} \int_o^\infty \langle D^k_{x_1} f \rangle^2_{\mu-k,\eta} \, dx_1 ,$$

and for $V=(u^+,u^-,\phi)$,

$$|||V|||^2_{\mu,\eta} = \langle\phi\rangle^2_{\mu+1,\eta} + \eta[\,|u^+|^2_{\mu,\eta} + |u^-|^2_{\mu,\eta}\,] + \langle u^+\rangle^2_{\mu,\eta} + \langle u^-\rangle^2_{\mu,\eta}.$$

In addition, we denote by B_s the space of functions $U(x',t,\xi_o,\xi',\eta)$, which belong to H^s with respect to x', t, and are positively homogeneous of degree 0, C^∞ smooth in R^{N+1} o with respect to ξ_o,ξ',η. Introducing the norm

$$\|U(x',t,\xi_o,\xi',\eta)\|_{B_s} = \sup_{\xi_o^2+|\xi'|^2+\eta^2=1} \|U(\cdot,\cdot,\xi_o,\xi',\eta)\|_{H^s} ,$$

we have the following lemmas:

<u>Lemma 6.1</u> Denote $e^+(x',t,\xi_o,\xi',\eta)=(i\xi_o+\eta)b_o+i\xi'b(=h+b_o\eta)$, If the uniform stable condition for a shock front mentioned in §2 is satisfied, then

$$\inf_{\substack{\xi_o \geqq o \\ \xi_o^2+|\xi'|^2+\eta^2=1}} |e^+(x',t,\xi_o,\xi',\eta)| \geqq \gamma. \tag{6.1}$$

Moreover, let $\pi(x',t,\xi_o,\xi',\eta)$ be 0 the projection matrix defined by $\pi v = v - \frac{(v,e^+)}{|e^+|^2} e^+$, then

$$\inf_{\substack{\xi_o \geqq o \\ \xi_o^2+|\xi'|^2+\eta^2=1}} |\pi(x',t,\xi_o,\xi',\eta)M\tilde{v}| \geqq \gamma|\tilde{v}|. \tag{6.2}$$

holds for any $\tilde{v}\in E^+(x',t,\xi_o,\xi',\eta)$.

<u>Lemma 6.2</u> There exist constants $\delta_1,\delta_2 > 0$ and a matrix $R(x_1,x',t_1,\xi_o,\xi',\eta)$, such that

i) R is continuous with respect to x_1, valued in $B_{s'}$, and $\frac{\partial R}{\partial x_1}$ is continuous with respect to x_1, valued in $B_{s'-1}$.

ii) RA_1 is a Hermitian matrix.

iii) $\mathrm{Re}(R((i\xi_o+\eta)I + \sum_{j=2}^{N} i\xi_j\tilde{A}_j) \geqq \delta_2 I\eta$

iv) $R\tilde{A}_1 + \delta_1^{-1}K^*K \geqq \delta_1 I$, where $K=\pi M$.

The proof of these two lemmas can be found in [8]. Next we apply them to the proof of the following fact.

<u>Lemma 6.3</u> If $V_k=(u_k^+,u_k^-,\phi_k)$ satisfies (5.9), (5.10), then for any integer μ, satisfying $0\leqq\mu\leqq s'$, and sufficiently large η,

$$|||V_k|||^2_{\mu,\eta} \leqq C_\mu\left(\frac{|\tilde{f}_k|^2_{\mu,\eta}}{\eta} + \langle g_k\rangle^2_{\mu,\eta}\right), \qquad (6.3)_\mu$$

<u>Proof</u>. Denoting by T_R the paradifferential operator corresponding to the symbol $R(x_1.x',t,\xi_0,\xi',\eta)$ with parameters x_1 and η applying it to both sides of (5.11), integrating with respect to x', t after multiplying w, we have

$$(\tilde{w},\ T_R T_{\tilde{p}_0}\frac{\partial\tilde{w}}{\partial x_1}) + (\tilde{w}, T_R T_{\tilde{p}+\eta}\tilde{w}) + (\tilde{w},\ T_R T_{\tilde{q}+b_0\eta}\psi)$$

$$= (w,\ T_R e^{-\eta t}\tilde{f}). \qquad (6.4)$$

By Lemma 4.4 we know that, if $s>\frac{N}{2}+\frac{3}{2}$, then $T_R T_{\tilde{p}_0} - T_{R\tilde{p}_0}$ is a 1-regular operator. In view of the fact that the boundary $x_1=0$ is non-characteristic, $\det|\tilde{p}_0|\neq 0$, hence

$$|T_R T_{\tilde{p}_0}\frac{\partial w}{\partial x_1} - T_{R\tilde{p}_0}\frac{\partial\tilde{w}}{\partial x_1}|_0 \leqq C(|\tilde{w}|_0 + \|\psi\|_0 + |e^{-\eta t}\tilde{f}|_0). \qquad (6.5)$$

Besides,

$$(\tilde{w},\ T_{R\tilde{p}_0}\frac{\partial\tilde{w}}{\partial x_1}) = \frac{\partial}{\partial x_1}(\tilde{w},T_{R\tilde{p}_0}\tilde{w}) - (\frac{\partial\tilde{w}}{\partial x_1},T_{R\tilde{p}_0}\tilde{w}) + r_2$$

$$= \frac{\partial}{\partial x_1}(\tilde{w},T_{R\tilde{p}_0}\tilde{w}) - (T_{R\tilde{p}_0}\frac{\partial\tilde{w}}{\partial x_1},\tilde{w}) + r_2 + r_3\ ,$$

where r_2, r_3 can be written as the form $(\tilde{w},R\frac{\partial\tilde{w}}{\partial x_1})$ with $|R\frac{\partial\tilde{w}}{\partial x_1}|_0$ satisfying the estimate of type (6.5). Hence,

$$\mathrm{Re}\int_0^\infty(\tilde{w},T_{R\tilde{p}_0}\frac{\partial\tilde{w}}{\partial x_1})dx_1 = (\tilde{w},T_{R\tilde{p}_0}\tilde{w})|_{x_1=0} + C|w|_0|R\frac{\partial w}{\partial x_1}|_0\ ,$$

By Lemma 6.2 and Lemma 4.5, if η is large enough, then

$$(\tilde{w},T_{R\tilde{p}_0}\tilde{w})|_{x_1=0} \geqq -\delta_1^{-1}(\tilde{w},T_{k*k}\tilde{w}) + \frac{\delta_1}{2}\|\tilde{w}\|^2_o$$

$$\geqq -C\|T_{k*\pi M}\tilde{w}\|^2_o + \frac{\delta_1}{3}\|\tilde{w}\|^2_o$$

$$\geqq -C\|T_{k*\pi}T_M\tilde{w}\|^2_o + \frac{\delta_1}{4}\|\tilde{w}\|^2_o\ ,$$

that is,

$$(\tilde{w},T_{R\tilde{p}_0}\tilde{w})|_{x_1=0} + C\|T_{k*\pi}T_M\tilde{w}\|^2_o \geqq \frac{\delta_1}{4}\|w\|^2_o\ . \qquad (6.6)$$

Furthermore, (5.12) shows that

$$T_M\tilde{w} = e^{-\eta t}g - T_{h+b_0\eta}\psi.$$

Since the leading term of $\pi(h+b_0\eta)$ is $\pi e^+=0$, we have

$$\|T_{k*\pi}T_M\tilde{w}\|_o^2 \leqq \|T_{k*\pi}e^{-\eta t}g\|_o^2 + \|T_{k*\pi}T_{h+b_o\eta}\psi\|_o^2$$

$$\leqq C(\|e^{-\eta t}g\|_o^2 + \|\psi\|_o^2). \tag{6.7}$$

Considering other terms in (6.4), by lemma 6.2 and Corollary 4.7 we know that the following inequality holds for large η,

$$\text{Re}\int_0^\infty (\tilde{w}, T_R T_{\tilde{p}+\eta}\tilde{w})dx_1 \geqq \frac{\delta_2}{2}\eta|\tilde{w}|_o^2, \tag{6.8}$$

and obviously,

$$\begin{aligned}&\int_o^\infty |(\tilde{w}, T_R T_{\tilde{q}+b_0\eta}\psi)|dx_1 \leqq \varepsilon\|\psi\|_{1(\eta)}^2 + \frac{C}{\varepsilon}|\tilde{w}|_o^2,\\ &\int_0^\infty |(\tilde{w}, T_R(e^{-\eta t}\tilde{f}))|dx_1 \leqq \varepsilon\eta|\tilde{w}|_o^2 + \frac{C}{\varepsilon\eta}|e^{-\eta t}\tilde{f}|_o\end{aligned} \tag{6.9}$$

Summarizing (6.4)-(6.9), we obtain

$$\frac{\delta_2}{4}\eta|\tilde{w}|_o^2 + \frac{\delta_1}{4}\|\tilde{w}\|_o^2 \leqq C(\|e^{-\eta t}g\|_o^2 + \frac{1}{\eta}|e^{-\eta t}\tilde{f}|_o^2) + \varepsilon\|\psi\|_{1(\eta)}^2. \tag{6.10}$$

In order to obtain the estimate of ψ, let $Q(x',t,\xi_o,\xi',\eta)$ satisfy $Qv=\frac{(v,e^+)}{|e^+|^2}$. Obviously, Q is positively homogeneous of degree -1 with respect to ξ_o, ξ', η, and $Qe^+=1$. Letting T_Q act on (5.12), we have

$$\begin{aligned}&T_{Q(h+b_0\eta)}\psi + T_Q T_M\tilde{w} = T_Q e^{-\eta t}g + r_3,\\ &\psi + T_Q T_M\tilde{w} = T_Q e^{-\eta t}g + r_3 + r_4,\end{aligned} \tag{6.11}$$

where r_3, r_4 have the form $R\psi$ with R as a 1-regular operator. Therefore, by (6.11)

$$\|\psi\|_{1(\eta)}^2 \leqq C(\|\tilde{w}\|_o^2 + \|e^{-\eta t}g\|_o^2) \tag{6.12}$$

holds for sufficiently large η. Hence $(6.3)_o$ is obtained from (6.10) and (6.12).

For the case $\mu\geqq 1$, we can obtain estimates of norms for derivatives of higher order by applying differential operators ∂_t and $\partial_{x'}$. For instance, letting ∂_t act on (5.11), by virtue of $s>\frac{N}{2}+\frac{3}{2}$, we have

$$\partial_t T_{\tilde{p}_o}\frac{\partial w}{\partial x_1} = T_{\tilde{p}_o}\frac{\partial}{\partial x_1}(\frac{\partial\tilde{w}}{\partial t}) + r,$$

where $\|r\|_o^2 \leqq C\|\frac{\partial\tilde{w}}{\partial x_1}\|_o^2 \leqq C(\|\tilde{w}\|_{1(\eta)}^2 + \|\psi\|_{1(\eta)}^2 + \|e^{-\eta t}\tilde{f}\|_o^2)$.

Similarly, by computing derivatives of other terms with respect to t, we obtain

$$T_{\tilde{p}_o} \frac{\partial}{\partial x_1}\left(\frac{\partial \tilde{w}}{\partial t}\right) + T_{\tilde{p}_o \eta}\frac{\partial \tilde{w}}{\partial t} + T_{\tilde{q}+q_o \eta} \frac{\partial \psi}{\partial t} = \frac{\partial}{\partial t}(e^{-\eta t}\tilde{f}) + r_a, \tag{6.13}$$

where $\| T_a \|_o^2 \leqq C(\| \tilde{w} \|^2_{1(\eta)} + \| \psi \|^2_{1(\eta)} + \| e^{-\eta t}\tilde{f} \|_o^2)$.

Moreover, letting ∂_t act on (5.12) gives

$$\partial_t \psi + T_Q T_M \partial_t w = T_Q \partial_t (e^{-\eta t} g) + r_b , \tag{6.14}$$

where $\| r_b \|_o^2 \leqq C(\| \tilde{w} \|_o^2 + \| \psi \|_o^2)$.

Now replacing (5.11), (5.12) by (6.13), (6.14), and by the same procedure as that in deriving $(6.3)_o$, we can obtain

$$||| \frac{\partial V}{\partial t} |||^2_{o,\eta} \leqq C\Big(\frac{|\partial_t e^{-\eta t}\tilde{f}|_o^2 + |\tilde{f}|_o^2 + |\tilde{w}|^2_{1(\eta)} + \| \psi \|^2_{1(\eta)}}{\eta}$$

$$+ \| \partial_t e^{-\eta t} g \|_o^2 + \| \tilde{w} \|_o^2 + \| \psi \|_o^2 \Big) .$$

For $||| \frac{\partial V}{\partial x'} |||^2_{o,\eta}$, we can estimate in a similar way. Then using the fact that the boundary $x_1 = 0$ is non-characteristic, we have

$$||| V |||^2_{1,\eta} \leqq C\Big(\frac{|f|^2_{1,\eta}}{\eta} + \langle g \rangle^2_{1,\eta} \Big),$$

that is $(6.3)_1$. The case $1 < \mu \leqq s'$ can be treated similarly.

In order to improve the smoothness of the solution, we have to estimate $||| V |||_{\mu,\eta,\delta}$ norm. Its meaning is given as follows:

$\| g \|_{\mu(\eta),\delta}$ is defined by (3.18),

$$\langle g \rangle_{\mu,\eta,\delta} = \| e^{-\eta t} g \|_{\mu(\eta),\delta} ,$$

$$|f|_{\mu(\eta),\delta} = \sum_{k=o}^{k} \int_o^\infty \| D^k_{x_1} f \|^2_{(\mu-k)(\eta),\delta} dx_1 ,$$

$$|f|^2_{\mu,\eta,\delta} = \sum_{k=o}^{\mu} \int_o^\infty \langle D^k_{x_1} f \rangle^2_{\mu-k,\eta,\delta} dx_1 ,$$

$$||| V |||^2_{\mu,\eta,\delta} = \langle \phi \rangle^2_{\mu+1,\eta,\delta} + \eta [|u^+|^2_{\mu,\eta,\delta} + |u^-|^2_{\mu,\eta,\delta}] +$$

$$+ \langle u^+ \rangle^2_{\mu,\eta,\delta} + \langle u^- \rangle^2_{\mu,\eta,\delta} ,$$

Assume that J_ε is an operator with $j(\varepsilon\xi_o, \varepsilon\xi_2, \ldots, \varepsilon\xi_N, \varepsilon\eta)$ as its Fourier multiplier, and that j satisfies (3.19) (3.20) with $\nu > \mu+1$, Then (3.21) implies

$$C_1 |||V|||^2_{\mu,\eta,\delta} \leqq |||V|||^2_{\mu,\eta} + \int_0^1 |||J_\varepsilon V|||^2_{o,\eta} dI \leqq C_2 |||V|||^2_{\mu,\eta,\delta} \tag{6.15}$$

where $dI=\varepsilon^{-2(\mu+1)}(1+\frac{\delta^2}{\varepsilon^2})^{-1}\frac{d\varepsilon}{\varepsilon}$.

<u>Lemma 6.4</u> Under the same assumptions as in Lemma 6.3.

$$|||V_k|||^2_{\mu,\eta,\delta} \leqq C'_\mu\left(\frac{|f_k|^2_{\mu,\eta,\delta}}{\eta} + \langle g_k\rangle^2_{\mu,\eta,\delta}\right), \tag{6.16$_\mu$}$$

if η is large enough.

<u>Proof</u>. By virtue of lemma 6.3 and (6.15) we only need to estimate $\int_0^1 |||J_\varepsilon V_k|||^2_{o,\eta} dI$. Letting J_ε act on (5.11), (5.12), we obtain

$$T_{\tilde{p}_0}\frac{\partial}{\partial x_1}(J_\varepsilon\tilde{w})+T_{\tilde{p}+\eta}J_\varepsilon\tilde{w}+T_{\tilde{q}+q_0\eta}J_\varepsilon\psi$$

$$=[T_{\tilde{p}_0},J_\varepsilon]\frac{\partial w}{\partial x_1}+[T_{\tilde{p}},J_\varepsilon]\tilde{w}+[T_{\tilde{q}+q_0\eta},J_\varepsilon]\psi+J_\varepsilon(\tilde{f}e^{-\eta t}), \tag{6.17}$$

$$T_{h+b_0\eta}J_\varepsilon\psi+T_M J_\varepsilon\tilde{w}=[T_{h+b_0\eta},J_\varepsilon]\psi+[T_M,J_\varepsilon]\tilde{w}+J_\varepsilon(ge^{-\eta t}). \tag{6,18}$$

Using lemma 3.9, we have

$$\int_0^1 \|[T_{\tilde{p}_0},J_\varepsilon]\frac{\partial\tilde{w}}{\partial x_1}\|_o^2\, dI \leqq C\|\frac{\partial\tilde{w}}{\partial x_1}\|^2_{(\mu-1)(\eta),\delta}$$

$$\leqq C(\|\tilde{w}\|^2_{\mu(\eta),\delta}+\|\psi\|^2_{\mu(\eta),\delta}+\|e^{-\eta t}f\|^2_{(\mu-1)(\eta),\delta})$$

$$\int_0^1 \|[T_{\tilde{p}},J_\varepsilon]w\|_o^2\, dI \leqq C\|\tilde{w}\|_{\mu(\eta),\delta}\ ,$$

$$\int_o^1 \|[T_{\tilde{q}+q_0\eta},J_\varepsilon]\psi\|_o^2\, dI \leqq C\|\psi\|^2_{\mu(\eta),\delta}\ ,$$

$$\int_0^1 \|[T_{h+b_0\eta},J_\varepsilon]\psi\|_o^2\, dI \leqq C\|\psi\|^2_{\mu(\eta),\delta}\ ,$$

$$\int_0^1 \|[T_M,J_\varepsilon]\tilde{w}\|_o^2\, dI \leqq C\|\tilde{w}\|^2_{(\mu-1)(\eta),\delta}\ .$$

Form the boundedness of J_ε,

$$\int_0^1 \|J_\varepsilon e^{-\eta t}\tilde{f}\|_o^2\, dI \leqq C\langle\tilde{f}\rangle_{\mu,\eta,\delta}\ ,$$

$$\int_0^1 \|J_\varepsilon e^{-\eta t}g\|_o^2\, dI \leqq C\langle g\rangle_{\mu,\eta,\delta}\ .$$

Applying Lemma 6.3 to (6.17), (6.18), we obtain

$$\int_0^1 |||J_\varepsilon V_k|||^2_{o,\eta}\, dI \leqq C(\frac{1}{\eta}(|\tilde{w}|^2_{\mu(\eta),\delta}+\|\psi\|^2_{\mu(\eta),\delta}+\langle f\rangle^2_{\mu,\eta,\delta}+$$

$$+\|\tilde{w}\|^2_{(\mu-1)(\eta)\delta}+\|\psi\|^2_{\mu(\eta),\delta}+\langle g\rangle^2_{\mu,\eta,\delta})$$

$$\leqq C(|||V_k|||^2_{\mu-1,\eta,\delta}+\frac{1}{\eta}\langle\tilde{f}\rangle^2_{\mu,\eta,\delta}+\langle g\rangle^2_{\mu,\eta,\delta}).$$

Combining it with Lemma 6.3, the estimate $(6.16)_\mu$ is obtained.

Having established the estimate of the solution of the localized system (5.9), (5.10), we may follow the standard fashion for boundary value problems of the hyperbolic system (see [9], [13]) to obtain estimates of the solution of the original system (5.2), (5.3) in the domain between the space-like surface $\pi_{1\Delta}$ and $\{t=0\}$. That is, for sufficiently large η,

$$|||V|||^2_{\mu,\eta,\delta;G_\Delta} \leq C_\mu\Big(\frac{|\tilde{r}_1|^2_{\mu,\eta,\delta;G_\Delta}}{\eta} + \langle r_2\rangle_{\mu,\eta,\delta;G_\Delta} + |||V|||_{\mu,\eta,\delta;G\cap\{t<o\}}\Big) \tag{6.19}$$

By (5.4), (5.5), we have for $\mu\leqq s'$

$$|||V|||^2_{\mu,\eta,\delta,G_\Delta} \leq C_\mu(|||V|||^2_{\mu-1,\eta,\delta;G} + |||V|||^2_{\mu,\eta,\delta;G\cap\{t<o\}}) \tag{6.20}$$

Starting with (6.20) we can prove the smoothness theorem mentioned in §2 by induction. In fact, by the property of $H^{s,\delta}$, if $\|u\|_{s,\delta}$ is uniformly bounded for $\delta\to 0$, then $u\varepsilon H^{s+1}$, and $\|u\|_{s+1}$ is dominated by the same constant (see [13]). Obviously, for any fixed η, a similar conclusion is valid for the norm $\|\cdot\|_{s(\eta),\delta}$.

The conditions in the main theorem imply that $(u^\pm,\phi)$ satisfies $(5.1)_{s'}$. Without loss of generality we may assume s' is an integer, thus for $\mu=s'$ and any η, $V\varepsilon H^\mu(G)$ and $|||V|||^2_{\mu,\eta;G}$ is bounded by a constant independent of η. Therefore, when $\delta\to 0$, $|||V|||^2_{\mu-1,\eta,\delta;G}$ is uniformly bounded with respect to δ and η. Besides, $u^\pm,\phi$ are in C^∞ for $t<0$, which implies $|||V|||^2_{\mu,\eta,\delta;G\cap\{T<O\}}$ is uniformly bounded. Thus, it is possible to obtain from (6.20) that $V\in H^{\mu+1}(G_\Delta)$ and $|||V|||^2_{\mu+1,\eta;G}$ is bounded by a constant independent of η. So we assert that $(u^\pm,\phi)$ satisfy condition $(5.1)_{s'+1}$, in other words, when we shrink the domain G to G_Δ, we may improve the smoothness of the solution by one order. Noting that Δ is arbitrarily small, by repeating the above mentioned procedure as many times as we need, we finally conclude that the C^∞ smoothness of the shock front S and the C^∞ smoothness of $u^\pm$ on both sides of S in $t<0$ can propagate to the whole domain G.

REFERENCES

1. J.M.Bony, Ann. Scien.de l'Ecole Norm. Sup. 14(1981), 209-246.
2. ---, Sem. Goulaouic-Meyer-Schwartz, exp. no.2 (1981-1982).
3. J.Rauch & M.Reed. Duke Math. J. 49(1982), 397-475.
4. M.Beals & G.Metivier, Duke Math. J. 53(1986), 125-137.
5. ---, Reflection of transversal progressing waves in nonlinear strictly hyperbolic mixed problems, preprint.
6. S.Alinhac, Paracomposition et operateurs paradifferentiels, Comm. in PDE, 11(1986), 87-121.
7. ---, Evolution d'une onde simple pour des equations non-lineaires generales, preprint.

8. A.Majda, Memoirs A.M.S., 275(1983).

9. ---, Memoirs A.M.S., 281(1983).

10. H.O.Kreiss, Comm. Pure Appl. Math. 23(1970), 277-298.

11. A.Cordoba & C.Fefferman, Comm. P.D.E. 3(1978).

12. R.Coifman & Y.Meyer, Asterisque 57(1978).

13. J.Chazarain & A.Piriou, Introduction to the theory on linear partial differential equations, North-Holland Publishing Company (1982).

14. L.Hörmander, The Analysis of linear partial differential operators, Springer-Verlag (1985).

15. P.Godin, Analytic regularity of uniformly stable shock fronts with analytic data, preprint.

On Degenerate Monge-Ampere Equations in Convex Domains

Chen Yazhe

Peking University

§1 Introduction

In [1] the author solved degenerate equations of Monge-Ampere type in $C^{1,1}(\bar{\Omega})$ with some restrictions on Dirichlet data or on degeneracy. N.Trudinger in [2] discussed degenerate Monge-Ampere equations in $C^{1,1}(B)\bigcap C^{0,1}(\bar{B})$ without these restrictions but only in balls B.

Enlightened by [2] we extend the results in [2] to the equations of Monge-Ampere type in a general convex domain. The idea comes from Krylov's work [3] in which he considered elliptic operator on manifolds and applied it to getting the boundary estimates for derivatives of solutions of Dirichlet problem with homogeneous boundary values. The method given in this paper is much simpler than those in [3], and it is used to find interior second derivative estimates for general Dirichlet data.

Let Ω be a convex domain in $\mathbf{R}^n$. In Ω we consider the Dirichlet problem

$$\det D^2 u = f(x,u,Du) \quad \text{in } \Omega, \tag{1.1}$$

$$u = \varphi \quad \text{on } \partial\Omega, \tag{1.2}$$

where Du and D^2u are the gradient and the Hessian matrix of u respectively.

Set $g(x,z,p) = [f(x,z.p)]^{1/n}$. Assume $f(x,z,p)$ satisfies

(F1) $g \geq 0$in $\Omega \times \mathbf{R} \times \mathbf{R}^n$ and $g \in C^{1,1}(\bar{\Omega} \times \mathbf{R} \times \mathbf{R}^n)$.

(F2) $g_z \geq 0$in $\Omega \times \mathbf{R} \times \mathbf{R}^n$.

(F3) There exists a constant N and $\tilde{g} \in C^1(\Omega)$, $h \in L^1_{loc}(\mathbf{R}^n)$ such that

$$f(x,-N,p) \leq \tilde{g}(x)/h(p), \quad \forall x \in \Omega, p \in \mathbf{R}^n,$$

where

$$\int_\Omega \tilde{g} < \int_{\mathbf{R}^n} h, \quad h^{-1} \geq \mu_0 > 0, \quad \forall p \in \mathbf{R}^n.$$

(F4) There exist non-negative constants μ, α and β such that

$$f(x,\varphi(x),p) \leq \mu d^\beta (1+|p|^2)^{\alpha/2}, \forall x \in \mathcal{N}, p \in \mathbf{R}^n.$$

where $d = \text{dist}\{x,\partial\Omega\}$, $\mathcal{N}$ is a neighborhood of $\partial\Omega$ and $\beta \geq \alpha - n - 1$.

(F5) $g(x,z,p)$ is convex with respect to p.

In this paper we shall prove the following existence theorem:

Theorem 1.1 Let Ω be a convex domain in $\mathbf{R}^n$ such that there exists a concave function $\psi \in C^{3,1}(\bar{\Omega})$ satisfying

$$\psi = 0, \quad |D\psi| \neq 0 \text{ on } \partial\Omega,$$

$$D^2\psi \leq -I, \quad \text{in } \Omega$$

where I is the unit matrix, and let f satisfy (F1)-(F5). Assume $\varphi \in C^{1,1}(\bar{\Omega})$. Then Dirichlet problem (1), (2) has the unique convex solution $u \in C^{1,1}_{loc}(\Omega)\bigcap C^{0,1}(\bar{\Omega})$.

§2 Elliptic operators on manifolds

Consider the elliptic operators

$$Lu = a^{ij}D_{ij}u + b^iD_iu$$

where summation convention is used and the matrix (a^{ij}) satisfies

$$(a^{ij}) \geq 0.$$

The following maximum principle of elliptic operators on manifolds is due to Krylov [3]. Here we state only a special case.

Lemma 1.2 Let $\Psi \in C^2_{loc}(\Omega)$ and $\mathcal{M} = \{x \in \Omega | \Psi(x) = 0\}$. Assume $|D\Psi| \neq 0$ on $\mathcal{M}$ and $L\Psi = L\Psi^2 = 0$ on $\mathcal{M}$. If $u \in C^2_{loc}(\Omega)$ takes its local maximum at $x_0 \in \mathcal{M}$ with respect to the topology of $\mathcal{M}$, then $Lu(x_0) \leq 0$.

Proof Consider the neighborhood V of x_0 in Ω. Make the transformation $x \to \tilde{x} = (\Psi(x), x_2, \cdots, x_n)$ which is non-degenerate in V. The manifold $\mathcal{M} \bigcap V$ is mapped into the hyperplane $\tilde{x}_1 = 0$ and by the hypotheses of the lemma the operator L is changed to

$$\tilde{L}u = \sum_{i,j=2}^{n} a^{ij}\tilde{D}_{ij}u + \sum_{i=2}^{n} b^i \tilde{D}_i u$$

where $\tilde{D}_i = \frac{\partial}{\partial \tilde{x}_i}$. The conclusion is obvious now.

Remark The conditions $L\Psi = L\Psi^2 = 0$ are equivalent to $L\Psi = a^{ij}D_i\Psi D_j\Psi = 0$.

§3 The proof of Theorem 1.1

In [1] we showed the proof of Theorem 1.1 can be referred to the estimation of D^2u. Without loss of generality we can suppose u has enough smoothness and it is strictly convex. Write the equation (1.1) in the form

$$\Phi(D^2u) = [\det D^2u]^{1/n} = g(x, u, Du) \tag{3.1}$$

By [1] we know

$$\Phi^{ij}(r) = \frac{\partial \Phi}{\partial r_{ij}}(r) = \frac{1}{n}\Phi r^{-1} \tag{3.2}$$

$$\frac{\partial^2 \Phi}{\partial r_{ij} \partial r_{kl}} = \frac{1}{n}\Phi[-r^{ik}r^{jl} + n^{-1}r^{ij}r^{kl}] \tag{3.3}$$

where $(r^{ij}) = r^{-1}$ and

$$\det(\Phi^{ij}) = n^{-n}, \quad Tr(\Phi^{ij}) \geq 1. \tag{3.4}$$

Consider $2n+1$-dimensional space $(x, \xi, t) \in \mathbf{R}^n \times \mathbf{R}^n \times \mathbf{R}$. Set

$$\Psi(x, \xi, t) = \psi_{(\xi)} - t\psi^{1/2}, \tag{3.5}$$

where ψ is the function defined in **Theorem** 1.1. Hereafter we denote

$$\psi_i = D_i\psi, \quad \psi_{(\xi)} = \psi_i \xi_i, \quad \psi_{(\xi)(\xi)} = \psi_{ij}\xi_i\xi_j.$$

Define the manifold

$$\mathcal{M} = \{(x, \xi, t) \in \Omega \times \mathbf{R}^n \times \mathbf{R} | \Psi(x, \xi, t) = 0\}. \tag{3.6}$$

We can observe that for $(x, \xi, t) \in \bar{\mathcal{M}}$ and $x \in \partial\Omega$ the direction ξ is tagential to $\partial\Omega$ at x.

We decompose (Φ^{ij}) into

$$\Phi^{ij} = A^{ik}A^{jk}. \tag{3.7}$$

Define the $2n+1$-dimensional vector σ^k $(k = 1, 2, \cdots, n)$ with the components as follows

$$\begin{aligned} &\sigma_i^k = A^{ik}, \sigma_{n+i}^k = \frac{1}{2}\psi^{-1/2}A^{ik}t \ (i = 1, 2, \cdots, n), \\ &\sigma_{2n+1}^k = \psi^{-1/2}A^{ik}\psi_{(\xi)i}. \end{aligned} \tag{3.8}$$

such that

$$\sigma^k \perp \text{ the normal vector to } \mathcal{M}(k = 1, \cdots, n). \tag{3.9}$$

Then we construct the $2n+1$-dimensional elliptic operator

$$Lw = w_{(\sigma^k)(\sigma^k)} - g_{p_i}D_i w + bD_t w \tag{3.10}$$

such that

$$L\Psi = 0 \tag{3.11}$$

Simple computation shows that it suffices to take

$$\begin{aligned} b = &\psi^{-1/2}\Phi^{ij}[\psi_{(\xi)ij} + \frac{1}{2}t\psi^{-1/2}\psi_{ij} + \frac{1}{4}t\psi^{-3/2}\psi_i\psi_j \\ &- \psi^{-1}\psi_{(\xi)i}\psi_j] - \psi^{-1/2}g_{p_i}(\psi_{(\xi)i} - \frac{1}{2}t\psi^{-1/2}\psi_i) \end{aligned} \tag{3.12}$$

By the remark of Lemma 2.1, (3.9) and (3.11) imply that

$$L\Psi = L\Psi^2 = 0$$

Thus Lemma 2.1 can be applied to the operator (3.10) on the manifold $\mathcal{M}$.

Now consider the function, on $\bar{\Omega} \times \mathbf{R}^n \times \mathbf{R}$,

$$w(x, \xi, t) = u_{(\xi)(\xi)} - K\varsigma^2 q^2 - K^2 v^\theta, \tag{3.13}$$

where $\theta = \frac{1}{2n}$ and

$$\begin{aligned} &v = 1 - e^{-M\psi}, \quad \varsigma = e^{Nv}(1 + v^\theta), \\ &q = -2\psi_{(\xi)(\xi)} + t^2 \geq 2|\xi|^2 + t^2. \end{aligned} \tag{3.14}$$

Here M, N and K are positive constants to be determined. Obviously $w(x, \xi, t)$ takes its maximum on $\bar{\mathcal{M}}$ at some point (x_0, ξ_0, t_0). If $(x_0, \xi_0, t_0) \in \bar{\mathcal{M}} \bigcap \partial\Omega \times \mathbf{R}^n \times \mathbf{R}$, since ξ_0 is tagential to $\partial\Omega$ at $x_0, w(x_0, \xi_0, t_0)$ must be bounded by a constant depending only on $n, |\varphi|_{1,1}$, $|\psi|_{1,1}$ and $|u|_{0,1}$.

Now suppose $(x_0, \xi_0, t_0) \in \mathcal{M}$. By Lemma 2.1

$$Lw \leq 0 \quad \text{at} \quad (x_0, \xi_0, t_0).$$

On the other hand, we shall prove $Lw > 0$ at (x_0, ξ_0, t_0) by the suitable choices of K, M and N. The contradiction implies the boundness of w.

First by the properties of conditional maximum we can find, at (x_0,ξ_0,t_0), that

$$
\begin{aligned}
&u_{(\xi)i} = 4K\varsigma^2 q[-\psi_{(\xi)i} + \frac{1}{2}\psi^{-1/2}t\psi_i],\\
&u_{(\xi)(\xi)} = 2K\varsigma^2 q^2,\\
&u_{(\xi)(\xi)i} = 2K\varsigma q^2\varsigma_i + \theta K^2 v^{\theta-1}v_i + t\psi^{-1/2}u_{(\xi)i},\\
&u_{(\xi)(\xi)(\xi)} = [2K\varsigma q^2\varsigma'(\psi) + \theta K^2 v^{\theta-1}v'(\psi)]t\psi^{1/2} + t\psi^{-1/2}u_{(\xi)(\xi)}.
\end{aligned} \tag{3.15}
$$

We have to compute Lw term by term now. We first have

$$
\begin{aligned}
Lv^\theta =&\theta v^{\theta-1}Me^{-M\psi}[\Phi^{ij}\psi_{ij} - M\Phi^{ij}\psi_i\psi_j - g_{p_i}\psi_i]\\
&-\theta(1-\theta)v^{\theta-2}\Phi^{ij}v_iv_j.
\end{aligned} \tag{3.16}
$$

Denote the eigenvalues of (Φ^{ij}) by $0<\lambda_1\le\lambda_2\le\cdots\le\lambda_n$. By (3.4) and Young inequality we find

$$
n^{-n} = [\det(\Phi^{ij})]^{1/n} \le (\lambda_1\lambda_n^{n-1})^{1/n} \le \varepsilon Tr(\Phi^{ij}) + C_\varepsilon\lambda_1, \tag{3.17}
$$

where ε is any positive constant. Using this inequality and noting $D^2\psi\le -I$ and $|g_p|\le C$, we get

$$
\begin{aligned}
Lv^\theta \le& -\theta v^{\theta-1}Me^{-M\psi}[(1-C\varepsilon)Tr(\Phi^{ij}) + (M-C_\varepsilon)\Phi^{ij}\psi_i\psi_j]\\
&-\theta(1-\theta)v^{\theta-2}\Phi^{ij}v_iv_j.
\end{aligned} \tag{3.18}
$$

Now turn to the key computation $L(\varsigma^2q^2)$. By the definition (3.10) of the operator L we have

$$
\begin{aligned}
Lq =& -2\Phi^{ij}\psi_{(\xi)(\xi)ij} - 2t\psi^{-1/2}\Phi^{ij}\psi_{(\xi)ij} - t^2\Phi^{ij}\psi_{ij}\psi^{-1}\\
&+2\psi^{-1}\Phi^{ij}\psi_{(\xi)i}\psi_{(\xi)j} + 2g_{p_i}\psi_{(\xi)(\xi)i} + 2bt
\end{aligned}
$$

Noting the expression (3.12) we get

$$
\begin{aligned}
Lq \le& Cq\psi^{-1/2}Tr(\Phi^{ij}) + C\psi^{-1}q\\
&+2\psi^{-1}\Phi^{ij}[\psi_{(\xi)i}\psi_{(\xi)j} - t\psi^{-1/2}\psi_{(\xi)i}\psi_j + \frac{1}{4}t^2\psi^{-1}\psi_i\psi_j].
\end{aligned}
$$

The equalities (3.15) yields, at (x_0,ξ_0,t_0),

$$
\begin{aligned}
&\Phi^{ij}[\psi_{(\xi)i}\psi_{(\xi)j} - t\psi^{-1/2}\psi_{(\xi)i}\psi_j + \frac{1}{4}t^2\psi^{-1}\psi_i\psi_j]\\
&= \frac{\Phi^{ij}u_{(\xi)i}u_{(\xi)j}}{(4K\varsigma^2q)^2} = \frac{\Phi u_{(\xi)(\xi)}}{n(4K\varsigma^2q)^2} \le C,
\end{aligned}
$$

where C is independent of K,N and M. Thus at (x_0,ξ_0,t_0)

$$
Lq \le Cq\psi^{-1/2}Tr(\Phi^{ij}) + C\psi^{-1}(q+1) \tag{3.19}
$$

In addition, at (x_0,ξ_0,t_0) we have

$$
\begin{aligned}
q_{(\sigma^k)} &= A^{ik}(-2\psi_{(\xi)(\xi)i} - 2t\psi^{-1/2}\psi_{(\xi)i} + 2t\psi^{-1/2}\psi_{(\xi)i})\\
&= -2A^{ik}\psi_{(\xi)(\xi)i}
\end{aligned} \tag{3.20}
$$

and

$$
(3.21)\quad \begin{aligned} L\varsigma^2 &= 2\varsigma L\varsigma + 2\Phi^{ij}\varsigma_i\varsigma_j \\ &= 2\Phi^{ij}\varsigma_i\varsigma_j + 2\varsigma e^{Nv}\{Lv^\theta + N(1+v^\theta)[\Phi^{ij}v_{ij} + N\Phi^{ij}v_iv_j] \\ &\quad + 2N\theta v^{\theta-1}\Phi^{ij}v_iv_j - g_{pi}[N(1+v^\theta) + \theta v^{\theta-1}]v_i\} \\ &\le 4\Phi^{ij}\varsigma_i\varsigma_j + 2\varsigma e^{Nv}Lv^\theta + 2N\varsigma^2\Phi^{ij}v_{ij} + C\varsigma e^{Nv}[N + \theta v^{\theta-1}]Me^{-M\psi} \end{aligned}
$$

Noting (3.18) and

$$
(3.22)\quad \Phi^{ij}v_{ij} \le Me^{-M\psi}(-Tr(\Phi^{ij}) - M\Phi^{ij}\psi_i\psi_j)
$$

and using the second term and the third one on the right-hand side of (3.21) to control the last term by means of inequality (3.17) if M is taken large enough, we get

$$
(3.23)\quad L\varsigma^2 \le 4\Phi^{ij}\varsigma_i\varsigma_j + \frac{3}{2}\varsigma e^{Nv}Lv^\theta + \frac{3}{2}N\varsigma^2\Phi^{ij}v_{ij}.
$$

By (3.20) we have the estimation, at (x_0, ξ_0, t_o),

$$
(3.24)\quad \begin{aligned} &8\varsigma q\varsigma_{(\sigma^k)}q_{(\sigma^k)} \\ &= 8\varsigma q e^{Nv}[(1+v^\theta)N + \theta v^{\theta-1}]Me^{-M\psi}\Phi^{ij}\psi_i(-2\psi_{(\xi)(\xi)j}) \\ &\le C\varsigma^2q^2[N + \theta v^{\theta-1}]Me^{-M\psi}[\varepsilon Tr(\Phi^{ij}) + \frac{1}{4\varepsilon}\Phi^{ij}\psi_i\psi_j]. \end{aligned}
$$

where ε is any positive constant.

By (3.19), (3.20), (3.23) and (3.24) we obtain, if taking ε small enough and M large enough (independent of N and K) in (3.24),

$$
\begin{aligned} L(\varsigma^2q^2) &= q^2L\varsigma^2 + 2\varsigma^2qLq + 2\varsigma^2q_{(\sigma^k)}q_{(\sigma^k)} + 8\varsigma q\varsigma_{(\sigma^k)}q_{(\sigma^k)} \\ &\le 4\Phi^{ij}\varsigma_i\varsigma_jq^2 + \varsigma e^{Nv}q^2Lv^\theta + N\varsigma^2q^2\Phi^{ij}v_{ij} \\ &\quad + C\varsigma^2q^2\psi^{-1/2}Tr(\Phi^{ij}) + C\varsigma^2\psi^{-1}(q^2+1). \end{aligned}
$$

Now fix M taken above. Noting $\theta = \frac{1}{2n}$ we have

$$
(3.25)\quad \begin{aligned} \psi^{-1} &\le n^n(\lambda_1\lambda_n^{n-1})^{1/n}\psi^{-1} \\ &\le C[\psi^{\frac{1}{n}-1}Tr(\Phi^{ij}) + \psi^{\frac{1}{n}-2}\Phi^{ij}\psi_i\psi_j] \le -C\psi^\theta Lv^\theta \end{aligned}
$$

and so, at (x_0, ξ_0, t_0),

$$
\begin{aligned} L(\varsigma^2q^2) &\le 4\Phi^{ij}\varsigma_i\varsigma_jq^2 + C\psi^{-1}\varsigma^2 + \varsigma e^{Nv}q^2Lv^\theta \\ &\quad + N\varsigma^2\Phi^{ij}v_{ij}q^2 + C\varsigma^2q^2\psi^{-1/2}Tr(\Phi^{ij}) - C\psi^\theta q^2Lv^\theta. \end{aligned}
$$

Consider two cases: $\psi \le \delta M^{-1}$ and $\psi > \delta M^{-1}$. The last two terms on the right-hand side can be controlled by the third or fourth term respectively in the above two cases if take first δ small enough and then take N sufficiently large. Thus at (x_0, ξ_0, t_0).

$$
(3.26)\quad \begin{aligned} L(\varsigma^2q^2) \le& 4\Phi^{ij}\varsigma_i\varsigma_jq^2 + C\varsigma^2\psi^{-1} \\ &+ \frac{1}{2}\varsigma e^{Nv}q^2Lv^\theta + \frac{1}{2}N\varsigma^2\Phi^{ij}v_{ij}q^2. \end{aligned}
$$

Finally we compute

$$
\begin{aligned}
&Lu_{(\xi)(\xi)} \\
(3.27)\qquad &=\Phi^{ij}u_{(\xi)(\xi)ij}-g_{p_i}u_{(\xi)(\xi)i}+t\psi^{-1/2}\Phi^{ij}u_{(\xi)ij}+\frac{t^2}{2}\psi^{-1}\Phi^{ij}u_{ij}.
\end{aligned}
$$

It is easy to get

$$
|\Phi^{ij}u_{(\xi)ij}-g_{p_i}u_{(\xi)i}|\le C|\xi|
$$

and

$$
\begin{aligned}
&\Phi^{ij}u_{(\xi)(\xi)ij}-g_{p_i}u_{(\xi)(\xi)i} \\
(3.28)\qquad &\ge-\frac{\partial^2\Phi}{\partial r_{ij}\partial r_{kl}}u_{ij(\xi)}u_{kl(\xi)}-C|\xi||Du_{(\xi)}|-C|\xi|^2,
\end{aligned}
$$

where C depends only on $|g|_{1,1}$ and $|u|_{0,1}$. Since $\Phi=[\det D^2u]$ is invariant under the transformation $y=Bx$ where $\det B=1$, for any fixed direction α we can, following Pogorelov's trick ([4] p.74), make a choice of coordinate system in such a way that the direction α coincides with one of the axes and we have $u_{ij}=0$ at x_0 for $i\ne j$. Thus (no summation for index α)

$$
\begin{aligned}
&-\frac{\partial^2\Phi}{\partial r_{ij}\partial r_{kl}}u_{ij\alpha}u_{kl\alpha}=\frac{1}{n}\Phi[\sum_{i,j=1}^{n}\frac{(u_{ij\alpha})^2}{u_{ii}u_{jj}}-\frac{1}{n}(\sum_{i=1}^{n}\frac{u_{ii\alpha}}{u_{ii}})^2] \\
\ge&\frac{1}{n}\Phi\sum_{i\ne j}\frac{(u_{ij\alpha})^2}{u_{ii}u_{jj}}\ge\frac{1}{n}\Phi[2\sum_{i=1}^{n}\frac{u_{i\alpha\alpha}^2}{u_{ii}u_{\alpha\alpha}}-\frac{2u_{\alpha\alpha\alpha}^2}{u_{\alpha\alpha}^2}].
\end{aligned}
$$

In our case we have

$$
-\frac{\partial^2\Phi}{\partial r_{ij}\partial r_{kl}}u_{ij(\xi)}u_{kl(\xi)}\ge\frac{2\Phi^{ij}u_{i(\xi)(\xi)}u_{j(\xi)(\xi)}}{u_{(\xi)(\xi)}}-\frac{2\Phi u_{(\xi)(\xi)(\xi)}^2}{nu_{(\xi)(\xi)}}.
$$

With the help of (3.15) it follows that at (x_0,ξ_0,t_0),

$$
\begin{aligned}
-\frac{\partial^2\Phi}{\partial r_{ij}\partial r_{kl}}u_{ij(\xi)}u_{kl(\xi)}\ge&(4-\delta N^{-1})Kq^2\Phi^{ij}\varsigma_i\varsigma_j \\
&-C\delta^{-1}Nq\psi^{-1}\frac{\Phi^{ij}u_{(\xi)i}u_{(\xi)j}}{u_{(\xi)(\xi)}}-C(q+K^2\psi^{2\theta-1}q^{-2}+\psi^{-1}q),
\end{aligned}
$$

where δ is any positive constant less than 1. In view of $\Phi^{ij}u_{(\xi)i}=\Phi\xi_i/n$, the above inequality reduces to

$$
\begin{aligned}
&-\frac{\partial^2\Phi}{\partial r_{ij}\partial r_{kl}}u_{ij(\xi)}u_{kl(\xi)} \\
\ge&(4-\delta N^{-1})Kq^2\Phi^{ij}\varsigma_i\varsigma_j-C(q+K^2\psi^{2\theta-1}q^{-2}+\delta^{-1}N\psi^{-1}q).
\end{aligned}
$$

Substituting it into (3.28) and using (3.15) again we find that

$$
\begin{aligned}
\Phi^{ij}u_{(\xi)(\xi)ij}-g_{p_i}u_{(\xi)(\xi)i}\ge&(4-\delta N^{-1})Kq^2\Phi^{ij}\varsigma_i\varsigma_j \\
&-C(q+K^2\psi^{2\theta-1}q^{-2}+\delta^{-1}N\psi^{-1}q)-C(1+K\psi^{-1/2}q^2).
\end{aligned}
$$

It follows from this and (3.27) that by the convexity of u,

$$Lu_{(\xi)(\xi)} \geq (4-\delta N^{-1})Kq^2\Phi^{ij}\varsigma_i\varsigma_j - C(q+K^2\psi^{2\theta-1}q^{-2}+\delta^{-1}N\psi^{-1}q) - C(1+K\psi^{-1}q^2) \tag{3.29}$$

at (x_0,ξ_0,t_0). Without loss of generality we may suppose $q \geq K$, otherwise we get the boundness of w at once. Summing up (3.18), (3.26) and (3.29) we find, at (x_0,ξ_0,t_0), that

$$\begin{aligned} Lw &= Lu_{(\xi)(\xi)} - KL(\varsigma^2 q^2) - K^2 Lv^\theta \\ &\geq -\delta N^{-1}Kq^2\Phi^{ij}\varsigma_i\varsigma_j - C\psi^{-1}(\delta^{-1}Nq + K\varsigma^2 + Kq^2) \\ &\quad -\frac{1}{2}K\varsigma e^{Nv}q^2 Lv^\theta - \frac{K}{2}N\varsigma^2 q^2\Phi^{ij}v_{ij} - K^2 Lv^\theta \\ &= \sum_{i=1}^{5} I_i. \end{aligned} \tag{3.30}$$

It is easy to get

$$\begin{aligned} |I_1| &\leq C\delta K\varsigma^2(N+N^{-1}\theta^2 v^{2\theta-2})\Phi^{ij}\psi_i\psi_j \\ &\leq \frac{1}{4}(I_3+I_4) \end{aligned}$$

if δ is small enough (independent of N and K). Then

$$|I_2| \leq C\psi^{-1}Kq^2 + C(K+K^{-1}\delta^{-2}N^2)\varsigma^2 = I_{21}+I_{22}.$$

The term I_{21} can be controlled by $I_3/4$ or $I_4/4$ respectively in the cases $\psi \leq \varepsilon M^{-1}$ and $\psi > \varepsilon M^{-1}$ by means of (3.25) if take first ε small enough and then take N sufficiently large. Now fix δ and N. Similarly, with the help of (3.25) I_{22} can be controlled by $I_5/4$ if K is large enough. For such a choice of M, N and K we finally get, at (x_0,ξ_0,t_0),

$$Lw > 0.$$

The contradiction implies that

$$w \leq C \quad \text{on} \quad \mathcal{M} \tag{3.31}$$

where C depends only on n, $|g|_{1,1}$, $|\varphi|_{1,1}$, $|\psi|_{3,1}$ and $|u|_{0,1}$.

For any fixed direction $\alpha \in \mathbf{R}^n$, $|\alpha| = 1$, take

$$\xi = \psi^{1/2}\alpha.$$

On $\mathcal{M}$ we have

$$|t| \leq C.$$

It follows from (3.31) that

$$\psi u_{\alpha\alpha} \leq C,$$

i.e.

$$u_{\alpha\alpha} \leq \frac{C}{\psi(x)} \quad \text{for any} \quad \alpha \in \mathbf{R}^n, |\alpha| = 1,$$

where C depends only on n, $|g|_{1,1}$, $|\varphi|_{1,1}$, $|\psi|_{3,1}$, and $|u|_{0,1}$. This is what we want to estimate.

References

[1] Chen Yazhe, On degenerate Monge-Ampere equations (to appear).

[2] N.S.Trudinger, On degenerate fully nonlinear elliptic equations in balls (preprint).

[3] N.V.Krylov, On degenerate nonlinear elliptic equations, Mat. Sb., 120, 3(1983), 311-330 (Russian).

[4] A.V.Pogorelov, The Minkowski multidimensional problems.

INITIAL AND BOUNDARY PROBLEMS FOR THE DEGENERATE OR SINGULAR SYSTEM OF THE FILTRATION TYPE

Fu Hong-Yuan

Institute of Applied Physics
and Computational Mathematics,
Beijing, China

The filtration equation is nonlinearly parabolic. Since the diffusion coefficient has zero points, the degenerate parabolic equation has solution, which are not smooth. The first paper about the existence of weak solutions for such filtration equations is given by O.A.Oleinik, A.C.Kalashnikov and Zhou Yu-lin in 1958[1]. The equations of this sort arise in many application, including heat flow in materials with a temperature dependent conductivety, flow in a porous medium, biological model and so on. Therefore many papers have appeared about degenerate parabolic equations[2-7]. Applying difference methods in [8,9], the existence of the weak solution is proved, as well as for the case, when the diffusion coefficients are singular at some points.

There are also degenerate parabolic systems appearing in some physical problems. In 1984 Zhou Yu-lin considered the system of the filtration type[10]

$$u_t = [\mathrm{grad}_u \phi(u)]_{xx}$$

where u is a vector and $\phi(u)$ a scalar function. He proved the existence and uniqueness of the weak solution for the periodic boundary problems and Cauchy problems. In his paper the regularization method is considered. In his approach it is assumed that $\phi \in C^3$.

In this paper we consider some initial and boundary problems for degenerate or singular systems in general form

$$u_t = f(u)_{xx}$$

where $u=u(x,t)$ and $f=f(u)$ are J-dimensional vector valued function. The convergence of difference solutions is used to obtain the existence of the weak solution. The uniqueness is also proved.

We assume that the nonlinear vector valued functions f(u) satisfy monotonic and continuous conditions. We do not have to suppose that f(u) are differentiable. It is degenerate, if the diffusion coefficients exist and the diffusion coefficient matrix f'(u) degenerates at some points. We say it is singular, if f'(u) exists almost everywhere and there are some singular points. There may be many degenerate points.

§1 Suppose the vector valued functions f(u) satisfy the following conditions:
(F_1) $f(u) \in C(R^J)$ and satisfy the monotonic condition, i.e., the scalar product

$(u_1-u_2)^T[f(u_1)-f(u_2)] > 0, \ \forall\ u_1 \neq u_2, \quad u_1, u_2 \in R^J$;

(F_2) there are positive numbers α and β such that the inner product $u^T f(u) \geqq \alpha|u|^{1+\beta}$.

THEOREM 1. (Existence of Inverse Functions) Suppose the vector valued functions $f(u)$ satisfy the conditions (F_1) and (F_2). Assume $v \in R^J$ is given, then the system $v = f(u)$ has an inverse function $u = f^{-1}(v)$.

Proof. First, let us prove the uniqueness of the solution for the system $v=f(u)$ for given $v \in R^J$. Suppose there are two solutions u_1 and $u_2 \in R^J$, $u_1 \neq u_2$, Since $f(u_1)=f(u_2)=v$, it follows that the inner product

$$(u_1-u_2)^T[f(u_1)-f(u_2)] = 0.$$

This equation is contracdictive to Condition (F_1). Hence the solution is unique.

In order to prove the existence of the inverse function, consider the following system:

$$u = v + \lambda w - \lambda f(w) \tag{1}$$

where $\lambda \in [0,1]$ is a parameter. As $v \in R^J$ is given, a map $u=T_\lambda w$ is determined by system (1). It is easy to prove that the map is continuous respect to w for every $\lambda \in [0,1]$ and it is uniformly continuous with respect to $\lambda \in [0,1]$ as w belongs to a bounded set. Next we have to prove that for every possible solution of $u=T_\lambda u$, there is a bound independing of $\lambda \in [0,1]$. Writting u in (1) instead of w, we have

$$v = (1-\lambda)u + \lambda f(u). \tag{2}$$

Make inner product with u, then

$$|u||v| \geqq u^T v = (1-\lambda)|u|^2 + \lambda u^T f(u).$$

Using the condition (F_2), there is

$$|u||v| \geqq (1-\lambda)|u|^2 + \lambda\alpha|u|^{1+\beta}.$$

Hence

$$|u||v| \geqq (1-\lambda+\lambda\alpha)\ |u|^{\beta_0}$$

$$|u||v| \geqq \alpha_0|u|^{\beta_0}, \ \forall\ \lambda \in [0,1],$$

Where

$$\beta_0 = \begin{cases} \max\ (\ 2, 1+\beta\ ), & \text{for } |u| \leqq 1; \\ \min\ (\ 2, 1+\beta\ ), & \text{for } |u| > 1; \end{cases}$$

$$\alpha_0 = \min\ (\ 1, \alpha\ ).$$

From the assumption $\alpha, \beta > 0$ we know that $\alpha_0 > 0$, $\beta_0 > 1$.

Hence

$$|u| \leqq \left[\frac{1}{\alpha_0}|v|\right]^{\frac{1}{\beta_0-1}}$$

This means that, Since v is bounded, u is uniformly bounded for $\lambda \in [0,1]$. As $\lambda=0$, (2) has unique solution. From the Leray-Schauder fixed point theorem[11] we know that $u=T_\lambda u$ has a solution for $\lambda=1$, i.e., $f(u)=v$ has a unique solution, denoted by $u=f^{-1}(v)$.

The conditions (F_1) and (F_2) do not require that $f(u)$ be differentiable.

If the diffusion coefficient $f'(u)$ exists almost everywhere, it may be degenerate or singular at some points. It is allowed that $f'(u)$ has many zero points, for example, $f(u) = u + \mathrm{Sin}\, u$.

For example $f(u) = |u|^{s-1}u$, where u is a vector. The following inequalities may be proved

$$(u-v)^T[f(u)-f(v)] \geqq \tfrac{1}{2}(|u|^{s-1} + |v|^{s-1})|u-v|^2, \text{ for } s\geqq 1;$$
$$(u-v)^T[f(u)-f(v)] \geqq \tfrac{1}{2}(|u|^{1-s} + |v|^{1-s})\left||u|^s\frac{u}{|u|} - |v|^s\frac{v}{|v|}\right|^2, \quad 0<S\leqq 1.$$

From these it follows that the above example $f(u)$ satisfies (F_1) and (F_2). For $s\geqq 1$, it is degenerate at $u=0$. For $0<s<1$, it has a singular point. The case $0<s<1$, called "fast diffusion", also occurs in plasma physics.

In order to investigate the continuity of the inverse function, we give the following additional condition.

(F_3) For every point $u_1 \in R^J$, there is a neighbourhood $B(u_1)$ and positive numbers α and β such that

$$(u-u_1)^T[f(u)-f(u_1)] \geqq \alpha|u-u_1|^{1+\beta}, \quad u\in B(u_1).$$

The condition (F_3) is a local property for $f(u)$.

THEOREM 2. (Continuity of Inverse Functions) Suppose the vector valued functions $f(u)$ satisfy the conditions (F_1), (F_2) and (F_3), then for $v=f(u)$ the inverse function $u=f^{-1}(u)$ is local $\frac{1}{\beta}$-Hölder continuous for $\beta>1$ and local Lipschitz continuous for $0<\beta<1$.

Proof. Under the conditions (F_1) and (F_2), we know that the inverse function exists and is unique. Let $v_1=f(u_1)$, $v=f(u)$, then

$$v-v_1= f(u)-f(u_1).$$

From the continuity of $f(u)$ it follows that

$$\lim_{u\to u_1} [f(u)-f(u_1)] = \lim_{u\to u_1} (v-v_1)=0.$$

Making inner product and considering the condition (F_3), we obtain

$$|u-u_1||v-v_1| \geqq (u-u_1)^T(v-v_1) \geqq \alpha|u-u_1|^{1+\beta}, \text{ for } u\in B(u_1).$$

Thus we have

$$|u-u_1| \leqq \left(\frac{1}{\alpha}|v-v_1|\right)^{\frac{1}{\beta}}.$$

This implies that $u=u(v)$ is local Hölder continuous.

§2 In this section we consider the first boundary problem

$$u_t = f(u)_{xx}, \quad 0<x<b, \ t>0,$$
$$u(x,0) = u^\circ(x), \quad 0\leqq x\leqq b, \qquad \text{(I)}$$
$$u(0,t) = u(b,t) = 0, \quad t>0$$

where $u=(x,t)$, $f=f(u)$ and $u^\circ(x)$ are vector valued functions.

Denote τ, h the step length of the net $(mh, n\tau)$, where $h=b/(M+1)$, $N\tau=T$ for positive number T. Let $\Delta_+f(u_m)=f(u_{m+1})-f(u_m)$, $\Delta_-f(u_m)=f(u_m)-f(u_{m-1})$, $\delta_+u^n=u^{n+1}-u^n$,

$\delta_- u^n = u^n - u^{n-1}$.

Consider the implicit difference scheme

$$u_m^{n+1} = u_m^n + \frac{\tau}{h^2}\Delta_+\Delta_- f(u_m^{n+1}), \quad m=1,2, \dots, M; \tag{3}$$

$$u_m^o = u^o(mh), \quad u_o^{n+1} = u_{M+1}^{n+1} = 0.$$

The existence of solutions for the nonlinear equations (3) will be proved.

LEMMA 1. Suppose two sets of numbers u_m, v_m, $m=0,1, \dots, M+1$, then

$$\sum_{m=o}^{M} u_m \Delta_+ v_m = -u_o v_o - \sum_{m=1}^{M+1} (\Delta_- u_m) v_m + u_{M+1} v_{M+1},$$

$$\sum_{m=1}^{M} u_m \Delta_+ \Delta_- v_m = -u_o \Delta_+ v_o - \sum_{m=o}^{M} (\Delta_+ u_m)(\Delta_+ v_m) + u_{M+1} \Delta_- v_{M+1}.$$

Proof. It may be checked directly.

THEOREM 3. Suppose the vector valued function f(u) satisfies the condition (F_1), then the difference equations (3) have solutions.

Proof. Suppose u_m^n are already solved. Let $v=[v_m]_{m=o}^{M+1}$ Satisfying the conditions

$$v_o = v_{M+1} = 0, \quad \sum_{m=1}^{M} |v_m|^2 < \infty$$

Consider the system

$$u_m = u_m^n + \frac{\tau}{h^2}\Delta_+\Delta_- f(v_m), \quad m = 1,2, \dots, M; \tag{4}$$

$$u_o = u_{M+1} = 0,$$

where the parameter $\lambda \in [0,1]$. Denote $u=[u_m]_{m=o}^{M+1}$. A map $u=T_\lambda v$ is defined by the system (4). It may be proved that for every $\lambda \in [0,1]$ the map is continuous with respect to v. Since v belongs to a bounded set, it is uniformly continuous with respect to $\lambda \in [0,1]$. It will be proved that every possible solution of $u=T_\lambda u$ is bounded independent of λ. Instead of v we write u in (4), then making the inner product with u and applying the lemma 1, we obtain

$$\sum_{m=1}^{M} |u_m|^2 = \Sigma u_m^T u_m^n + \lambda \frac{\tau}{h^2} \Sigma\, u_m^T \Delta_+\Delta_- f(u_m)$$

$$= \sum_{m=1}^{M} u_m^T u_m^n - \frac{\lambda\tau}{h^2}[u_o^T \Delta_+ f(u_o) + \sum_{m=o}^{M} (\Delta_+ u_m)^T \Delta_+ f(u_m) - u_{M+1}^T \Delta_- f(u_{M+1})].$$

Using the boundary conditions $u_o = u_{M+1} = 0$ and the monotone condition (F_1), we obtain

$$\sum_{m=1}^{M} |u_m|^2 \leq \sum_{m=1}^{M} u_m^T u_m^n \quad .$$

From the Cauchy inequality it follows that

$$\sum_{m=1}^{M} |u_m|^2 \leq \sum_{m=1}^{M} |u_m^n|^2 \tag{5}$$

This inequality means that u is uniformuly bounded and the bound is independent of λ. For $\lambda=0$ there is obviously a unique solution for $u=T_\lambda u$. From the fixed point theorem of Leray-Schauder it follows that there exists a solution also for $\lambda=1$.

COROLLARY. Suppose $u^o(x) \in C[0,b]$, f(u) satisfy the condition (F_1), then the solutions of (3) are uniformly bounded in L_2-norm, i.e.,

$$\sum_{m=1}^{M} |u_m^{n+1}|^2 h \leq \sum_{m=1}^{M} |u_m^o|^2 h, \quad \forall\ n \geq 0.$$

Proof. Multiplying (5) by h, then the conclusion follows.

LEMMA 2. Suppose the sequence $[v_m]_0^{M+1}$, $v_\circ=0$ and $h=b/(M+1)$. If the following estimate holds

$$\sum_{m=0}^{M} \left|\frac{\Delta_+ V_m}{h}\right|^2 h \leq C ,$$

then $|V_m| \leq (b+c)/2$.

Proof. From the following equality

$$V_m = V_o + \sum_{\ell=0}^{m-1} (V_{\ell+1} - V_\ell) = V_o + \sum_{\ell=0}^{m-1} \frac{\Delta_+ V_\ell}{h} h$$

it follows that

$$|V_m| \leq \sum_{\ell=0}^{M} \left(h + \left|\frac{\Delta_+ V_\ell}{h}\right|^2 h\right)/2$$
$$\leq (b+c)/2$$

THEOREM 4. Suppose f(u) satiafies the condition (F); $f(0)=0$; $u^\circ(x) \in c[0,b]$, $\int_\circ^b \left|\frac{d}{dx} f[u^\circ(x)]\right|^2 dx < \infty$, then there is a subsequence of the difference solutions of (3), which is convergent and the limit function

$$u(x,t) \in L_\infty([0,T],\ C^{(0,\frac{1}{2\beta})}[0,b]) \cap \frac{\partial f(u(x,t))}{\partial x} \in L_\infty([0,T],L_2)$$

is a weak solution of the problem (I):

$$\int_\circ^T \int_\circ^b [u(x,t)^T \Psi_t(x,t) - f_x[u(x,t)]^T \Psi_x(x,t)]dxdt + \int_\circ^b u^\circ(x)^T \Psi(x,0)dx = 0 ,$$

where the vector valued test functions $\Psi(x,t)$ are such that $\Psi(\cdot,t) \in (\overset{\circ}{W}{}_2^{(1)}[0,b])^J$, $\Psi(x,T)=0$.

Proof. As $|u_m^{n+1} - u_m^n| \neq 0$, let A_m be the matrix defined

$$A_m = \frac{[f(u_m^{n+1}) - f(u_m^n)][u_m^{n+1} - u_m^n]^T}{|u_m^{n+1} - u_m^n|^2} . \tag{6}$$

Multiplying the difference equations (3) by the matrix (6), then

$$\frac{[\delta_- f(u_m^{n+1})][\delta_- u_m^{n+1}]^T [\delta_- u_m^{n+1}]}{|\delta_- u^{n+1}|^2} = \frac{\tau}{h^2} A_m \Delta_+ \Delta_- f(u_m^{n+1}) . \tag{7}$$

Note that, as $|\delta_- u_m^{n+1}| = 0$, both sides of the system

$$\delta_- u_m^{n+1} = \frac{\tau}{h^2} \Delta_- \Delta_+ f(u_m^{n+1})$$

are equal to zero. Therefore, as $|\delta_- u_m^{n+1}| = 0$, let $A_m=0$, so that the system (7) holds also for $u_m^{n+1} = u_m^n$. (7) may be written as

$$\delta_- f(u_m^{n+1}) = \frac{\tau}{h^2} A_m \Delta_+ \Delta_- f(u_m^{n+1}) .$$

Making inner product with $\Delta_- \Delta_+ f(u_m^{n+1})$ and summation with respect to m, we have

$$\sum_{m=1}^{M}[\Delta_-\Delta_+ f(u_m^{n+1})]^T f(u_m^{n+1}) = \sum_{m=1}^{M}[\Delta_-\Delta_+ f(u_m^{n+1})]^T f(u_m^n) + $$
$$+ \frac{\tau}{h^2}\sum_{m=1}^{M}[\Delta_-\Delta_+ f(u_m^{n+1})]^T A_m[\Delta_-\Delta_+ f(u_m^{n+1})].$$

Using Lemma 1, the boundary condition, and (3), we get

$$\sum_{m=o}^{M}|\Delta_+ f(u_m^{n+1})|^2 = \sum_{m=o}^{M}[\Delta_+ f(u_m^{n+1})]^T[\Delta_+ f(u_m^n)] -$$
$$- \frac{\tau}{h^2}\sum_{m=1}^{M}[\frac{h^2}{\tau}\delta_- u_m^{n+1}]^T A_m[\frac{h^2}{\tau}\delta_- u_m^{n+1}] . \tag{8}$$

From the definition of A_m and the monotone condition for $f(u)$, it follows that

$$[\delta_- u_m^{n+1}]^T A_m[\delta_- u_m^{n+1}] = [u_m^{n+1} - u_m^n]^T[f(u_m^{n+1}) - f(u_m^n)] \geqq 0 .$$

Using the Cauchy inequality, from (8) it follows that

$$\sum_{m=o}^{M}|\Delta_+ f(u_m^{n+1})|^2 \leqq \sum_{m=o}^{M}|\Delta_+ f(u_m^n)|^2 .$$

Put $v=f(u)$, then we get

$$\sum_{m=o}^{M}\left|\frac{\Delta_+ V_m^{n+1}}{h}\right|^2 h \leqq \sum_{m=o}^{M}\left|\frac{\Delta_+ f(u_m^\circ)}{h}\right|^2 h \leqq C, \quad \forall\ n\geqq 0 . \tag{9}$$

Hence we have the conclusion: $\sum_{m=o}^{M}\left|\frac{\Delta_+ V_m^{n+1}}{h}\right|^2 h$, $\forall\ n\geqq 0$ are uniformly bounded.

From the assumption $f(0)=0$, we know that $v_o^{n+1}=0$. From Lemma 2 we get $|V_m^{n+1}| \leqq (b+c)/2$, $\forall\ n\geqq 0$, i.e., $|V_m^{n+1}|$, $m=o,1,\dots,M+1$; $n=o,1,\dots$ are uniformly bounded. Using the theorem of inverse functions we know that u_m^n are uniformly bounded, and so are $f(u_m^{n+1})$.

For the domain $t>0$, $0\leqq x\leqq b$ we define the following functions

$$u_{h\tau}(x,t) = u_m^n \text{ for } x\in[mh, (m+1)h),\ t\in[n\tau,(n+1)\tau),$$
$$n=1,2,\dots;\ m=o,1,\dots,M;$$
$$u_{h\tau}(b,t) = u_{M+1}^n = 0, \text{ for } t\in[n\tau,(n+1)\tau),\ n=1,2,\dots .$$

There are piecewise constants, where u_m^n are the solutions to the difference system(3). Correspondently $V_{h\tau}(x,t) = f[u_{h\tau}(x,t)]$ are also functions with piecewise constants. It is already proved that $V_{h\tau}(x,t)$ are uniformly bounded and the bound is independent of h,τ and n. It will be proved that for every fixed t, the total variation of the functions $V_{h\tau}(\cdot,t)$ are uniformly bounded, denoted by $V_{h\tau}(\cdot,t)\in BV$.

Suppose $t\in[n\tau,(n+1)\tau)$, then for any partition $0=x_o<x_1<\dots<x_p = b$ the total variation may be estimated

$$\sum_{i=1}^{p}|V_{h\tau}(x_i,t) - V_{h\tau}(x_{i-1},t)| \leqq \sum_{m=1}^{M+1}|V_m^n - V_{m-1}^n|$$
$$\leqq (b + \sum_{m=1}^{M+1}\left|\frac{\Delta_- V_m^n}{h}\right|^2 h)/2 .$$

From (9) we know that the right-hand side of the inequality above is uniformly bounded and the bound is independent of h,τ and n, i.e., the total variations are uniformly bounded. Applying the Helly's theorem there is a subset $V_\gamma(x,t)$, $\gamma=1,2,\dots$ from the set of functions $[V_{h\tau}]$, such that $V_\gamma(x,t)$ converges in L_1 for given fixed t, as $\gamma\to\infty$ $(\tau,h\to 0)$

Applying the diagonal choosing method, for rational times $t=p/q$, there is a subsequence, denoted also by $v_\gamma(x,t)$ $\gamma=1,2,\ldots$, such that $v_\gamma(x,t)$ converges for numerable rational times. From Theorem 1 it follows that $u_\gamma(x,t)$ $\gamma=1,2,\ldots$ are convergent for numerable rational times. We have now to prove that for any t, as $\gamma\to\infty$, the set of functions $v_\gamma(x,t)$ is convergent.

For any t_1, t_2, without loss of generality, we suppose $t_1 > t_2$, there exist n and s such that

$$u_\gamma(x,t_1) - u_\gamma(x,t_2) = u_\gamma(x,(n+s)\tau) - u_\gamma(x,n\tau) .$$

Let $\Phi(x)\in \overset{o}{W}{}_2^{(1)}[0,b]$, consider

$$\begin{aligned}
&\int_0^b \Phi(x)[u_\gamma(x,t_1) - u_\gamma(x,t_2)]dx = \\
&= \int_0^b \Phi(x) \sum_{\alpha=1}^{S} [u_\gamma(x,(n+\alpha)\tau) - u_\gamma(x,(n+\alpha-1)\tau)]dx \\
&= \sum_{m=o}^{M} \sum_{a=1}^{S} \int_{mh}^{(m+1)h} \Phi(x)[u_\gamma(x,(n+\alpha)\tau) - u_\gamma(x,(n+\alpha-1)\tau)]dx \\
&= \sum_{m=1}^{M} \sum_{\alpha=1}^{S} \int_{mh}^{(m+1)h} \Phi(x)dx[u_m^{n+\alpha} - u_m^{n+\alpha-1}] \\
&= \sum_{m=1}^{M} \sum_{\alpha=1}^{S} [\psi((m+1)h) - \psi(mh)][\tfrac{\tau}{h^2}\Delta_-\Delta_+ f(u_m^{n+\alpha})] \\
&= -\sum_{\alpha=1}^{S} \sum_{m=o}^{M} [\Delta_+^2\Psi_m][\Delta_+ f(u_m^{n+\alpha})]\tfrac{\tau}{h^2}
\end{aligned}$$

where $\Psi'(x)=\phi(x)$, $\Psi_m=\psi(mh)$. Hence

$$\left|\int_0^b \Phi(x)[u_\gamma(x,t_1) - u_\gamma(x,t_2)]dx\right| \le \sum_{\alpha=1}^{S} [\sum_m |\frac{\Delta_+^2\psi_m}{h^2}|^2 h]^{\frac{1}{2}} [\sum_m |\frac{\Delta_+ V_m^{n+\alpha}}{h}|^2 h]^{\frac{1}{2}} \tau .$$

Using (9), it leads to

$$\left|\int_0^b \phi(x)[u_\gamma(x,t_1) - u_\gamma(x,t_2)]dx\right| \le C(|t_1 - t_2| + \tau) . \tag{10}$$

For any t we have the following estimation

$$\begin{aligned}
&\left|\int_0^b \phi(x)[u_\gamma(x,t) - u_{\gamma'}(x,t)]dx\right| \le \left|\int_0^b \phi(x)[u_\gamma(x,t) - u_\gamma(x,p/q)]dx\right. + \\
&+ \left|\int_0^b \phi(x)[u_\gamma(x,p/q) - u_{\gamma'}(x,p/q)]dx\right| + \left|\int_0^b \phi(x)[u_{\gamma'}(s,p/q) - u_{\gamma'}(x,t)]dx\right| \\
&\le \left|\int_0^b \phi(x)[u_\gamma(x,p/q) - u_{\gamma'}(x,p/q)]dx\right| + C(|t-p/q| + \tau) .
\end{aligned}$$

Hence $u_\gamma(x,t)$ is convergent for any t. The limit functions are denoted by $u(x,t)$. Then for any t the sequence $v_\gamma(x,t)$ is uniformly bounded and as are the total variations. Therefore for any t the sequence $v_\gamma(x,t)$ is convergent to $v(x,t)=f[u(x,t)]$ and $v(\cdot,t)\in BV$.

It will be proved that $v(\cdot,t)$ are continuous functions with respect to x. Let ξ_1, $\xi_2 \in [0,b]$ and suppose $\xi_1 < \xi_2$. From the definition of $v_{h\tau}(x,t)$ for given ξ_1 and ξ_2 there are n, m and s, such that

$$\begin{aligned}
v_\gamma(\xi_2,t) - v_\gamma(\xi_1,t) &= v_{m+s}^n - v_m^n \\
&= \sum_{\alpha=1}^{S}(v_{m+\alpha}^n - v_{m+\alpha-1}^n) .
\end{aligned}$$

Then

$$|v_\gamma(\xi_2,t) - v_\gamma(\xi_1,t)| \le (\sum_\alpha h)^{\frac{1}{2}} (\sum_\alpha |\frac{\Delta_- V_{m+\alpha}^n}{h}|^2 h)^{\frac{1}{2}}$$

Applying (9) we obtain

$$|v_\gamma(\xi_2,t) - v_\gamma(\xi_1,t)| \leq C(|\xi_2 - \xi_1| + h)^{\frac{1}{2}} \quad ,$$

$$|v(\xi_2,t) - v(\xi_1,t)| \leq C(|\xi_2 - \xi_1|)^{\frac{1}{2}} \quad ,$$

as $\gamma\to\infty$, $h\to 0$, i.e. $v(\cdot,t)\in C^{(0,\frac{1}{2})}[0,b]$.

From the theorem 2 it follows that $u(\cdot,t)\in C^{(0,\frac{1}{2\beta})}[0,b]$.

We define

$$w_{h\tau}(x,t) = \frac{\Delta_+ v_m^n}{h} \text{ , for } x\in[mh,(n+1)h), \quad t\in[n\tau,(n+1)\tau) \text{ ,}$$

$$n=1,2,\ldots;\ m=0,1,\ldots,\ M \text{ .}$$

It will be proved that there exists a subsequence, which converges weakly to $f[u(x,t)]_x$ in L_2-norm. From (9) we know that $w_{h\tau}(\cdot,t)\in L_2$, and $\|w_{h\tau}(\cdot,t)\|_{L_2(0,b)}\leq C$, $\forall\ t\in[0,T]$ are uniformly bounded. Then there is a subsequence $w_{r'}(x,t)$, which is weakly convergent to $w(x,t)\in L_\infty([0,T],L_2(0,b))$. Suppose $[r']\subseteq[r]$ and write also $[w_r]$.

Let the test function $\eta(x)\in\mathring{H}'(0,b)$. Consider the equality

$$\sum_{m=o}^{M}[\eta_m^T\frac{\Delta_+ v_m^n}{h}]h = -\sum_{m=1}^{M+1}[(\frac{\Delta_-\eta_m}{h})^T v_m^n]h \text{ .}$$

Hence

$$\int\eta(x)^T w(x,t)dx = -\int\eta'(x)^T v(x,t)dx \text{ .}$$

This means that $w(x,t)$ is the generalized derivative of $v(x,t)$ with respect to x and $w = \partial v/\partial x = f[u(x,t)]_x$ a.e. .

Now we will prove that $u(x,t)$ is a weak solution of (I).

Let $\psi(x,t)$ be test functions, $\psi_m^n=\psi(mh,n\tau)$, $\psi(mh,T)=0$. Multiplying (3) by $\psi_m^n\tau h$ and making summation with respect to m and n, we have

$$0 = \sum_{n=o}^{N}\sum_{m=1}^{M}(\psi_m^n)^T\frac{\delta_+ u_m^n}{\tau}\tau h - \sum_{n=o}^{N}\sum_{m=1}^{M}(\psi_m^n)^T\frac{\Delta_-\Delta_+ f(u_m^{n+1})}{h^2}\tau h$$

$$= -\sum_{n=1}^{N}\sum_{m=1}^{M}(\frac{\delta_-\psi_m^n}{\tau})^T u_m^n\tau h - \sum_{m=1}^{M}(\psi^\circ)^T u_m^\circ h + \sum_{n=o}^{N}\sum_{m=1}^{M+1}(\frac{\Delta_-\psi_m^n}{h})^T\frac{\Delta_- f(u_m^{n+1})}{h}\tau h \quad .$$

Putting $h,\tau\to 0$, we obtain the integral expression of the weak solution. The existence of the solution is proved.

<u>THEOREM 5</u>. The weak solution of Theorem 4 for (I) is unique.

<u>Proof</u>. Suppose there are two solutions $u_1(x,t)$ and $u_2(x,t)$ satisfying the

same initial data. Then we have

$$\iint[(u_1 - u_2)^T\psi_t(x,t) - [f(u_1)_x - f(u_2)_x]^T\psi_x]dxdt = 0 \quad .$$

Take

$$\psi(x,t) = \int_T^t[f(u_1(x,\tau)) - f(u_2(x,\tau))]d\tau \; .$$

Substituting it into the above equations, we get

$$\iint(u_1 - u_2)^T[f(u_1) - f(u_2)]dxdt = \iint[f(u_1)_x - f(u_2)_x]^T\int_T^t[f(u_1(x,\tau))_x - f(u_2(x,\tau))_x]d\tau dxdt$$

$$= \frac{1}{2}\int_0^b\int_0^T\frac{d}{dt}\left|\int_T^t[f(u_1(x,\tau))_x - f(u_2(x,\tau))_x]d\tau\right|^2dxdt$$

$$= -\frac{1}{2}\int_0^b\left|\int_0^T[f(u_1(x,\tau))_x - f(u_2(x,\tau))_xd\tau\right|^2dx \leqq 0 \quad .$$

Since the weak solutions u_1 and u_2 are continuous with respect to x, if there is a point $\xi\in(0,b)$, such that $u_1(\xi,t) \neq u_2(\xi,t)$ for some $t>0$ then there is a neighbourhood $d=(\xi-\delta,\xi+\delta)$ such that $u_1(x,t)\neq u_2(x,t)$ $\forall$ $x\in d$. According to (F_1) if $u_1\neq u_2$, then

$$(u_1-u_2)^T[f(u_1) - f(u_2)] > 0 \; .$$

Hence the integral

$$\int_0^b\int_0^T(u_1-u_2)^T[f(u_1)-f(u_2)]dxdt > 0 \; .$$

From this contradiction it follows $u_1 \equiv u_2$, thus proving the uniqueness.

§3 In this section we consider the case on the interval $x\in[0,\infty)$. The initial and boundary problem is

$$\begin{aligned} &u_t = f(u)_{xx} , \quad 0<x<\infty , \quad t>0 , \\ &u(0,t) = 0, \quad t\geqq 0 \; ; \; u(x,0) = u^\circ(x), \quad x\in[0,\infty) \end{aligned} \qquad \text{(II)}$$

where u and f are J-dimensional vectors.

Consider the implicit difference scheme

$$\begin{aligned} &u_m^{n+1} = u_m^n + \frac{\tau}{h^2}\Delta_+\Delta_-f(u_m^{n+1}), \quad m=1,2,\ldots \; ; \; n=0,1,\ldots \; ; \\ &u_o^{n+1} = 0, \quad n=0,1,\ldots \; ; \; u_m^\circ = u^\circ(mh), \quad m=0,1,\ldots \quad . \end{aligned} \qquad (12)$$

(12) is a system with infinite number of nonlinear equations, the existence of which will be proved.

LEMMA 3. Suppose $u=[u_m]_{m=o}^{\infty}$ and $v=[v_m]_{m=o}^{\infty}$ satisfy the condition $u\in l_2$, $v\in l_2$, i.e. $\Sigma|u_m|^2<\infty$, $\Sigma|v_m|^2<\infty$,

then the following equalities hold

$$\sum_{m=0}^{\infty} u_m \Delta_+ v_m = -u_0 v_0 - \sum_{m=1}^{\infty} (\Delta_- u_m) v_m ,$$

$$\sum_{m=1}^{\infty} u_m \Delta_+ \Delta_- v_m = -u_0 \Delta_+ v_0 - \sum_{m=0}^{\infty} (\Delta_+ u_m)(\Delta_+ v_m) . \tag{13}$$

Proof. Since $u \in l_2$ and $v \in l_2$, $\lim_{m\to\infty} u_m = \lim_{m\to\infty} v_m = 0$. From Lemma 1, let $m\to\infty$, then we get (13).

THEOREM 6. Suppose f satisfies (F_1) and $\sum_{m=0}^{\infty} |u_m^0|^2 < \infty$, then the system (12) has a solution $\forall\ n \geq 0$.

Proof. Suppose we have already solved u_m^n. Consider the system

$$u_m = u_m^n + \lambda \frac{\tau}{h^2} \Delta_+ \Delta_- f(v_m), \quad m=1,2,\ldots ;$$

$$u_0 = 0$$

where the parameter $\lambda \in [0,1]$. This system defines a map $u = T_\lambda v$. From the fixed point theorem as Theorem 3 it follows that $u = T_\lambda u$ has a solution for $\lambda = 1$.

COROLLARY. suppose $u^0(x) \in C[0,\infty) \cap u^0(x) \in L_2(0,\infty)$, $f(u)$ satisfy (F_1), then the solutions of (12) are uniformly bounded in L_2-norm, i.e., $\sum_{n=0}^{\infty} |u_m^{n+1}|^2 h \leq \sum_{m=0}^{\infty} |u_m^0|^2 h < C$, $\forall\ n \geq 0$.

THEOREM 7. Suppose f satisies (F), $f(0)=0$, $u^0(x) \in C[0,\infty) \cap L_2(0,\infty)$, $\int_0^\infty |\frac{d}{dx} f(u^0(x))|^2 dx < \infty$, then there is a subsequence of the solutions of (12), which is convergent and the limit function $u(x,t) \in L_\infty[(0,T), C^{(0,\frac{1}{2\beta})}[0,\infty)] \cap \frac{\partial f(u(x,t))}{\partial x} \in L_\infty([0,T], L_2(0,\infty))$ is a weak solution of (II):

$$\int_0^T \int_0^\infty [u(x,t)^T \psi_t(x,t) - f(u(x,t))_x^T \psi_x(x,t)] dxdt + \int_0^\infty u^0(x)^T \psi(x,0) dx = 0,$$

where $\psi(x,t)$ is the vector valued test function with compact support. $\psi(\cdot,t) \in w_2^{(1)}(0,\infty)$, $\psi(0,t) = \psi(x,T) = 0$. The problem, therefore, (II) has a weak solution.

Proof. The proof is similar to that of Theorem 4. Let the matrix A be

$$A_m = \begin{cases} \dfrac{[f(u_m^{n+1}) - f(u_m^n)][u_m^{n+1} - u_m^n]^T}{|u_m^{n+1} - u_m^n|^2}, & |u_m^{n+1} - u_m^n| \neq 0 \\ 0, & |u_m^{n+1} - u_m^n| = 0 \end{cases}$$

Multiplying (12) by A_m, yields

$$\delta_- f(u_m^{n+1}) = \frac{\tau}{h^2} A_m \Delta_+ \Delta_- f(u_m^{n+1}) .$$

Making inner product and sumation, we have

$$\sum_{m=0}^{M}[\Delta_-\Delta_+ f_m^{n+1}]^T f_m^{n+1} = \sum_{m=1}^{M}[\Delta_-\Delta_+ f_m^{n+1}]^T f_m^n + \frac{\tau}{h^2}\sum_{m=1}^{M}(\Delta_-\Delta_+ f_m^{n+1})^T A_m(\Delta_-\Delta_+ f_m^{n+1}) ,$$

Using Lemma 3 and the boundary condition, considering the system (12), the definition of A_m and (F_1) we obtain

$$\sum_{m=0}^{M}(\Delta_+ f_m^{n+1})^T(\Delta_+ f_m^{n+1}) - (f_{M+1}^{n+1})^T\Delta_- f_{M+1}^{n+1} = \sum_{m=0}^{M}(\Delta_+ f_m^{n+1})^T(\Delta_+ f_m^n) -$$
$$- (f_{M+1}^{n+1})^T\Delta_- f_{M+1}^n - \frac{\tau}{h^2}\sum_{m=1}^{M}[\frac{h^2}{\tau}\delta_- u_m^{n+1}]^T A_m[\frac{h^2}{\tau}\delta_- u_m^{n+1}] , \tag{14}$$

$$[\delta_- u_m^{n+1}]^T A_m[\delta_- u_m^{n+1}] = [u_m^{n+1}-u_m^n]^T[f(u_m^{n+1})-f(u_m^n)] \geq 0 .$$

Considering the Cauchy inequality, it follows from (14) that

$$\sum_{m=0}^{M}|\Delta_+ f_m^{n+1}|^2 \leq \sum_{m=0}^{M}|\Delta_+ f_m^n|^2 + 2(f_{M+1}^{n+1})^T\Delta_- f_{M+1}^{n+1} - 2(f_{M+1}^{n+1})^T\Delta_- f_{M+1}^n$$

Note that $\lim_{M\to\infty} u_M^n = \lim_{M\to\infty} u_M^{n+1} = 0$, $f(0) = 0$.

Suppose $\sum_{m=0}^{\infty}|\Delta_+ f_m^n|^2 < \infty$, put $M\to\infty$, then

$$\sum_{m=0}^{\infty}|\Delta_+ f_m^{n+1}|^2 \leq \sum_{m=0}^{\infty}|\Delta_+ f_m^n|^2 .$$

Let $v=f(u)$, from the inequality above it follows that

$$\sum_{m=0}^{\infty}\left|\frac{\Delta_+ v_m^{n+1}}{h}\right|^2 h \leq \sum_{m=0}^{\infty}\left|\frac{\Delta_+ f(u_m^{\circ})}{h}\right|^2 h \leq C, \ \forall \ n\geq 0 \tag{15}$$

This means $\sum_{m=0}^{\infty}\left|\frac{\Delta_+ v_m^{n+1}}{h}\right|^2 h$, $n\geq 0$ are uniformly bounded.

From the assumption $f(0)=0$, $u_0^{n+1}=0$ we know that $v_0^{n+1}=0$. Let $L>0$ be a finite number. Applying Lemma 2 we know that $|v_m^{n+1}|\leq\frac{1}{2}(L+C)$, $n\geq 0$, $0\leq mh\leq L$. It implies v_m^{n+1}, $n\geq 0$, $0\leq mh\leq L$ are uniformly bounded. From the theorem of inverse function we know that u_m^n, $\forall$ $n\geq 0$, $0\leq mh\leq L$ are also uniformly bounded.

For the domain $t>0$, $0\leq x<\infty$ we define the functions

$$u_{h\tau}(x,t)=u_m^n, \quad x\in[mh,(m+1)h), \quad t\in[n\tau,(n+1)\tau),$$
$$n=1,2,\dots ; \ m=1,2,\dots .$$

Then there are

$$v_{h\tau}(x,t)=f(u_{h\tau}(x,t)), \quad t\geq 0, \quad x\geq 0.$$

From the corollary of Theorem 7 it follows that there is a subsequence $u_r(x,t)$

as $r\to\infty$ is convergent weakly to $u(x,t)$, $u(\cdot,t)\in L_2(0,\infty)$ and $\|u(\cdot,t)\|_{L_2(0,\infty)}$ uniformly bounded $\forall\, t\in[0,T]$. From the proof of Theorem 4 we know that there is a subsequence, denoted also by $[v_r(x,t)]$, $[u_r(x,t)]$ converging to $v(x,t)$ and $u(x,t)$ respectively on finite interval $0\leq x\leq L$, $0\leq t\leq T$. And $v(\cdot,t)\in C^{(0,\frac{1}{2})}[0,L]$, $u(\cdot,t)\in C^{(0,\frac{1}{2\beta})}[0,L]$.

Define the functions with constants by the piece

$$w_{h\tau}(x,t)=\frac{\Delta_+ v_m^n}{h}\,,\quad x\in[mh,(m+1)h),\quad t\in[n\tau,(n+1)\tau),$$
$$n=1,2,\dots\,;\ m=0,1,\dots\ .$$

From (15) we know $\|w_{h\tau}(\cdot,t)\|_{L_2(0,\infty)}$ is uniformly bounded.

Therefore, there is a subequence $w_{r'}(x,t)$ as $r'\to\infty$, converging weakly to $w(x,t)\in L_\infty([0,T],\ L_2(0,\infty))$. Let $[r']\subseteq[r]$. It is denoted also by w_r.

Let $\eta(x)\in H'[0,\infty)$, $\eta(0)=0$ with compact support. Consider

$$\sum_{m=0}^{\infty}\eta_m^T\frac{\Delta_+ v_m^n}{h}h=-\sum_{m=0}^{\infty}\left(\frac{\Delta_-\eta_m}{h}\right)^T v_m^n\,h\ .$$

Noting that $v_r(x,t)$ is convergent to $v(x,t)$ for any finite interval $[0,L]$ and $\eta(x)$ has compact support, we have

$$\int\eta(x)^T w(x,t)\,dx=-\int(\eta')^T v(x,t)\,dx\ .$$

That is to say, $w(x,t)$ is the generalized derivative with respect to x, i.e. $\frac{\partial v}{\partial x}=w$. Hence $\frac{\partial v(\cdot,t)}{\partial x}\in L_2(0,\infty)$ i.e. $f[u(\cdot,t)]_x\in L_2(0,\infty)$ and $\|f[u(\cdot,t)]_x\|_{L_2(0,\infty)}$, $\forall\, t\in[0,T]$ is uniformly bounded.

Next it will be proved that $u(x,t)$ is a weak solution of (II). Let $\psi(x,t)$ be a vector valued test function with compact support and $\psi(0,t)=\psi(x,T)=0$, $\psi(\cdot,t)\in H^1(0,\infty)$.

Making inner product, we have

$$0=\sum_{n=0}^{N}\sum_{m=1}^{\infty}(\psi_m^n)^T\frac{\delta_+u_m^n}{\tau}\tau h-\sum_{n=0}^{N}\sum_{m=1}^{\infty}(\psi_m^n)^T\frac{\Delta_-\Delta_+f(u_m^{n+1})}{h^2}\tau h$$

$$=-\sum_{n=1}^{N}\sum_{m=1}^{\infty}\left(\frac{\delta_-\psi_m^n}{\tau}\right)^T u_m^n\tau h-\sum_{m=1}^{\infty}(\psi^o)^T u_m^o h+\sum_{n=0}^{N}\sum_{m=1}^{\infty}\left(\frac{\Delta_-\psi_m^n}{h}\right)^T\left(\frac{\Delta_-f_m^{n+1}}{h}\right)\tau h$$

Let $h,\tau\to 0$, then we get the expression of the weak solution.

<u>THEOREM 8</u>. The weak solution satisfying the conditions of the Theorem 7 is unique.

Proof. It is similar to that of Theorem 5. The only difference is that the test functions are taken differently.

Suppose there are two solutions u_1 and u_2 of (II). Then

$$\int_o^T\int_o^\infty\{(u_1,u_2)^T\psi_t(x,t) - [f(u_1)_x - f(u_2)_x]^T\psi_x(x,t)\}dxdt = 0.$$

Let

$$\phi(x,t) = f[u_1(x,t)] - f[u_2(x,t)] ,$$

$$\eta_z(x) = 1, \quad 0\leq x\leq z; \quad \eta_z(x)=0, \quad x\geq z+1; \quad 0\leq\eta_z(x)\leq 1, \quad z\leq x\leq z+1 .$$

Let $\eta_z(x)$ be continuous differentiable and uniformly bounded with respect to x. Put the test functions $\psi(x,t)=\eta_z(x)\int_T^t\phi(x,\tau)d\tau$. It is easy to see that $\psi(0,t)=0$, $\forall\ t\geq 0$; supp. $\psi(x,t)<\infty$, the generalized derivatives are

$$\psi_t(x,t) = \eta_z(x)\ \phi(x,t) ,$$

$$\psi_x(x,t) = \eta_z'(x)\int_T^t\phi(x,\tau)d\tau+\eta_z(x)\int_T^t\phi(x,\tau)_x d\tau$$

and that $\psi(x,t)$ satisfy the conditions of test functions.

Substituting it into the integral relations of weak solution, we have

$$\int_o^T\int_o^\infty\eta_z(x)(u_1-u_2)^T\phi(x,t)dxdt = \int_o^T\int_o^\infty[\eta_z'(x)+\eta_z(x)]\ \phi(x,t)_x^T\ \int_T^t\phi(x,\tau)_x d\tau dxdt$$

$$= \int_o^T\int_z^{z+1}\eta_z'(x)\phi(x,t)_x^T\ \int_T^t\phi(x,\tau)d\tau dxdt - \frac{1}{2}\int_o^\infty\eta_z(x)\left|\int_o^T\phi(x,\tau)_x d\tau\right|^2dx. \quad (16)$$

If there is a point (ξ,t), $\xi>0$, $t>0$ such that $u_1(\xi,t)\neq u_2(\xi,t)$, according to (F_1) there is

$$[u_1(\xi,t) - u_2(\xi,t)]^T[f(u_1(\xi,t)) - f(u_2(\xi,t))] = \varepsilon > 0 ,$$

then there is an interval $d=(\xi-\delta,\xi+\delta)$, where $\delta>0$, such that

$$[u_1(x,t) - u_2(x,t)]^T[f(u_1(x,t) - f(u_2(x,t))] \geq \frac{\varepsilon}{2} , \ \forall\ x\in d .$$

Assume $z>\varepsilon+1$, then

$$\int_o^T\int_o^\infty\eta_z(x)(u_1-u_2)^T[f(u_1)-f(u_2)]dxdt \geq \ \varepsilon T/2.$$

Since $\phi(x,t)_x \in L_2(0,\infty)$; $\phi(x,t)\to 0$ for $x\to\infty$.

Take z sufficiently big, such that

$$\left|\int_o^T\int_z^{z+1}\eta_z'(x)\phi(x,t)_x^T\int_T^t\phi(x,\tau)d\tau dxdt\right| < \varepsilon T/2.$$

Substituting it into (16) leads to a contradiction.

From this it follows that $u_1\equiv u_2$, thus proving the uniqueness.

§4. The difference methods may be used to proved the existence of solutions for another degenerate parabolic system under weaker conditions. We have the following results.

THEOREM 9. Suppose $f(v)$ satisfies (F), $f(0)=0$, $u^\circ(x)\in C'[0,b]$ and $\int_0^b\left|\frac{d}{dx}f[u_x^\circ(x)]\right|^2dx<\infty$. Then there exists a unique weak solution in global for the problem

$$u_t = f(u_x)_x,\ 0<x<b,\ t>0,$$
$$u_x(0,t) = u_x(b,t) = 0;\ u(x,0) = u^\circ(x) \qquad \text{(III)}$$
$$u(x,t)\in L_\infty([0,T],C^{(1,\frac{1}{2\beta})}[0,b]) \cap f[u_x(x,t)]_x\in L_\infty([0,T],L_2(0,b)) .$$

THEROEM 10. Suppose f satisfies the monotone and continuous conditions (F), $f(0)=0$, $u^\circ(x)\in C^2[0,b]$ and $\int_0^b\left|\frac{d}{dx}f[u_{xx}^\circ(x)]\right|^2dx<\infty$. Then there exists a unique weak solution in global for the problem

$$u_t = f(u_{xx}),\ 0<x<b,\ t>0,$$
$$u(x,0) = u^\circ(x);\ u(0,t) = u(b,t) = 0, \qquad \text{(IV)}$$
$$u(x,t)\in L_\infty([0,T],C^{(2,\frac{1}{2\beta})}[0,b]) \cap f[u_{xx}(x,t)]_x \in L_\infty([0,T],L_2(0,b)) .$$

The proofs will be given in another paper.

REFERENCES

1 Oleinik, O.A., Klashnikow, A.S. and Zhou Yu-lin, The Cauchy problem and boundary problems for equations of the type of nonstationary filtration, Izv. Akad. Nauk. SSSR, ser. mat. 22(1958) 667-704.

2 Aronson, D.G., Regularity proerties of flows through porous media, SIAM J. Appl. Math., 17(1969), 461-467.

3 Gilding, B.H., Peletier, L.A., The Cauchy problem for an equation in the theory of filtration, Arch. Rat. Mech. Anal. 61(1976), 127-140.

4 Gilding, B.H., Properties of solutions of an equation in the theory of filtration, Arch. Rat. Mech. Anal., 65(1977), 203-225.

5 Knerr, B.F., The porous medium equation in one dimension, Trans. American Math. Soc. 234(1977), 381-414.

6 Aronson, D.G., Regularity properties of flows through pororus media: A counter-example, SIAM J. Appl. Math. 19(1970), 299-307.

7 Kamin, S., Source-type solutions for equations of nonstationary filtration, J. Math. Anal. Appl., 64(1978), 263-276.

8 Fu Hong-Yuan, Convergence of difference solutions for filtration equations, Math. Numer. Sinica, 3(1985), 302-308.

9 Fu Hong-Yuan, Convergence of difference solutions for degenerate or singular parabolic equations,to appear.

10 Zhou Yu-lin, Initial value problems for nonlinear degenerate systems of filtration type, Chin. Ann. of Math. 5B(1984) 632-652.

11 Friedman, A., Partial Differential Equation of Parabolic Type, Prentice Hall, Inc. 1964.

On Interior Regularity of Solutions
of a Class of Hypoelliptic Equations

Huang Yumin
Dept. of Math., Nankai Univ.
Tianjin, China

1. Introduction. As one knows, in 1967 Hörmander proved a famous theorem on "Sum of Squares"[1]. He considered operators of the form

$$P = \sum_{j=1}^{k} X_j^2 + X_0$$

where X_0 and $X_1, \ldots, X_k$ are smooth real vector fields. It was proved that such operators are hypoelliptic, if the Lie algebra generated by $X_0, X_1, \ldots, X_k$ gives a basis for all vector fields at any point. A simpler proof of this theorem was given by Kohn[2]. Olejnik and Radkevic gave an alternative proof and extended this theorem to a class of operators of second order with nonnegative principal symbol (see [3], Thm.2.6.2). Using the Fefferman-Phong inequality, Hormander obtained a similar result for more general pseudodifferential operators (see [4]. Thm.22.2.6). But in the results above, the estimates on regularity of solutions are less precise. For operators of the form $\sum X_j^2 + X_0$, the best possible such estimates were proved via analysis on nilpotent Lie groups by Rothschild and Stein[5].

It is natural to ask can the estimates of Rothschild and Stein be extended to more general operators? In [6], Bolley, Camus and Nourrigat gave a positive answer for operators of the form $\sum A_j^* A_j$, where A_j are pseudodifferential operators with real principal symbol. In this paper, we discuss operators of general form with consideration of effect of lower terms and get precise conditions to ensure hypoellipticity with loss of 4/3 derivatives.

2. Main result. We consider a linear partial differential operator P(X,D) defined in an open subset Ω of R^n with C^∞ coefficients. Let

$$P(x,D) = P_m(x,D)+P_{m-1}(x,D)+\ldots+P_1(x,D)+P_0(x),$$

where $D=(D_1,\dots,D_n)$, $D_j=-i\frac{\partial}{\partial x_j}$ and $P_{m-j}(x,D)$ is a homogeneous operator of degree m-j. We assume

$$P_m(x,\xi)\geq 0,\quad \operatorname{Re} P_{m-1}(x,\xi)=0$$

Set

$$Q_j(x,\xi)=\begin{cases}\frac{\partial P_m}{\partial\xi_j}(x,\xi), & j=1,\dots,n\\ \langle\xi\rangle^{-1}\frac{\partial P_m}{\partial x_{j-n}}(x,\xi), & j=n+1,\dots,2n\end{cases}$$

$$Q_o(x,\xi)=\text{the principal symbol of } \frac{1}{2i}(P-P^*)$$

where $\langle\xi\rangle=(1+|\xi|^2)$. In fact, $Q_o(x,\xi)$ is the imaginary part of sub-principal symbol of P. We denote the properly supported pseudodifferential operator with symbol $Q_j(x,\xi)$ by $Q_j(x,D)$. For a multiple index $I=(\alpha_1,\dots,\alpha_k)$, where $\alpha_j\in\{0,1,2,\dots,2n\}$, we write

$$|I|=\sum_{j=1}^{k}\lambda_j\ ,\quad \text{here}\quad \lambda_j=\begin{cases}1, & \alpha_j\neq 0\\ 2, & \alpha_j=0\end{cases}$$

and $Q_I=[Q_{\alpha_1},[Q_{\alpha_2},\dots[Q_{k-1},Q_{\alpha_k}]\dots]$

Obviously Q_I is a pseudodifferential operator of $k(m-1)-k+1$ th order. The main result in this paper is the following.

Theorem. If for any $(x,\xi)\in\Omega\times\mathbb{R}^n\backslash\{o\}$, there is Q_I with $|I|\leq 3$ and Q_I being elliptic at (x,ξ), then for any $s\in R^1$ we have

$$u\in\mathcal{D}'(\Omega),\qquad Pu\in H_s^{loc}(\Omega)\Rightarrow u\in H^{loc}_{m-\frac{4}{3}+s}(\Omega)$$

Remark.1. It is easy to extend this theorem to pseudodifferential operator under similar hypotheses.

2. Comparing this theorem with Theorem 22.2.6 in [4], one can see that under the same assumptions, Theorem 22.2.6 gives only

$$u\in\mathcal{D}'(\Omega),\quad Pu\in H_s^{loc}(\Omega)\Rightarrow\quad u\in H^{loc}_{m-\frac{7}{4}+s}(\Omega)$$

Hence the result in this paper improves Theorem 22.2.6 partially.

The proof of the main result will be given in following sections. We shall prove three lemmas first, then get interior estimates, and finally complete the proof of the theorem.

3. Three lemmas.

Lemma 1. For any compact set $K \subseteq \Omega$ and $s \in R^1$ there is $C_{k,s} > 0$ such that

$$\sum_{j=i}^{2n} \|Q_j u\|_s^2 \leq C_{k,s}(\mathrm{Re}(Pu,u)_{s+\frac{m}{2}-1} + \|u\|_{s+m-2}^2), \quad u \in C_o^\infty(K) .$$

Proof. First we quote a well-known proposition: Let $f(t) \in C^2(R^1)$ and $f(t) \geq 0$, $\sup_t |f''(t)| < +\infty$, then

$$|f'(t)|^2 \leq 2(\sup_s |f''(s)|)f(t) . \tag{1}$$

From $P_m(x,\xi) \geq 0$ and (1), we can easily conclude that for any compact set $K \subseteq \Omega$ there is a constant $C_k > 0$ such that

$$|<\xi>^{-m+2}Q_j(x,\xi)|^2 \leq C_k <\xi>^{-m+2}P_m(x,\xi),\ x \in K,\ j=1,\ldots,2n$$

Since $\mathrm{Re}\, P_{m-1}(x,\xi)=0$, we have

$$|<\xi>^{-m+2}Q_j(x,\xi)|^2 \leq C_k<\xi>^{-m+2}\mathrm{Re}P(x,\xi)+C_k<\xi>^{-m+2}[-\mathrm{Re}(P_{m-2}+\ldots+P_o)]$$

Obviously $<\xi>^{-m+2}[-\mathrm{Re}(P_{m-2}+\ldots+P_o)] \in S^o_{1,0}$. Using the Fefferman-Phong enequality we can get

$$\|Q_j u\|_{-m+2}^2 \leq C_k(\mathrm{Re}(Pu,u)_{-\frac{m}{2}+1} + \|u\|_o^2), \quad u \in C_o^\infty(K) \tag{2}$$

where C_k is a constant depending on K.

For any real number s, we denote the pseudodifferential operator with symbol $<\xi>^s$ by $<D>^s$. Taking $\phi \in C_o^\infty(\Omega)$ and $\phi \equiv 1$ in a neighborhood of K, we have for $u \in C_o^\infty(K)$

$$\|Q_j u\|_{-m+2+s} = \|<D>^s Q_j u\|_{-m+2} \leq \|\phi<D>^s Q_j u\|_{-m+2} + \|(1-\phi)<D>^s Q_j u\|_{-m+2}$$

$$\leq \|Q_j\phi<D>^s u\|_{-m+2} + \|[\phi<D>^s, Q_j]u\|_{-m+2} + \|(1-\phi)<D>^s Q_j u\|_{-m+2}$$

Since $\mathrm{Supp}(1-\phi) \cap \mathrm{Supp}\, u=\Phi$ and $[\phi<D>^s, Q_j]$ is a pseudodifferential operator of m-2+s th order, there exists $C_{k,s} > 0$ such that

$$\|Q_j u\|_{-m+2+s} \leq \|Q_j\phi<D>^s u\|_{-m+2} + C_{k,s}\|u\|_s$$

According to (2), we can find positive constants C_k and $C_{k,s}$ such that

$$\|Q_j u\|^2_{m+2+s} \leq C_k \mathrm{Re}(P\phi\langle D\rangle^s u,\ \phi\langle D\rangle^s u)_{-\frac{m}{2}+1} + C_{k,s}\ \|u\|^2_s\ , \qquad (3)$$

where $u \in C_o^\infty(K)$ and $j=1,2,\ldots,2n$. Using the symbol calculus of pseudo-differential operators, we can prove easily that

$$[P,\phi\langle D\rangle^s] = \sum_{\ell=1}^{2n} \alpha_\ell(x,D)Q_\ell(x,D)+\beta(x,D)$$

where $\alpha_\ell(x,D)$ are operators of s th order and $\beta(x,D)$ of s+m-2 th order;

$$[\phi,\langle D\rangle^s] \sim 0 \quad \text{in a neighborhood of } K.$$

Hence there is $C_{k,s} > 0$ and $C_{k,\varepsilon} > 0$ such that for $u \in C_o^\infty(K)$

$$\mathrm{Re}(P\phi\langle D\rangle^s u,\phi\langle D\rangle^s u)_{-\frac{m}{2}+1} \leq \mathrm{Re}(Pu,u)_{-\frac{m}{2}+1+s} + C_{k,s}\Big(\sum_{\ell=1}^{2n} \|Q_\ell u\|_{-m+2+s}\|u\|_s + \|u\|^2_s\Big)$$

Using (3), we can get a constant $C_{k,s} > 0$ such that

$$\sum_{j=1}^{2n} \|Q_j u\|^2_{-m+2+s} \leq C_{k,s}(\mathrm{Re}(Pu,u)_{-\frac{m}{2}+1+s} + \|u\|^2_s),\ u \in C_o^\infty(K)$$

In view of arbitrariness of s, the proof of Lemma 1 is complete.

Lemma 2. For any compact set $K \subseteq \Omega$ and $\varepsilon > 0$, there is a constant $C_{k,\varepsilon} > 0$ such that $\|Q_o u\|_o \leq \varepsilon\|u\|_{m-\frac{4}{3}} + C_{k,\varepsilon}(\|Pu\|_o + \|u\|_{m-\frac{2}{3}})$, $u \in C_o(K)$.

Proof. Since $Q_o(x,D)$ is the principal part of $\frac{1}{2i}(P-P^*)$, we can write

$$Q_o(x,D) = \frac{1}{2i}(P(x,D)-P^*(x,D))+R_1(x,D)$$

$$Q^*(x,D) = Q_o(x,D)+R_2(x,D)$$

where R_1 and R_2 are differential operators of m-2 th order. With $u \in C_o^\infty(K)$, we consider

$$I_m(Pu,Q_o u)_o = \frac{1}{2i}((Pu,Q_o u)_o-(Q_o u,Pu)_o)$$

$$= \frac{1}{2i}((Q_o^* P-P^*Q_o)u,u)_o$$

In view of $\quad Q_o^*P-P^*Q_o = [Q_oP]+2iQ_o^2+R_2P-2iR_1Q_o$

$$= [Q_oP]+2iQ_o^*Q_o+R_2P-2i(R_1+R_2)Q_o,$$

hence $\text{Im}(Pu,Q_0 u)_0 = \|Q_0 u\|_0^2 + \frac{1}{2i}([Q_0,P]u,u)_0 + (Gu,u)_0$,

where $G = \frac{1}{2i} R_2 P - (R_1+R_2)Q_0$. Therefore

$$\|Q_0 u\|_0^2 \leq |\text{Im}(Pu,Q_0 u)_0| + |\frac{1}{2i}([Q_0,P]u,u)_0| + |(Gu,u)_0|$$

$$\leq |\frac{1}{2i}([Q_0,P]u,u)_0| + \delta\|Q_0 u\|_0^2 + C_{k,\delta}(\|Pu\|_0^2 + \|u\|_{m-2}^2) ,$$

where $\delta > 0$ is arbitrarily given and $C_{k,\delta}$ is a constant depending on K and δ. Taking $\delta = \frac{1}{2}$, we can find a constant $C_k > 0$ such that

$$\|Q_0 u\|_0^2 \leq |\frac{1}{2i}([Q_0, P]u,u)_0| + C_k(\|Pu\|_0^2 + \|u\|_{m-2}^2). \tag{4}$$

By symbol calculus, we have

$$[Q_0,P] = \sum_{\ell=1}^{2n} \alpha_\ell(x,D)Q_\ell(x,D) + \beta(x,D)$$

where α_ℓ are pseudodifferential operators of m-1 th order and $\beta(x,D)$ of 2m-3 th order. Hence there is $C_k > 0$ such that

$$|\frac{1}{2i}([Q_0,P]u,u)_0| \leq C_k(\|u\|_{m-\frac{4}{3}} \sum_{\ell=1}^{2n} \|Q\ u\|_{\frac{1}{3}} + \|u\|_{m-\frac{3}{2}}^2),\ u \in C_0^\infty(K)$$

Using Lemma 1, we can find a constant $C_k > 0$ such that

$$\sum_{\ell=1}^{2n} \|Q_\ell u\|^2 \leq C_k(\text{Re}(Pu,u)_{\frac{m}{2}-\frac{2}{3}} + \|u\|_{m-\frac{5}{3}}^2)$$

$$\leq C_k(\|Pu\|_0 \|u\|_{m-\frac{4}{3}} + \|u\|_{m-\frac{5}{3}}^2),\quad u \in C_0^\infty(K). \tag{5}$$

Therefore for any $\varepsilon > 0$ there is $C_{k,\varepsilon} > 0$ such that

$$|\frac{1}{2i}([Q_0,P]u,u)_0| \leq \varepsilon\|u\|_{m-\frac{4}{3}}^2 + C_{k,\varepsilon}(\|Pu\|_0^2 + \|u\|_{m-\frac{3}{2}}^2) . \tag{6}$$

Combining (4) and (6), we get the conclusion of this lemma.

Lemma 3. For any compact subset k of Ω and $\varepsilon > 0$, there is $C_{k,\varepsilon} > 0$ such that

$$\sum_{j=1}^{2n} \|[Q_0,Q_j]u\|_{-m+\frac{5}{3}} \leq \varepsilon\|u\|_{m-\frac{4}{3}} + C_{k,\varepsilon}(\|Pu\|_0 + \|u\|_{m-\frac{3}{2}}),\ u \in C_0^\infty(K)$$

Proof. The set consisting of all properly suported pseudodifferential operators defined in Ω with symbol $S_{1,0}^m$ is denoted by $\psi_{1,0}^m$. Obviously $[Q_0,Q_j] \in \psi_{1,0}^{2m-3}$. Hence there exists $R(x,D) \in \psi_{1,0}^{1/3}$ such that

$$\|[Q_o,Q_j]u\|^2_{-m+\frac{5}{3}} = ([Q_o,Q_j]u,[Q_o,Q_j]u)_{-m+\frac{5}{3}}$$

$$= ([Q_o,Q_j]u,Ru)_o \;, \; u \in C_o^\infty(K) \qquad (7)$$

Since $(Q_oQ_ju,Ru)_o = (Q_ju,Q_o^*Ru)_o = (Q_ju,RQ_o^*u)_o+(Q_ju,[Q_o^*,R]u)_o$

$$= (Q_ju,RQ_ou)_o+(Q_ju,Gu)_o \;,$$

where $G=R(Q^*-Q_o)+[Q^*,R] \in \Psi^{m-\frac{5}{3}}_{1,0}$, hence there is $C_k > 0$ such that

$$|(Q_oQ_ju,Ru)_o| \leq C_k\|Q_ju\|_{\frac{1}{3}} (\|Q_ou\|_o+\|u\|_{m-2}), \; u \in C_o^\infty(K)$$

By Lemma 2 and (5), it can be easily proved that for any $\varepsilon > 0$, there is $C_{k,\varepsilon} > 0$ such that

$$|(Q_oQ_ju,Ru)_o| \leq \varepsilon\|u\|^2_{m-\frac{4}{3}} +C_{k,\varepsilon}(\|Pu\|^2_o+\|u\|^2_{m-\frac{3}{2}}) \;. \qquad (8)$$

Similarly way we can estimate $|(Q_jQ_ou,Ru)_o|$. The result of Lemma 3 follows immediately from (7) and (8).

4. Interior estimates. In this section we shall prove that under the hypotheses of the theorem, for any compact set $k \subseteq \Omega$ and $s \in R^1$ there is a constant $C_{k,s} > 0$ such that

$$\|u\|_{s+m-\frac{4}{3}} \leq C_{k,s}(\|Pu\|_s+\|u\|_{s+m-2}), \; u \in C_o^\infty(K) \;. \qquad (9)$$

To prove (9), we need following proposition: Let $A_j \in \Psi^{m_j}_{1,0}$, $j=1, \ldots, \ell$, $m_j \in R^1$ and $A_j-A_j^* \in \Psi^{m_j-1}_{1,0}$. If for $(x_o,\xi_o) \in T^*\Omega\setminus\{o\}$ there exists $[A_{i_1},[A_{i_2},\ldots[A_{i_{r-1}},A_{i_r}]\ldots]$ which is a elliptic operator of $m_{i_1}+\ldots+m_{i_r}-r+1$ th order at (x_o,ξ_o), then for any $s \in R^1$ we have

$$u \in \mathscr{D}'(\Omega), A_ju \in H_{s-m_j} \text{ at}(x_o,\xi_o), j=1,\ldots,\ell \Rightarrow u \in H_{s-1+\frac{1}{r}} \text{ at}(x_o\xi_o). \quad (10)$$

This proposition is the main result in [6]. The reader can find its proof in that paper.

By the hypotheses of the theorem, we know that for any $(x_o,\xi_o) \in T^*\Omega\setminus\{o\}$ there exists Q_I with $|I| \leq 3$ and Q_I is elliptic at (x_o,ξ_o). According to definitions of Q_I and $|I|$, there are only four cases

(i) There exists Q_j $(j=0,1,2,\ldots,2n)$ which is a elliptic operator of m-1 th order at (x_o,ξ_o) . Then

$$Q_j u \in H_{-\frac{1}{3}} \text{ at } (x_o,\xi_o) \Longrightarrow u \in H_{m-\frac{4}{3}} \text{ at } (x_o,\xi_o)$$

(ii) There exists $[Q_j,Q_k]$ $(j,k=1,\ldots,2n)$ which is elliptic of 2m-3 th order at (x_o,ξ_o). By (10), we have

$$Q_j u, Q_k u \in H_{\frac{1}{6}} \text{ at } (x_o,\xi_o) \Longrightarrow u \in H_{m-\frac{4}{3}} \text{ at } (x_o,\xi_o)$$

(iii) There exists $[Q_j,[Q\ ,Q_k]]$ $(j,\ ,k=1,\ldots,2n)$ which is elliptic of 3m-5 th order at (x_o,ξ_o). By (10), we have

$$Q_j u, Q\ u, Q_k u \in H_{\frac{1}{3}} \text{ at } (x_o,\xi_o) \Longrightarrow u \in H_{m-\frac{4}{3}} \text{ at } (x_o,\xi_o).$$

(iv) There exists $[Q_o,Q_j]$ $(j=1,2,\ldots,2n)$ which is elliptic of 2m-3 th order at (x_o,ξ_o). Then

$$[Q_o,Q_j]u \in H_{-m+\frac{5}{3}} \text{ at } (x_o,\xi_o) \Longrightarrow u \in H_{m-\frac{4}{3}} \text{ at } (x_o,\xi_o).$$

In view of arbitrariness of (x_o,ξ_o), we obtain that

$$u \in \mathcal{D}'(\Omega) \text{ and } \left.\begin{array}{l} Q_j u \in H^{loc}_{\frac{1}{3}}(\Omega),\ j=1,2,\ldots,2n \\ Q_o u \in H^{loc}_{-\frac{1}{3}}(\Omega), \\ [Q_o,Q_j]u \in H^{loc}_{-m+\frac{5}{3}}(\Omega), j=1,\ldots,2n \end{array}\right\} \Longrightarrow u \in H^{loc}_{m-\frac{4}{3}}(\Omega)$$

Using the closed graph theorem and a standard method, we can prove easily that for any compact set $k \subseteq \Omega$ there is $C_k > 0$ such that

$$\|u\|_{m-\frac{4}{3}} \leq C_k\left(\sum_{j=1}^{2n} \|Q_j u\|_{\frac{1}{3}} + \|Q_o u\|_{-\frac{1}{3}} + \sum_{j=1}^{2n} \|[Q_o,Q_j]u\|_{-m+\frac{5}{3}} + \|u\|_{m-2}\right)$$

$$\leq C_k\left(\sum_{j=1}^{2n} \|Q_j u\|_{\frac{1}{3}} + \|Q_o u\|_0 + \sum_{j=1}^{2n} \|[Q_o,Q_j]u\|_{-m+\frac{5}{3}} + \|u\|_{m-2}\right)$$

where $u \in C_o^\infty(k)$. From Lemma 2, Lemma 3 and (5), we conclude that for any compact set $k \subseteq \Omega$ and $\varepsilon > 0$ there is $C_{k,\varepsilon} > 0$ such that

$$\|u\|_{m-\frac{4}{3}} \leq \varepsilon\|u\|_{m-\frac{4}{3}} + C_{k,\varepsilon}(\|Pu\|_0 + \|u\|_{m-\frac{5}{3}} + \|u\|_{m-\frac{3}{2}} + \|u\|_{m-2})$$

where $u \in C_o^\infty(k)$. We note that for any $\delta > 0$ there is $C_\delta > 0$ such that

$$\|u\|_{m-\frac{5}{3}} \leq \delta\|u\|_{m-\frac{4}{3}} + C_\delta\|u\|_{m-2}$$

$$u \in C_o^\infty(\Omega)$$

$$\|u\|_{m-\frac{3}{2}} \leq \delta\|u\|_{m-\frac{4}{3}} + C_\delta\|u\|_{m-2}$$

Therefore there is $C_k > 0$ such that

$$\|u\|_{m-\frac{4}{3}} \leq C_k(\|Pu\|_o + \|u\|_{m-2}), \quad u \in C_o^\infty(K) \tag{11}$$

(9) is a direct consequence of (11) and Lemma 1. The rest of the proof is omitted.

5. The proof of the theorem. From Lemma 1 we have

$$\sum_{j=1}^{2n} \|Q_j u\|^2_{s+\frac{1}{3}} \leq C_{k,s}(\mathrm{Re}(Pu,u)_{s+\frac{m}{2}-\frac{2}{3}} + \|u\|^2_{s+m-\frac{5}{3}})$$

$$\leq C_{k,s}\|Pu\|_s\|u\|_{s+m-\frac{4}{3}} + \|u\|^2_{s+m-\frac{5}{3}}), \quad u \in C_o^\infty(K)$$

Hence by (9) it can be easily proved that

$$\|u\|_{s+m-\frac{4}{3}} + \sum_{j=1}^{2n} \|Q_j u\|_{s+\frac{1}{3}} \leq C_{k,s}(\|Pu\|_s + \|u\|^2_{s+m-2}), \quad u \in C_o^\infty(K) \tag{12}$$

Using (12), for any $(x_o,\xi_o) \in T^*\Omega\setminus\{o\}$ we can prove that

$$u \in \mathcal{D}'(\Omega),\ Pu \in H_s \text{ at } (x_o,\xi_o) \Rightarrow u \in H_{s+m-\frac{4}{3}} \text{ at } (x_o,\xi_o),$$

$$Q_j u \in H_{s+\frac{1}{3}} \text{ at } (x_o,\xi_o),$$

$$j=1,\dots,2n.$$

In fact, for $u \in \mathcal{D}'(\Omega)$ we can choose a positive integer N such that

$$u \in H_{t+m-\frac{4}{3}} \text{ at } (x_o,\xi_o),\ Q_j u \in H_{t+\frac{1}{3}} \text{ at } (x_o,\xi_o),\ j=1,2,\dots,2n$$

where $t=s-\frac{N}{3}$. By standard method (see, for example, the proof of Lemma 22.2.5 in [4]), we may obtain that

$$u \in H_{t+m-1} \text{ at } (x_o,\xi_o),\quad Q_j u \in H_{t+\frac{2}{3}} \text{ at } (x_o,\xi_o),\ j=1,2,\dots,2n$$

Repeating this process N times yields

$$u \in H_{s+m-\frac{4}{3}} \text{ at } (x_0,\xi_0),\ Q_j u \in H_{s+\frac{1}{3}} \text{ at } (x_0,\xi_0),\ j=1,2,\dots,2n$$

The proof of the theorem is complete.

References

[1] Hörmander, L., Hypoelliptic second order differential equations, Acta Math. 119, 147-171 (1967).

[2] Kohn, J.J., Pseudo-differential operators and hypoellipticity, Proc. Symp. Pure Math., 23, 61-69 (1973).

[3] Oleinik, O.A. and Radkevitch, E.M., Second order equations with nonnegative characteristic form, Itogi Nauk, Moscow (1971).

[4] Hörmander, L., The analysis of linear partial differential operators III, Springer-Verlag (1985).

[5] Rothchild. L.P. and Stein, E.M., Hypoelliptic differential operators and nilpotent groups, Acta Math., 137, 247-320 (1976).

[6] Bolley, P., Camus, J. et Nourrigat, J., La Condition de Hörmander-Kohn Pour les operateurs pseudo-differentiels, Comm. in P.D.E., 7(2), 197-221 (1982).

A COUNTEREXAMPLE TO THE YAMABE PROBLEM FOR COMPLETE NONCOMPACT MANIFOLDS

Jin Zhiren
Institute of Mathematics,
Academia Sinica, Beijing, China

1. Introduction

In this paper we consider conformal deformation of metrics on a class of complete noncompact Riemannian Manifolds. In particular we consider the noncompact version of the famous Yamabe problem. In this problem one is given a complete noncompact Riemannian manifold (M^n,g) of dimension n, $n>3$, and seeks a pointwise conformal metric $g'=ug$, with $u>0$, so that g' is complete and has constant scalar curvature.

Our main result is a counter example to this problem. The simplest example is a complete metric on R^n, $n\geq 3$, which is not pointwise conformal to a complete metric with constant scalar curvature. More generally, we can construct similar examples on any noncompact manifold M obtained by deleting a finite number of points from a compact manifold. These examples are surprising because we now know that, for compact manifolds, the Yamabe problem can always be solved. This simply means that there are great differences between the conformal deformation of metrics on compact manifolds and those on complete noncompact manifolds.

The history of the problem can be briefly stated as follows. After Yamabe's original paper (9), N.S. Trüdinger (3) found a serious error in Yamabe's work and also showed how to solve the problem in some cases. The general case for a compact manifold was very recently solved by R. Schoen (5), who used earlier results of T. Aubin (7). The solvability of the noncompact Yamabe problem was raised in (6), problem 32, and also in (2).

The Yamabe problem may be viewed as a special case of a more general problem of studying the pointwise conformal deformation of metrics on manifolds. For noncompact manifolds, there are some results, such as those of W.N. Ni (8) who studied the pointwise conformal deformation of standard flat metric $g_\circ$ on R^n with prescribed scalar curvature. In a defferent way, we consider the conformal deformation of metrics that are quite arbitrary on a class of noncompact manifolds.

Let $(M^n_\circ,g_\circ)$ be a compact Riemannian manifold, $n\geq 3$, $p_1,p_2,\cdots,p_k$ be k points in $M^n_\circ$. $M=M^n_\circ\backslash\{p_1,\cdots, p_k\}$. It is easy to find a function $\phi\in C^\infty(M)$, $\phi>0$, so that with the metric $g=\phi^{4/(n-2)}g_\circ$, the manifold (M,g) is complete.

Now we consider the pointwise conformal deformation of g. Suppose $g_1=u^{4/(n-2)}g$, $u>0$, $u\in C^\infty(M)$, then the requirement that g_1 be complete and have scalar curvature S is equivalent to u being a smooth solution of the following problem:

$$
(\mathrm{I})\begin{cases} -\gamma_n \Delta_g u + S_g u = S u^{\frac{n+2}{n-2}} \text{ on } M, \\ u>0 \text{ on } M, \\ u^{4/(n-2)} g \text{ is complete on } M. \end{cases}
$$

where $\gamma_n = \frac{4(n-1)}{(n-2)}$, S_g is the scalar curvature of g, Δ_g is the Laplacian of g.

On the other hand, since $g=\phi^{\frac{4}{n-2}} g_\circ$, for $v=u\phi$, we have $g_1=v^{\frac{4}{n-2}} g_\circ$, and (I) is equivalent to (II):

$$
(\mathrm{II})\begin{cases} -\gamma_n \Delta_{g_\circ} v + s_{g_\circ} v = s v^{\frac{n+2}{n-2}} \text{ on } M, & (\mathrm{II})_1 \\ v>0 \text{ on } M, & (\mathrm{II})_2 \\ v^{4/(n-2)} g_\circ \text{ is complete on } M. & (\mathrm{II})_3 \end{cases}
$$

To consider our problem, it will be convenient to use the first eigenvalue $\lambda_1(g_\circ)$ of the linear problem

$$L\psi \equiv -\gamma_n \Delta_{g_\circ} \psi + s_{g_\circ} \psi = \mu\psi \quad \text{on } M_o^n ,$$

where μ is a constant. It is well known that, in the compact case, $\lambda_1(g_\circ)$ is useful not only because its sign is invariant under conformal deformation but also because one can always find a conformal metric whose scalar curvature has the same sign as $\lambda_1(g_\circ)$ (see(2)). Thus, for example, if $\lambda_1(g_\circ)<0$, then there is a conformal metric g' with scalar curvature $S'<0$ everywhere. For the noncompact manifolds that we consider, the eigenvalue $\lambda_1(g_\circ)$ also gives crucial information to the problem.

Here are our conclusions.

Theorem 1. Assume $\lambda_1(g_\circ)\leqq 0$; if either 1) or 2) occurs, then (II) has no solutions:

1) For some i, $1\leqq i\leqq k$, there is an open set $U\subset M_o^n$ and positive constants c_1 and c_2, so that $p_i\in U$, $-c_1\geqq S\geqq -c_2$ on $U\setminus\{p_i\}$.

2) $S\geqq 0$ on M.

Remark Actually we will prove a slightly stronger conclusion that in case 2) we can delete the completeness assumption, so there are no solutions for $(\mathrm{II})_1$, $(\mathrm{II})_2$ if either $\lambda_1(g_\circ)<0$ or $\lambda_1(g_\circ)=0$, $s\not\equiv 0$.

Theorem 2. Assume $\lambda_1(g_\circ)>0$.

a) If $S=0$, then (II) has a solution.

b) If S satisfies 1) of Theorem 1, then (II) has no solutions.

c) If $S>0$, then sometimes (II) has a solution and sometimes it has no solutions for different S.

Theorem 3. Let M_o^n be a compact differential manifold, $n\geq 3$, $p_i \in M_o^n$, $i=1,2,\cdots,k$. $M=M_o^n\setminus\{p_1,p_2,\cdots, p_k\}$. Then there is a complete Riemannian metric g on M so that g can not be pointwise conformal to any complete metric g' with constant scalar curvature.

We will see that Theorem 1 - Theorem 3 are obvious consequences of the results in the following sections; Therefore their proof will not be formulated again.

Acknowledgement The author would like to thank Prof. Ding Wei-Yue for bringing the problem to his attention and recommending the paper (4) to him which is used to simplify the original proof of Theorem 1. Discussions with Ding are always helpful and fruitful. Also, the author would like to express his deep thanks to prof. Jerry Kazdan for his help and kindness in the completion of the paper.

2. The Case $\lambda_1(g_\circ)<0$

First of all, we quote a special case of a local result about the removable singularities of solutions to certain elliptic equations from Theorem 2.2 in (4).

Lemma 2.1 (P. Aviles (4)) Let $p\in M_o^n$, g_o is a metric on M_o^n. Let $U\subset M_o^n$ be a neighbourhood of p, c_1 and c_2 are positive constants, so that for $f\in L^\infty(U)$, $-c_1\geqq f\geqq -c_2$ on $U\setminus\{p\}$.

If $u\in W^{2,n}_{loc}(U\setminus\{p\})$ satisfies

$$-\Delta_{g_\circ} u+hu=fu^\alpha \quad \text{on } U\setminus\{p\}$$

where $h\in C^\beta(U)$, $0<\beta<1$, $\frac{n}{n-2}\leqq\alpha\leqq\frac{n+2}{n-2}$, then u can be extended to a C^1 function on U, here we do not restrict the sign of $\lambda_1(g_\circ)$.

Now we have

Theorem 2.1 If for some i, $1\leqq i\leqq k$, there is an open set $U\subset M_o^n$ and positive constants c_1 and c_2, so that $p_i\in U$, $-c_1\geqq S\geqq -c_2$ on $U\setminus\{p_i\}$, then (II) has no solutions, where the sign of $\lambda_1(g_\circ)$ is not restricted.

Proof. Combining the assumption and Lemma 2.1, we know that a solution v of (II) can be extended to a C^1 function on U. But On the other hand, however, we know $g_\circ$ is complete on M_o^n, hence $v^{4/(n-2)}g_\circ$ is not complete on $\bar{U}'\setminus\{p_i\}$ for any neighbourhood U' of p_i. This is a contradiction to the property that $v^{4/(n-2)}g_\circ$ is complete on M. Q.E.D.

Let $\bar{g}$ be a metric on M_o^n, $B_\rho(p_i)$ is a geodesic ball centered at p with radirs ρ, $1\leq i\leq k$. Set $\Omega_\rho = M_o^n \setminus \bigcup_{i=1}^k B_\rho(p_i)$. Assume that $\lambda_{\rho,1}(\bar{g})$ is the first eigenvalue of the linear Dirichlet problem

$$(*)\qquad \begin{cases} -\gamma_n \Delta_{\bar{g}}\psi + S_{\bar{g}}\psi = \mu\psi & \text{on } \Omega_\rho \\ \psi = 0 & \text{on } \partial\Omega_\rho,\ \mu \text{ is a constant.} \end{cases}$$

Lemma 2.2 If $n\geq 3$, then $\lim_{\rho\to 0}\lambda_{\rho,1}(\bar{g})=\lambda_1(\bar{g})$. In particular, if $\lambda_1(\bar{g})<0$, then there is a $\rho_\circ>0$, so that $\lambda_{\rho,1}(\bar{g})<0$, for $\rho\in(0,\rho_\circ)$.

Proof. Assume that $\bar{u}$ is the eigenfunction for $\lambda_1(\bar{g})$. Then

$$\lambda_1(\bar{g}) = \frac{\gamma_n \int_{M_o^n} |\nabla_{\bar{g}}\bar{u}|^2 dv_{\bar{g}} + \int_{M_o^n} S_{\bar{g}}\bar{u}^2 dv_{\bar{g}}}{\int_{M_o^n} \bar{u}^2 dv_{\bar{g}}} \tag{2.1}$$

Let $\bar{\psi}\in C^2(M_o^n)$ and

$$\bar{\psi} = \begin{cases} 0 & \text{on } \bigcup_{i=1}^k B_\rho(p_i) \\ 1 & \text{on } M_o^n\setminus\bigcup_{i=1}^k B_{2\rho}(p_i) \end{cases}$$

$$|\nabla\bar{\psi}| \leq C/\rho$$

Set $v=\bar{u}\bar{\psi}$, then it is easy to verify that

$$\left|\int_{M_o^n} |\nabla_{\bar{g}}\bar{u}|^2 dv_{\bar{g}} - \int_{M_o^n} |\nabla_{\bar{g}}v|^2 dv_{\bar{g}}\right| \leq C\rho^{n-2} \tag{2.2}$$

$$\left|\int_{M_o^n} S_{\bar{g}}\bar{u}^2 dv_{\bar{g}} - \int_{M_o^n} S_{\bar{g}}v^2 dv_{\bar{g}}\right| \leq C\rho^n \tag{2.3}$$

$$\left|\int_{M_o^n} \bar{u}^2 dv_{\bar{g}} - \int_{M_o^n} v^2 dv_{\bar{g}}\right| \leq C\rho^n \tag{2.4}$$

Combining (2.1)-(2.4), we see that for $\varepsilon>0$, there is a $\delta>0$, so that for $0<\rho<\delta$

$$\frac{\gamma_n \int_{M_o^n} |\nabla_{\bar{g}}v|^2 dv_{\bar{g}} + \int_{M_o^n} S_{\bar{g}}v^2 dv_{\bar{g}}}{\int_{M_o^n} v^2 dv_{\bar{g}}} < \lambda_1(\bar{g}) + \varepsilon$$

But $v=0$ on $M_o^n\setminus\Omega_\rho$, therefore

$$\frac{\gamma_n \int_{\Omega\rho} |\nabla_{\bar{g}}v|^2 dv_{\bar{g}} + \int_{\Omega\rho} S_{\bar{g}}v^2 dv_{\bar{g}}}{\int_{\Omega\rho} v^2 dv_{\bar{g}}} < \lambda_1(\bar{g}) + \varepsilon \quad \text{for } \delta>\rho>0,$$

that is

$$\lambda_{\rho,1}(\bar{g}) \leq \lambda_1(\bar{g}) + \varepsilon \quad \text{for } \delta>\rho>0.$$

On the other hand, it is easy to see that

$$\lambda_1(\bar{g}) \leqq \lambda_{\rho,1}(\bar{g}) \quad \text{for any } \rho>0.$$

Then $\lambda_1(\bar{g}) \leqq \lambda_{\rho,1}(\bar{g}) \leqq \lambda_1(\bar{g}) + \varepsilon$ for $\delta>\rho>0$. This gives the desired conclusion. Q.E.D.

Theorem 2.2 If $\lambda_1(g_\circ)<0$, $S \geqq 0$ on M, then there is no function $v \in C^2(M)$ satisfying $(II)_1$, $(II)_2$.

Proof. Suppose that there is a $v \in C^2(M)$ satisfying $(II)_1$, $(II)_2$. We take a small positive constant ρ, so that for $1 \leqq i \leqq k$, $B_\rho(P_i)$ are in the normal coordinate system at p_i with respect to $g_\circ$, $\rho<\rho_\circ$, where $\rho_\circ$ is the number prescribed by Lemma 2.2 for $\bar{g}=g_\circ$. Let u be the eigenfunction of $\lambda_{\rho,1}(g_\circ)$, u>0 on Ω_ρ.

Now by $(II)_1$, (*) and an integration by parts, we have

$$\int_{\Omega\rho} S\, u\, v^{\frac{n+2}{n-2}}\, dv_{g_\circ} = \int_{\Omega\rho} u(-\gamma_n \Delta_{g_\circ} v + S_{g_\circ} v)\, dv_{g_\circ}$$

$$= \int_{\Omega\rho} v(-\gamma_n \Delta_{g_\circ} u + S_{g_\circ} u)\, dv_{g_\circ} - \gamma_n \sum_{i=1}^{k} \int_{\partial B_\rho(p_i)} v \frac{\partial u}{\partial x^h} g^{hl} \nu^l \, dS$$

$$= \int_{\Omega\rho} \lambda_{\rho,1}(g_\circ) uv \, dV_{g_\circ} - \gamma_n \sum_{i=1}^{k} \int_{\partial B_\rho(p_i)} v \frac{\partial u}{\partial \nu} \, dS \qquad (2.5)$$

where ν is the outer normal on $\partial B_\rho(p_i)$. Since u>0 on Ω_ρ, u=0 on $\partial\Omega_\rho$, $\frac{\partial u}{\partial \nu} \geqq 0$ on $\partial B_\rho(p_i)$, $i=1,2,\cdots,k$. Then $\lambda_{\rho,1}(g_\circ)<0$, u>0, v>0 imply the right-hand side of (2.5) is negative while the left-hand side of (2.5) is nonnegative. The contradiction gives the conclusion. Q.E.D.

Hitherto, we have proved Theorem 1 in the case $\lambda_1(g_\circ)<0$. Theorem 3 follows by combining this with the following proposition.

Proposition 2.1 For a compact differential manifold $M_\circ^n$, $n \geqq 3$, there is a Riemannian metric $g_\circ$ with $S_{g_\circ}<0$ and $\lambda_1(g_\circ)<0$.

Proof. It is well known that there is a Riemannian metric $\bar{g}$ on $M_\circ^n$. For $(M_\circ^n, \bar{g})$, by Theorem 3 in (7), we know that there is a $g_\circ$ with $S_{g_\circ}<0$. Let u be the eigenfunction for $\lambda_1(g_\circ)$, u>0 on $M_\circ^n$, then

$$-\gamma_n \Delta_{g_\circ} u + S_{g_\circ} u = \lambda_1(g_\circ) u \qquad \text{on } M_\circ^n .$$

Now if $u(x_\circ) = \operatorname{Min}_{x \in M_\circ^n} u(x)$, $x_\circ \in M_\circ^n$, then

$$-\Delta_{g_\circ} u(x_\circ) \leqq 0, \quad S_{g_\circ}(x_\circ)<0, \quad u(x_\circ)>0,$$

hence $\lambda_1(g_\circ)<0$. Q.E.D.

3. The Case $\lambda_1(g_\circ) = 0$

The conclusion in this section is

Theorem 3.1 If $\lambda_1(g_\circ) = 0$, $S\geqq 0$, then there is no $v\in C^2(M)$ satisfying $(II)_1, (II)_2$ unless v is a constant. In particular, if $S\geqq 0$, $S\not\equiv 0$, then $(II)_1$, $(II)_2$ has no solutions.

Proof. Let $\phi_1\in C^\infty(M_\circ^n)$ be the eigenfunction for $\lambda_1(g_\circ)$, $\phi_1>0$ on $M_\circ^n$. Then $\lambda_1(g_\circ)=0$ implies that for $g_2=\phi_1^{\frac{4}{n-2}} g_\circ$, $S_{g_2}=0$ on $M_\circ^n$. Now if $(II)_1$, $(II)_2$ has a solution v, then $\bar{v}=v\phi_1^{-1}$ satisfies

$$-\gamma_n\Delta_{g_2}\bar{v} = S\bar{v}^{\frac{n+2}{n-2}} \quad \text{on } M, \tag{3.1}$$

$$\bar{v} > 0 \quad \text{on } M.$$

Since $S\geqq 0$,

$$\Delta_{g_2}\bar{v} \leqq 0 \quad \text{on } M. \tag{3.2}$$

Then by the maximum principle, we see that $\bar{v}$ can not achieve local minima in the interior of M unless $\bar{v}$ is a constant.

But we claim that $\bar{v}$ will achieve a local minimum in the interior of M. This gives the conclusion of the theorem.

Fix a point p_i, $1\leqq i\leqq k$, let ρ_i be a small positive constant, such that $B_{\rho_i}(p_i)$ is a geodesic ball with radius ρ_i. Now we choose a function $G(x)\in C^\infty(B_{\rho_i}(p_i)\setminus\{p_i\})$, satisfying

$$-\Delta_{g_2}G = 0 \quad \text{on } B_{\rho_i}(p_i)\setminus\{p_i\}\ , \tag{3.3}$$

$$G > 0 \quad \text{on } B_{\rho_i}(p_i)\setminus\{p_i\}\ .$$

$$G(x)\to+\infty \quad \text{when } x\to p_i\ . \tag{3.4}$$

The existence of G can be easily seen, for example, one may take a Green function G of the following problem and G has a singularity at p_i:

$$-\Delta_{g_2} u = 0 \quad \text{on } B_{\rho_i}(p_i)$$

$$u = 0 \quad \text{on } \partial B_{\rho_i}(p_i)\ .$$

For $\alpha\in(0,1)$, set

$$w_\alpha = \alpha\bar{v} + (1-\alpha)G \tag{3.5}$$

By (3.4), we see that for any $\alpha\to(0,1)$, there is a $\rho_\alpha\in(0,\rho_i)$ so that

$$W_\alpha\Big|_{\partial B_{\rho_i}(p_i)} \leq W_\alpha\Big|_{\partial B_{\rho_\alpha}(p_i)} \tag{3.6}$$

Without loss of generality, we may assume $\rho_\alpha \to 0$ as $\alpha \to 1$. Then there is a $\delta>0$, so that

$$\tfrac{1}{2}\rho_i > \rho_\alpha \quad \text{for} \quad \alpha \in (1-\delta, 1) \tag{3.7}$$

Now combining (3.2)-(3.7), we have

$$\Delta_{g_2} W_\alpha \leq 0 \qquad \text{on } B_{\rho_i}(p_i) \setminus B_{\rho_\alpha}(p_i)$$

$$W_\alpha\Big|_{\partial B_{\rho_i}} \leq W_\alpha\Big|_{\partial B_{\rho_\alpha}}$$

$$\tfrac{1}{2}\rho_i > \rho_\alpha, \ \alpha(1-\delta,1), \quad \delta>0.$$

Then the maximum principle implies that

$$\inf_{\partial B_{\rho_i}} W_\alpha \leq W_\alpha\Big|_{\partial B_{\frac{1}{2}\rho_i}}, \quad \alpha \in (1-\delta, 1)$$

Letting $\alpha \to 1$, we get

$$\inf_{\partial B_{\rho_i}} \bar{v} \leq \bar{v}\Big|_{\partial B_{\frac{1}{2}\rho_i}}$$

By analogous reasoning, we find $\rho_1, \rho_2, \cdots, \rho_k$, so that

$$\inf_{\partial B_{\rho_i}(p_i)} \bar{v} \leq \bar{v}\Big|_{\partial B_{\frac{1}{2}\rho_i}(p_i)}, \quad i=1,2,\cdots,k.$$

This simply means that $\bar{v}$ takes the minimum on Ω in the interior of Ω where $\Omega = M \setminus \bigcup_{i=1}^{k} B_{\frac{1}{2}\rho_i}(p_i)$. Therefore $\bar{v}$ has a local minimum in the interior of M. Q.E.D.

4. The Case $\lambda_1(g_\circ)>0$

<u>Theorem 4.1</u> If $\lambda_1(g_\circ)>0$, $S=0$, then (II) has a solution.

<u>Proof.</u> Since $\lambda_1(g_\circ)>0$, there are Green functions $G_i = G_i(x,p_i) \in C^\infty(M_0^n \setminus \{p_i\})$, $1 \leq i \leq k$, satisfying

$$-\gamma_n \Delta_{g_\circ} G_i + S_{g_\circ} G_i = 0 \quad \text{on } M_0^n \setminus \{p_i\} \tag{4.1}$$

$$G_i > 0 \qquad \text{on } M_0^n \setminus \{p_i\}. \tag{4.2}$$

$$|x-p_i|^{n-2} G_i(x) \to 1 \quad \text{as } x \to p_i. \tag{4.3}$$

Now set

$$v = \sum_{i=1}^{k} G_i \qquad \text{on M.}$$

Then using (4.1)-(4.3), we can verify that v is a solution of (II). Q.E.D.

For the case $S>0$, the situation may be complicated. Here we only give two examples to illustrate the situation.

<u>Example 4.1</u> Let $S^n \hookrightarrow R^{n+1}$ be the n-dimension unit sphere, g_0 is the standard

metric on S^n. $M=S^n\setminus\{p\}$ where p is the north pole of S^n. Now we consider the solvability of (II) with S=1

If (II) has a solution u, then $g_1=u^{4/(n-2)}g_\circ$ is complete on M and $S_{g_1}=1$.

On the other hand, there is a conformal diffeomorphism F: $S^n\setminus\{p\}\to R^n$, so that for the Euclidean metric $\bar{g}_\circ$ on R^n, there is a $\phi\in C^\infty(M)$, $\phi>0$ and

$$g_\circ = \phi^{4/(n-2)}F^*\bar{g} \quad .$$

Hence

$$g_1 = (u\phi)^{4/(n-2)}F^*\bar{g}_\circ = F^*((u\circ F^{-1}\cdot\phi\circ F^{-1})^{4/(n-2)}\bar{g}_\circ)$$

Since g_1 and $\bar{g}_1=(u\circ F^{-1}\cdot\phi\circ F^{-1})^{4/(n-2)}\bar{g}_\circ$ are isometric, $S_{\bar{g}_1}=1$ and $\bar{g}_1$ is complete on R^n.

This simply means, for $v=u\circ F^{-1}\cdot\phi\circ F^{-1}$

$$-\gamma_n\Delta_{\bar{g}_\circ} v = v^{(n+2)/(n-2)} \quad \text{on } R^n \tag{4.4}$$

$$v^{4/(n-2)}\bar{g}_\circ \text{ is complete on } R^n \quad . \tag{4.5}$$

Theorem 3 in (1) claims that if v satisfies (4.4) and v>0, then

$$v = \frac{c(n)\lambda^{\frac{n-2}{2}}}{(\lambda^2 + |x-a|^2)^{\frac{n-2}{2}}} \qquad \text{on } R^n$$

for some constant c(n), $\lambda\in R$, $a\in R$, $\lambda>0$. Hence we have a contradiction to (4.5), this means that (II) has no solutions.

<u>Example 4.2</u> Let $M=S^n\setminus\{p,\bar{p}\}$, where p and $\bar{p}$ are the north pole and south pole of S^n respectively. Then (II) has a solution for S=1, where g is the standard metric on S^n.

In fact, it is easy to verify that $v(y)=\gamma_n^{\frac{n-2}{4}} m^m|y|^{-m}$, $m=\frac{1}{2}(n-2)$, satisfies

$$-\gamma_n\Delta_{\bar{g}_\circ} v = v^{\frac{n+2}{n-2}} \qquad \text{on } R^n\setminus\{0\},$$

$$v > 0 \qquad \text{on } R^n\setminus\{0\},$$

$$G = v^{4/(n-2)}\bar{g}_\circ \quad \text{is complete on } R^n\setminus\{0\}.$$

where $\bar{g}_\circ$ is the Euclidean metric on R^n.

Therefore, for the conformal diffeomorphism F: $S^n\to R^n$ $F(p)=\infty$, $F(\bar{p})=0$, we have $F^*\bar{g}_\circ=\phi^{-4/(n-2)}g_\circ$ and

$$g_1 = F^*(v^{4/(n-2)}\bar{g}_\circ)=(v\circ F)^{4/(n-2)}F^*\bar{g}_\circ=((v\circ F)\phi^{-1})^{4/(n-2)}g_\circ \quad \text{is complete on } M,$$

$$S_{g_1} = F^*S_G = 1.$$

This simply means that $u=(v\circ F)\phi^{-1}$ is a solution of (II).

REFERENCES

(1) B. Gidas, Symmetry and Isolated Singularities of Conformally Flat Metrics and of Solutions of the Yang-Mills Equations. Seminar on Diff. Geo., 423-442, edited by S.T. Yau, 1982 Princeton Press.

(2) J. Kazdan, Prescribing the Curvature of a Riemannian Manifold, Expository Lectures from the CBMS Regional Conference held at Polytechnic Institute of New York 1984.

(3) N. Trüdinger, Remarks Concerning the Conformal Deformation of Riemannian Structures of Compact Manifolds, Ann. Scuola Norm. Sup. Pisa 3, 265-274 (1968)

(4) P. Aviles, A Study of the Singularities of Solutions of a Class of Nonlinear Elliptic Partial Differential Equations. Comm. in PDE, 7(6), 609-643, (1982)

(5) R. Schoen, Conformal Deformation of a Riemannian Metric to Constant Scalar Curvature, J. Diff. Geo. 20(1984), 479-495.

(6) S.T. Yau, Problem Section, Seminar on Diff. Geo. 669-706, edited by S.T. Yau, 1982, Princeton Press.

(7) T. Aubin, Best Constants in the Sobolev Imbedding Theorem; The Yamabe Problem, Siminar on Diff. Geo. 173-194, edited by S.T. Yau, 1982, Princeton Press.

(8) W.N. Ni, On the Elliptic Equation $\Delta u+Ku^{(n+2)/(n-2)}=0$, its Generalization and Application in Geometry, India Uni. Math. Journal, vol 31, 4(1982), 493-529.

(9) H. Yamabe, On a Deformation of Riemannian Structures on Compact Manifolds, Osaka Math. J. 12, 21-37(1960).

FREE BOUNDARY PROBLEMS FOR DEGENERATE PARABOLIC EQUATIONS*

Li Huilai

Institute of Mathematics,
Jilin University, China.

§1. Introduction

During the last three decades a great deal of progress has been made on the research of free boundary problems for quasilinear parabolic equations, which arise from mechanics, biochemistry, filtration theory and other fields.

Wu Zhuoqun [24] studied a kind of free boundary problem related to the investigation of the structure of discontinuous solutions of degenerate quasilinear parabolic equations: to find functions u, λ such that

$$\begin{aligned} &u_t=A(u)_{xx} && \lambda(t)<x<\infty, && 0<t<T, \\ &u(x,0)=\phi(x) && 0<x<\infty, && \\ &u(\lambda(t),t)=0 && 0<t<T, && \\ &A(u)_x(\lambda(t),t)=\psi(\lambda(t))\lambda'(t) && 0<t<T, && \end{aligned} \tag{1.1}$$

where $A'(u)=a(u)=0$ for $u<0$ and $a(u)>0$ for $u>0$, $\phi(x)\geqq 0$ for $x>0$ and $\psi(x)<0$ for $x<0$.

A thorough treatment of the problem was given by Wu for the following special case:

$$A(u)=0 \text{ for } u<0 \text{ and } A(u)=u^m \text{ for } u>0 \tag{1.2}$$

$$\phi(x)=\alpha>0 \text{ for } x>0 \text{ and } \psi(x)=\beta<0 \text{ for } x<0 \tag{1.3}$$

where $m>1$ and α, β are constants. In this special case the problem can be reduced to a problem in ordinary differential equations upon introduction of the appropriate similarity variable. Beyond its inherent interest, the solvability of the resulting problem shows that an analysis by Vol'pert and Hudjaev of jump conditions satisfied by solutions of degenegate parabolic equations is not correct in general, see [23] [24].

In this paper we are concerned with the problem (1.1) for general bounded measurable functions $\phi\geq 0$ and $\psi<0$. For simplicity, we confine our discussion to the special case (1.2). The similar arguement for general A(u) will be made in another paper.

The problem (1.1) is different from other kinds of free boundary problems studied by many authers. One kind is that arising in the study of the porous medium equation

$$u_t = u^m_{xx} \qquad (x,t) \in R^1 x\ (0,T) \tag{1.4}$$

* This work is partly supported by the Science Fund of the National Education Committee.

where m>1 is a constant. It is well known that if the initial data

$$u(x,0) = u_o(x) \begin{cases} >0 & \text{for } x>0, \\ =0 & \text{for } x<0, \end{cases} \tag{1.5}$$

then for the solution u the interface of $\{u>0\}$ and $\{u=0\}$ is determined by a curve, say $x=\lambda(t)$, and u satisfies

$$u_x^{m-1}(\lambda(t),t) = -\frac{m-1}{m}\lambda'(t) \tag{1.6}$$

see [1],[2],[3],[5],[6],[12],[17],[20],[21],[22].

Thus the problem (1.4),(1.5) is equivalent to the following free boundary one

$$\begin{aligned} &u_t = u^m_{xx} && \lambda(t)<x<\infty,\ 0<t<T, \\ &u(x,0) = u_o(x) && 0<x<\infty, \\ &u(\lambda(t),t) = 0 && 0<t<T, \\ &u_x^{m-1}(\lambda(t),t) = -\frac{m-1}{m}\lambda'(t) && 0<t<T, \end{aligned} \tag{1.7}$$

which differs from our problem only in the second free boundary condition.

Another kind is the so-called Stefan problem: to find functions u,λ such that

$$\begin{aligned} &u_t = A(u)_{xx} && \lambda(t)<x<\infty,\ 0<t<T, \\ &u(x,0) = u_o(x) && 0<x<\infty, \\ &u(\lambda(t),t) = 0 && 0<t<T, \\ &A(u)_x(\lambda(t),t) = -k\lambda'(t) && 0<t<T, \end{aligned} \tag{1.8}$$

where $A'(u)>0$. See [8],[9],[13] [15]. The last condition in our problem is the same as in the problem (1.8). The essential difference is that in our case the equation is degenerate in the neighbourhood of the free boundary, while in (1.8) the equation is strictly parabolic.

The present paper is constructed as follows. We first define the weak sulutions of the problem and then establish the existence and uniqueness of the weak solutions in §2. The main idea is to reduce the free boundary conditions to a weak form.

Subsequently, in §3 we discuss the regularity properties of solutions. We prove that the free boundary conditions is satisfied almost everywhere and that the solutions are classical if the initial data are positive.

In the last section, §4, we are devoted to the regularity properties of free boundaries and give a necessary and sufficient condition under which the free boundaries are in C^1. Our conclusion is stronger than that obtained for the porous medium equation.

The author would like deeply to thank Prof. Wu Zhuoqun for the selection of the subject and the elaborate instruction during the preparation of the paper, and to thank Dr. Zhao Junning for several helpful discussions.

§2. Definition, Existence and Uniqueness

Consider the free boundary problem

$$u_t = u^m_{xx} \qquad \lambda(t)<x<\infty,\ 0<t<T, \tag{2.1}$$

$$u(x.0) = \phi(x) \qquad 0<x<\infty, \tag{2.2}$$

$$u(\lambda(t),t) = 0 \qquad 0<t<T, \tag{2.3}$$

$$u^m_x(\lambda(t),t) = \psi(\lambda(t))\lambda'(t) \qquad 0<t<T, \tag{2.4}$$

where $\phi(x)\geqq 0$, $\psi(x)\leqq\varepsilon_o<0$ are bounded measurable functions and $\varepsilon_o>0$, $m>1$ are constants. Suppose (u,λ) is a classical solution of (2.1) — (2.4). Let $v(x,t)=u(x+\lambda(t),t)$. Then v satisfies

$$v_t = v^m_{xx} + \lambda'(t)v_x \qquad \text{in}\quad Q_{o,T} = (0,\infty)x(0,T), \tag{2.5}$$

$$v(x,0) = \phi(x) \qquad \text{on}\quad R^1_+ = (0,\infty), \tag{2.6}$$

$$v(0,t) = 0 \qquad \text{on}\quad (0,T), \tag{2.7}$$

$$v^m_x(0,t) = \psi(\lambda(t))\lambda'(t) \qquad \text{on}\quad (0,T). \tag{2.8}$$

Integrating (2.5) and using (2.6) — (2.8) we have

$$\Psi(\lambda(t))=\int_0^x \frac{x-y}{x}(v(y,t)-\phi(y))dy - \frac{1}{x}\int_0^x v^m(x,s)ds - \frac{1}{x}\int_0^t\int_0^x \lambda'(s)v(y,s)dyds \tag{2.9}$$

where $\Psi(s)=-\int_0^s\psi(t)dt$ and $x>0$ is arbitrary. Therefore, if (u,λ) is a classical solution of (2.1) — (2.4), then (v,λ) must satisfy (2.5), (2.6), (2.7) and (2.9).

Now we have the following

DEFINITION 2.1 A pair of functions (u,λ) is said to be a weak solution of (2.1) — (2.4) if for $v(x,t)=u(x+\lambda(t),t)$ we have

i) $v(x,t) \in C^\alpha(\bar{Q}_{o,T})\cap L^\infty(\bar{Q}_{o,T})$ for some $\alpha\in(0,1]$ and $v\geqq 0$,

ii) $\lambda(t)$ is nonincreasing and Lipschitz continuous on $[0,T]$,

iii) (v,λ) satisfies

$$\int_0^T\int_0^r(vf_t+v^mf_{xx}-\lambda'(t)vf_x)dxdt - \int_0^r\phi(x)f(x,0)dx+\int_0^T v^mf_x\Big|_0^r dt \tag{2.10}$$

for any $f \in C^{2,1}(\bar{Q}_{o,r,T})$ with $f(x,T)=f(0,t)=f(r,t)=0$, where $Q_{o,r,T}=(0,r)x(0,T)$ and $r>0$ is arbitrary.

iv) (v,λ) satisfies (2.9).

If, instead of (iv), $v^m_x(0,t)=\psi(\lambda(t))\lambda'(t)$ a.e., then we say (u,λ) is a strong solution.

To discuss the existence of solutions, we begin by assuming

$$(H): -m_0 \leq \psi(x) \leq -\varepsilon_0 < 0,\ \phi^m(x) \in C^1[0,\infty),\ |\tfrac{d}{dx}\phi^m| \leq K_2,\ 0 \leq \phi(x) \leq K_1,\ \phi(0)=0.$$

where $\varepsilon_0>0$, $M_0>0$, K_1, K_2 are constants.

Consider the first boundary value problem

$$\begin{cases} v_t = v^m_{xx} + g(t)v_x & \text{in } Q_{o,T} \\ v(x,0)=\phi(x) & \text{on } R^1_+ \\ v(0,t)=0 & \text{on } (0,T) \end{cases} \tag{2.11}$$

where $g \in L^\infty$ and $-k_3 \leq g(t) \leq 0$, $K_3>0$ is a constant.

In order to solve (2.11), we first consider its regularized problem

$$(IVP)_n \begin{cases} v_t=(a_n(v)v_x)_x + g_n(t)v_x & \text{in } Q_{o,n,T}, \\ v(x,0)=\phi_n(x) & \text{on } (0,n), \\ v(0,t)=\varepsilon_n & \text{on } (0,T), \\ v(n,t)=\phi_n(n) & \text{On } (0,T). \end{cases}$$

where

$$\begin{cases} a_n(v)=\begin{cases} mv^{m-1} & \text{if } v>\varepsilon_n \\ \text{smoothly connected} & \text{if } \varepsilon_n/2 \leq v \leq \varepsilon_n \\ \varepsilon_n/2 & \text{if } v<\varepsilon_n/2 \end{cases} \\ g_n(t) \in C^\infty[0,T],\ -k_3 \leq g_n \leq 0,\ g_n \to g \text{ in } L^1, \\ \phi_n^{(k)}(0)=\phi_n^{(k)}(n)=0\ (k=1,2,3,\ldots),\ \varepsilon_n \downarrow 0,\ \phi_n(0)=\varepsilon_n,\ \varepsilon_n \leq \phi_n(x) \leq K_1, \\ \phi_n \in C^\infty[0,T], \text{ and } \phi_n \to \phi \text{ uniformly in any compact subset of } [0,\infty) \end{cases} \tag{2.12}$$

By virtue of [7] or [19] it follows that

<u>Lemma 2.1</u>: Under the hypotheses and (2.12), $(IVP)_n$ admits a unique solution v_n satisfying

$$v_n \in C^\infty(\bar{Q}_{o,nT}),\ \varepsilon_n \leq v_n \leq K_1+\varepsilon_n \text{ and } v_{nt}=v^m_{nxx}+g_n(t)v_{nx} \text{ in } \bar{Q}_{o,n,T}.$$

Moreover, we have

<u>Lemma 2.2</u>:

$$|v^m_{nx}| \leq C_1, \qquad 0 \leq v^m_{nx}(0,t) \leq K_2 \tag{2.13}$$

$$|v^m_n(x,t) - v^m_n(y,s)| \leq C_2(|x-y|^2 + |t-s|)^{\frac{1}{2}} \tag{2.14}$$

where C_1 and C_2 depend only on K_1, K_2 and K_1,K_2,K_3, respectively.

<u>Proof</u>: Let $L=\frac{\partial}{\partial t} - mv_n^{m-1}\frac{\partial^2}{\partial x^2} - g_n\frac{\partial}{\partial x}$ and $V=v^m_n$. Then $LV=0$ and $V(x,0)=\phi^m_n(x)$,

$V(0,t)=\varepsilon_n^m$, $V(n,t)=\phi_n^m(n)$. Using (H) and (2.11) and the comparison functions

$$V^+=\varepsilon_n^m+K_2x, \text{ and } V^-=\varepsilon_n^m - K_2x$$

we get $0\leqq V_x(0,t)\leqq K_2$. In the same way we can obtain

$$|V_x(n,t)| \leqq K_2 \tag{2.15}$$

Subsequently, set

$$w = \int_0^{v_n} \frac{ms^{m-1}}{\log(2+s)} ds$$

Then

$$w_t = mv_n^{m-1}w_{xx} + \frac{w_x^2}{2+v_n} + g_n w_x \qquad \text{in } Q_{o,n,T}$$

Put $p=w_x$. We obtain

$$\frac{1}{2}(p^2)_t-mv_n^{m-1}pp_{xx} = -\frac{p^4}{(2+v_n)^2 mv_n^{m-1}}\log(2+v_n)+\frac{m-1}{m}p^2p_x\log(2+v_n)+g_npp_x+\frac{2p^2p_x}{2+v_n} \tag{2.16}$$

Let $z=p^2$. If z takes its positive maximum in $Q_{o,n,T}$, then at its maximum point, say (x_o,t_o), one has

$$z_x=0, \quad z_t-mv_n^{m-1}z_{xx} \geqq 0$$

i.e.,

$$p_x=0, \quad \frac{1}{2}(p^2)_t - mv_n^{m-1}pp_{xx} \geqq 0 \qquad \text{at } (x_o,t_o).$$

From this and (2.16) it follows that

$$0\leqq-\frac{z^2(x_o,t_o)}{(2+v_n(x_o,t_o))^2mv_n^{m-1}(x_o,t_o)}\log(2+v_n(x_o,t_o))<0$$

a contradiction, which shows that z must take its maximum on the parabolic boundary of $Q_{o,n,T}$. Thus $|v_{nx}^m|\leqq 2K_2(1+K_1)\equiv C_1$ and (2.13) is proved. (2.14) then follows from (2.13) and [11] or [18].

Lemmas 2.1 and 2.2 show that $\{v_n\}$ is compact on any $\bar{Q}_{o,n,T}$ $(n=1,2,\ldots)$. Thus we can find a subsequence of $\{v_n\}$, we still write it as $\{v_n\}$, which converges to a continuous function v in $\bar{Q}_{o,T}$ satisfying (2.14), and the convergence is uniform on every compact subset of $\bar{Q}_{o,T}$.

On the other hand, we have

$$\iint_{Q_{o,r,T}} (v_nf_t+v_n^mf_{xx}-g_nv_nf_x)dxdt = -\int_0^r\phi_n(x)f(x,0)dx-\varepsilon_n^m\int_0^T f_x(0,t)dt +$$

$$+\phi_n^m(r)\int_0^T f_x(r.t)dt$$

for any r>0, n>r, and any f as in Definition 2.1. Therefore

$$\iint_{Q_{o,r,T}} (vf_t+v^m f_{xx}-gvf_x)dxdt = -\int_0^r \phi(x)f(x,0)dx+\int_0^T v^m f_x\Big|_o^r dt$$

Obviously $v(x,0)=\phi(x)$ and $v(0,t)=0$. Thus v is a solution of (2.11).

Moreover we have

Theorem 2.1 Suppose that (H) is satisfied. Then the problem (2.11) admits a unique solution such that $v^m \in C^{1,\frac{1}{2}}(\bar{Q}_{o,T})$.

Proof: We only prove the uniqueness. Let v^* be another solution of (2.11). We want to prove

$$\iint_{Q_{o,T}} (v^*-v)h = 0 \tag{2.17}$$

for any $h \in C_o^\infty(Q_{o,T})$.

Let $\{v_n\}$ be the solution of $(IVP)_n$ which converges to v. Then for any r>0, when n>r, one has

$$\iint_{Q_{o,r,T}} (v_n f_t+v_n^m f_{xx}-g_n v_n f_x)dxdt = -\int_0^r \phi_n(x)f(x,o)dx+\int_0^T v_n^m f_x\Big|_o^r dt$$

for any f as in Definition 2.1. For such f, one also has

$$\iint_{Q_{o,r,T}} (v^* f_t+v^{*m} f_{xx}-gv^* f_x)dxdt = -\int_0^r \phi(x)f(x,o)dx+\int_0^T v^{*m} f_x\Big|_o^r dt$$

It follows that

$$\iint_{Q_{o,r,T}} w_n(f_t+a_n(x,t)f_{xx}-g_n f_x)dxdt = \iint_{Q_{o,r,T}} (g_n-g)v^* f_x dxdt +$$

$$+\int_0^r(\phi_n(x)-\phi(x))f(x,0)dx + \int_0^T(v_n^m(r,t)-v^{*m}(r,t))f_x(r,t)dt -$$

$$-\int_0^T \varepsilon_n^m f_x(0,t)dt \tag{2.18}$$

where $w_n=v_n-v^*$, $a_n(x,t)=m\int_0^1 (\theta v^*+(1-\theta)v_n)^{m-1}d\theta$ and $a_n(x,t)\geq\varepsilon_n^{m-1}>0$.

Set $\quad L_n=\frac{\partial}{\partial t} + a_n(x,t)\frac{\partial^2}{\partial x^2} - g_n\frac{\partial}{\partial t}$.

For any $h \in C_o^\infty$, the problem $L_n f=h, f|_{t=T}=f|_{x=0,r}=0$ has a unique solution

$f_{n,r} \in C^{2,1}(\bar{Q}_{o,r,T})$ such that $|f_{n,r}| \leq M$, $|f_{n,r}(r,t)| \leq Mr^{-2}$ and $|f_{n,r,x}| \leq M_1$, where M is a constant independent of n,r and M_1 depends only on M,T and $\|h_x\|_{L^\infty}$. Substituting these $f_{n,r}$ into (2.18), letting $n\to\infty$ first and then $r\to\infty$, it is easy to see that (2.17) is valid.

Remark 2.1 The uniqueness of the solution of (2.11) implies that the whole sequence $\{v_n\}$ converges and the limit is just the solution of (2.11).

Suppose that v_n is the solution of $(IVP)_n$. Then for $0<x<n$, we have

$$F_n(t) = -\frac{1}{x}\int_0^t v_n^m(x,s)ds - \frac{1}{x}\int_0^t\int_0^x g_n(s)v_n(y,s)dyds + \frac{1}{x}\int_0^x (x-y)(v_n(y,t)-\phi_n(y))dy$$
$$+\frac{\varepsilon_n t}{x} + \frac{\varepsilon_n}{x}\int_0^t g_n(s)ds \qquad (2.19)$$

where $F_n(t) = -\int_0^t v^m_{n,x}(0,s)ds$. From Lemma 2.2 it follows that

$F_n(t)$ is nonincreasing on $(0,T)$ and independent of $x \in (0,n)$

$$|F_n(t_1)-F_n(t_2)| \leq K_2|t_1-t_2| \qquad \text{on } [0,T] \qquad (2.20)$$

Theorem 2.2 Suppose that (H) is satisfied. Then the problem (2.5)-(2.8) admits a weak solution (v,λ) with $v \in C^{1/m,1/2m}(\bar{Q}_{0,T})$, $\lambda\in \mathrm{Lip}[0,T]$.

Proof: Set

$$V = \{\lambda(t);\ \lambda(0) = 0,\ \lambda(t) \text{ is nonincreasing and } |\lambda(t_1)-\lambda(t_2)| \leq \frac{K_2}{\varepsilon_o}|t_1-t_2|\}.$$

Then V is compact in $C^o[0,T]$. Take $\lambda\in V$ and set $g(t)=\lambda'(t)$ in (2.11). By Theorem 2.1, the problem (2.11) has a unique solution v. Let v_n be the corresponding solution of $(IVP)_n$. We have

$$v(x,t) = \lim v_n(x,t)$$

uniformly on any compact subset in $\bar{Q}_{0,T}$. Define the operator $T: V\to C^o[0,T]$ as follows

$$\Psi(T\lambda)(t) - \int_0^x \frac{x-y}{x}(v(y,t)-\phi(y))dy - \frac{1}{x}\int_0^t v^m(x,s)ds - \int_0^t\int_0^x \lambda'(s)v(y,s)dyds \qquad x>0,\ t\in[0,T].$$

By virtue of (2.19) and (2.20), the operator T is well defined. Also (2.19) implies $\lim_{n\to\infty} F_n(t)=\psi(T\lambda)$. From (H) we see that Ψ^{-1} exists and $T\lambda=\Psi^{-1}(\lim_{n\to\infty} F_n(t))$ is nonincreasing and Lipschtz continuous by (2.20). Moreover we have

$$|\Psi^{-1}(\lim_{n\to\infty} F_n(t_1)) - \Psi^{-1}(\lim_{n\to\infty} F_n(t_2))| \leq K_2|t_1-t_2|/\varepsilon_o$$

on $[0,T]$, i.e., $T\lambda \in V$. Obviously $T\lambda(0)=0$. Therefore $TV\subset V$. In order to prove that T has a fixed point, it remains to prove that $T: V\to V$ is continuous in $C^o[0,T]$.

Let $\lambda_n, \lambda \in V$, $\lambda_n \to \lambda$ in $C^o[0,T]$. Let $\bar{v}_n$ and $\bar{v}$ be the solutions couresponding to $g=\lambda_n'$ and $g=\lambda'$ in (2.11), respectively. It is easily seen from Theorem 2.1 and Lemma 2.2 that

$$\lim_{n\to\infty} \bar{v}_n(x,t) = \bar{v}(x,t) \tag{2.21}$$

uniformly on any compact subset in $\bar{Q}_{o,T}$(Notice λ_n, $\lambda \in V$!). Also we have

$$\lambda_n' \longrightarrow \lambda' \text{ in } L^1[0,T] \text{ weakly} \tag{2.22}$$

since $\lambda_n' \leq 0$, $\lambda' \leq 0$. Therefore

$$|T\lambda_n - T\lambda| \leq \frac{1}{\varepsilon_o} \|\psi(T\lambda_n)-\psi(T\lambda)\| \leq \frac{T}{\varepsilon_o x} \sup_{[o,T]} |\bar{v}_n^m(x,\cdot)-\bar{v}^m(x,\cdot)| +$$
$$+ \frac{K_2 T+x}{0^x} \int_0^x \sup|\bar{v}_n(y,\cdot)-\bar{v}(y,\cdot)|dy + \frac{1}{\varepsilon_o x}\left|\int_0^x\int_0^T (\lambda_n'-\lambda)\bar{v}(y,s)dyds\right|$$

It follows from (2.21) and (2.22) that

$$\lim_{n\to\infty} \|T\lambda_n - T\lambda\|_{C^o[0,T]} = 0.$$

Thus T has a fixed point by a theorem in [10], this λ and the corresponding v yield a solution of (2.5) - (2.8).

Theorem 2.3 Suppose that (H) is satisfied. Then the solution of (2.5) - (2.8) is unique.

Proof: Let (v_1,λ_1) and (v_2,λ_2) be two solutions of (2.5) - (2.8) corresponding to the same initial data and free boundary conditions. We will prove $(v_1,\lambda_1)=(v_2,\lambda_2)$.

Since (v_i,λ_i) satisfies (2.5) - (2.7) in which $g=\lambda_i'(i=1,2)$, as the proof of Theorem 2.1, v_i can be approached by its corresponding solutions $v_{i,n}$ of $(IVP)_n$ in which $g_i^n \to \lambda_i'(i=1,2)$.

For any $f \in C^{2,1}([-r,r]x[0,T])$, where $r>0$, with $f(x,T)=f(\pm r,t)=0$. Let $n>r-\lambda_i(T)$ $(i=1,2)$ and use $(IVP)_n$. We get

$$\int_0^T \int_0^{r-\lambda_i(t)} (v_{i,n}\bar{f}_{it} + v_{i,n}^m \bar{f}_{ixx} - g_i^n v_{i,n}\bar{f}_{ix})dxdt$$

$$= - \int_0^r \phi_n(x)\bar{f}_i(x,0)dx - \int_0^T v_{i,n}^m(r-\lambda_i(t),t)\bar{f}_{ix}(r-\lambda_i(t),t)dt +$$

$$+ \int_0^T v_{i,n\,x}^m(0,t)\bar{f}_i(0,t)dt \tag{2.23}$$

where $\bar{f}_i(x,t) = f(x+\lambda_i(t),t)$. Notice that

$$\int_0^T v^m_{i,n\,x}(0,t)\bar{f}_i(0,t)dt = -\int_0^T F'_{i,n}(t)\bar{f}_i(0,t)dt \qquad \text{(see (2.19))}$$

$$= \frac{1}{x}\int_0^T v^m_{i,n}(x,t)\bar{f}_i(0,t)dt + \frac{1}{x}\iint_{oo}^{xT} g^n_i(t)v_{i,n}(y,t)\bar{f}_i(0,t)dydt$$

$$-\frac{1}{x}\iint_{oo}^{xT}(x-y)v_{i,n\,t}(y,t)\bar{f}_i(0,t)dtdy - \frac{\varepsilon_n}{x}\int_0^T \bar{f}_i(0,t)dt - \frac{\varepsilon_n}{x}\int_0^T g^n_i(t)\bar{f}_i(0,t)$$

$$= \frac{1}{x}\int_0^T v^m_{i,n}(x,t)\bar{f}_i(0,t)dt + \frac{1}{x}\iint_{oo}^{xT} g^n_i(t)v_{i,n}(y,t)\bar{f}_i(0,t)dydt$$

$$+\frac{1}{x}\iint_{oo}^{xT}(x-y)v_{i,m}(y,t)\bar{f}_{it}(0,t)dt + \frac{1}{x}\int_0^x (x-y)\phi_n(y)\bar{f}_i(0,0)dy$$

$$-\frac{\varepsilon_n}{x}\int_0^T \bar{f}_i(0,t)dt - \frac{\varepsilon_n}{x}\int_0^T g^n_i(t)\bar{f}_i(0,t)dt$$

Therefore

$$\int_0^T v^m_{i,n\,x}(0,t)\bar{f}_i(0,t)dt \longrightarrow$$

$$\frac{1}{x}\int_0^T v^m_i(x,t)\bar{f}_i(0,t)dt + \frac{1}{x}\iint_{oo}^{xT}\lambda'_i(t)v_i(y,t)\bar{f}_i(0,t)dtdy$$

$$+\frac{1}{x}\iint_{oo}^{xT}(x-y)v_i(y,t)\bar{f}_{it}(0,t)dydt + \frac{1}{x}\int_0^x (x-y)\phi(y)\bar{f}_i(0,0)dy \quad (n\to\infty).$$

On the other hand, by (2.9), $\forall\varepsilon>0$, choosing n large enough, we have

$$n\int_\varepsilon^{T-\varepsilon}(\Psi(\lambda_i(t+\tfrac{1}{n})) - \Psi(\lambda_i(t)))\bar{f}_i(0,t)dt$$

$$= \frac{n}{x}\int_0^x\int_\varepsilon^{T-\varepsilon}(x-y)(v_i(y,\tfrac{1}{n}+t) - v_i(y,t))\bar{f}_i(0,t)dydt$$

$$-\frac{n}{x}\int_\varepsilon^{T-\varepsilon}\int_t^{t+\frac{1}{n}} v^m_i(x,s)\bar{f}_i(0,t)dsdt - \frac{n}{x}\int_\varepsilon^{T-\varepsilon}\int_t^{t+\frac{1}{n}}\int_0^x \lambda'_i(s)v_i(y,s)\bar{f}_i(0,t)dydsdt$$

$$= \frac{n}{x}\int_0^x\int_\varepsilon^{T-\varepsilon}(x-y)v_i(y,t)(\bar{f}_i(0,t-\tfrac{1}{n}) - \bar{f}_i(0,t))dydt$$

$$-\frac{n}{x}\int_0^x\int_\varepsilon^{\frac{1}{n}+\varepsilon}(x-y)v_i(y,t)\bar{f}_i(0,t-\tfrac{1}{n})dydt + \frac{n}{x}\int_0^x\int_{T-\varepsilon}^{T-\varepsilon+\frac{1}{n}}(x-y)v_i(y,t)\bar{f}_i(0,t-\tfrac{1}{n})dydt$$

$$-\frac{n}{x}\int_\varepsilon^{T-\varepsilon}\int_t^{t+\frac{1}{n}} v^m_i(x,s)\bar{f}_i(0,t)dsdt - \frac{n}{x}\int_\varepsilon^{T-\varepsilon}\int_t^{t+\frac{1}{n}}\int_0^x \lambda'_i(s)v_i(y,s)\bar{f}_i(0,t)dydsdt$$

Letting $n\to\infty$ and then $\varepsilon\to 0$, it follows that

$$\int_0^T \psi(\lambda_i(t))\lambda_i'(t)\bar f_i(0,t)dt =$$

$$= -\frac{1}{x}\int_0^x\int_0^T (x-y)v_i(y,t)\bar f_{it}(0,t)dydt - \frac{1}{x}\int_0^x (x-y)\phi(y)\bar f_i(0,0)dy$$

$$-\frac{1}{x}\int_0^T v_i^m(x,t)\bar f_i(0,t)dt - \frac{1}{x}\int_0^T\int_0^x \lambda_i'(t)v_i(y,t)\bar f_i(0,t)dydt$$

Therefore, letting $n\to\infty$ in (2.40), we obtain that

$$\int_0^T\int_0^{r-\lambda_i(t)} (v_i\bar f_{it} + v_i^m\bar f_{ixx} - \lambda_i' v_i\bar f_{ix})dxdt$$

$$= -\int_0^r \phi(x)\bar f_i(x,0)dx - \int_0^T v_i^m(r-\lambda_i(t),t)\bar f_{ix}(r-\lambda_i(t),t)dt$$

$$-\int_0^T\psi(\lambda_i(t))\lambda_i'(t)\bar f_i(0,t)dt$$

in all of the above expressions $i=1,2$. Set $u_i(x,t)=v_i(x-\lambda_i(t),t)$. Changing the variables in the above identity with (x,t) by $(x-\lambda_i(t),t)$, we get

$$\int_0^T\int_{\lambda_i(t)}^r (u_if_t + u_i^m f_{xx})dxdt = -\int_0^r \phi(x)f(x,0)dx -$$

$$-\int_0^T u_i^m(r,t)f_x(r,t)dt - \int_0^T \psi(\lambda_i(t))\lambda_i'(t)f(\lambda_i(t),t)dt \qquad (i=1,2) \tag{2.24}$$

Also

$$\int_0^T\int_{-r}^{\lambda_i(t)}\psi(x)f_t dxdt = -\int_{-r}^0 \psi(x)f(x,0)dx + \int_0^T\lambda_i'(t)\psi(\lambda_i(g))f(\lambda_i(t),t)dt .$$

Adding this to (2.24) and setting

$$U_i = \begin{cases} u_i & x>\lambda_i \\ \psi(x) & x<\lambda_i \end{cases}, \quad A(u) = \begin{cases} u^m & u>0 \\ 0 & u<0 \end{cases}, \quad \Phi(x) = \begin{cases} \phi(x) & x>0 \\ \psi(x) & x<0 \end{cases},$$

one has

$$\int_0^T\int_{-r}^r (U_if_t + A(U_i)f_{xx})dxdt = -\int_{-r}^r \Phi(x)f(x,0)dx - \int_0^T u_i^m(r,t)f_x(r,t)dt$$

Set $W=U_1-U_2$ and

$$a(x,t) = \int_0^1 A'(\theta U_1+(1-\theta)U_2)d\theta.$$

Then

$$\int_0^T\int_{-r}^r W(f_t+a(x,t)f_{xx})dxdt = -\int_0^T (u_1^m(r,t)-u_2^m(r,t))f_x(r,t)dt \tag{2.25}$$

$h \in C_o^\infty((-\infty,\infty)x(0,T))$. Choose r large enough such that supp $h\subset(-r,r)x(0,T)$. Consider the problem

$$\begin{aligned} &f_t+a_n(x,t)f_{xx}=h \qquad \text{in } (-r,r)x(0,T) \\ &f|_{t=T} = f|_{x=\pm r} = 0 \end{aligned} \tag{2.26}$$

where $a_n \in C^\infty([-r,r]x[0,T])$ satisfying $\|a_n-a\|_{L^2(\bar{Q}_{o,r,T})}\leq C(r)/n$ and $a_n\geq 1/n$.

As in the proof of Theorem 2.2, we know that (2.26) has a unique $C^{2,1}$ solution $f_{n,r}$ such that $|f_{n,r}|<M$, $|f_{n,r\,x}(\pm r,t)|\leq Mr^{-2}$, $|f_{n,r\,x}(x,t)|\leq M$ where M is a constant independent of n,r. Moreover we have

$$\|f_{n,r\,xx}\|_{L^2(Q_{o,r,G})}\leq M_1n^{\frac{1}{2}} + Mr^{\frac{1}{2}} \tag{2.27}$$

Indeed, setting $V=f_{n,r\,x}$, then V satisfies

$$V_t + (a_nV_x)_x = h_x \tag{2.28}$$

Since $h \in C_o^\infty$ and (2.26), we get

$$a_n(\pm r,t)V_x(\pm r,t)=0 \text{ i.e. } V_x(\pm r,t)=0 \qquad \text{on } [0,T].$$

Also $\qquad V_x(x,T)=0 \qquad$ on $[-r,r]$.

Therefore integrating (2.28) one has (2.27).

Substituting $f_{n,r}$ into (2.25) yields

$$\begin{aligned} \left|\int_o^T\int_{-r}^r Wh\right| &\leq \|a_n-a\|_{L^2}\|f_{n,r\,xx}\|_{L^2}+\int_o^T|u_1^m(r,t)-u_2^m(r,t)||f_{n,r\,x}(r,t)|dt \\ &\leq C(r)M/n^{\frac{1}{2}} + 2K_1^mTM/r^{\frac{3}{2}} \end{aligned}$$

Letting $n\to\infty$ and then $r\to\infty$, we get

$$\int_o^T\int_{-\infty}^\infty Whdxdt = 0$$

which shows W=0, i.e., $U_1=U_2$. Thus follow $u_1=u_2$ and $\lambda_1=\lambda_2$.

<u>Corollary 2.1</u> Let ϕ_i and ψ_i satisfy (H)(i=1,2). Let (u_i,λ_i) be the weak solutions of the problem (2.1) - (2.4) corresponding to ϕ_i and ψ_i(i=1,2), respectively. If $\phi_1\geq\phi_2$ and $\psi_1\geq\psi_2$, then $u_1\geq u_2$, $\lambda_1\leq\lambda_2$.

To sum up, we obtain

<u>Theorem 2.4</u> Suppose that (H) is satisfied. Then the problem (2.1) - (2.4) admits a unique weak solution.

<u>Remark 2.2</u> In the proof of Theorem 2.3, if a(x,t) is symmetric in x about $x=x_o$,

then we may take a_n to be also symmetric in x about $x=x_o$. Thus the solution of (2.25) is symmetric in x about $x=x_o$ if h is. Moreover, the first derivative w.r.t. x of the solution is zero at x_o. This remark will be used in §4.

§3. Regularity of Weak Solutions

Recalling the proofs of Theorem 2.2 and 2.3, we see that the solution of (2.1)-(2.4) can be approached by the solution U_n of the following problem

$$(IVP)'_n \qquad \begin{array}{ll} U_t = (a_n(U)U_x)_x + g_n(t)U_x & \text{in } Q_{o,T} \\ U(x,0) = \phi_n(x) & \text{on } R^1_+ \\ U(0,t) = 0 & \text{on } [0,T] \end{array}$$

where a_n, g_n, ϕ_n and ε_n still satisfy (2.12). And U_n have the corresponding properties. All estimates for v_n obtained before are still valid for U_n. Moreover we first have

<u>Proposition 3.1</u> Suppose that v is a solution of the problem (2.5) - (2.7), Then for any bounded set $A=(a,b)X(c,d)\subset\subset Q_{o,T}$ where $A\subset\subset Q_{o,T}$ means $\bar{A}\subset Q_{o,T}$, there is a constant C depending only on b-a, d-c, c, and K_1 such that

$$v_x^{m-1} \text{ exists and } |v_x^{m-1}| \leqq C \text{ in } A. \tag{3.1}$$

If, moreover, $|\phi_x^{m-1}|\leqq K_4$, then (3.1) is true for $A=(a,b)X(0,d)\subset Q_{o,T}$ when C is independent of c but depends on K_4. Thus $v \in C^{1/(m-1),1/2(m-1)}_{loc}(Q_{o,T})$.

<u>Proposition 3.2</u> Suppose that v is a solution of (2.5) - (2.7). Then $v_x^m \in C(Q_{0,T})$ and $v_x^m(x,t)=0$ if $v(x,t)=0$. If, moreover, ϕ_x^{m-1} is continuous, the v_x^m is continuous up to the boundary $\{x>0,\ t=0\}$.

The proof of Proposition 3.1 is similar to that in [1] and the proof of Proposition 3.2 to that in [12].

<u>Proposition 3.3</u> Suppose that (v,λ) satisfies (2.5) - (2.8) and that $(\phi^m)''\geqq-C$. Then $\lim_{x\to o^+} V_x^m(x,t)$ exists everywhere and

$\lim_{x\to o^+} v_x^m(x,t)=\psi(\lambda(t))\lambda'(t)$ a.e. on [0,T].

<u>Proof</u>: Suppose U_n is the solution of $(IVP)'_n$ where $(\phi_n^m)''\geqq-C$. Set $W=U_n^m$. Then W satisfies

$$W_t = mW^\alpha W_{xx} + g_n(t)W_x \qquad \text{in } Q_{o,T} \tag{3.2}$$

$$W(x,0) = \phi_n^m(x) \qquad \text{on } R_+^1,\ W(0,t) = \varepsilon_n^m \qquad \text{on } [0,T] \tag{3.3}$$

where $\alpha=(m-1)/m$. Differentiating (3.2) w.r.t. x twice time, one has

$$W_{xxt} = mW^\alpha W_{xxxx} + (2m\alpha W^{\alpha-1}W_x + g_n(t))W_{xxx} + m\alpha(\alpha-1)W^{\alpha-2}(W_x)^2W_{xx} + mW^{\alpha-1}(W_{xx})^2$$

Define

$$LV = \frac{\partial V}{\partial t} - mW^\alpha\frac{\partial^2 V}{\partial x^2} - (2m\alpha W^{\alpha-1}W_x + g_n(t))\frac{\partial V}{\partial x} - m\alpha(\alpha-1)W^{\alpha-2}(W_x)^2V.$$

Then $L(W_{xx}) = mW^{\alpha-1}(W_{xx})^2 \geqq 0$. Notice $W_{xx}(x,0) = (\phi_n^m)'' \geqq -C$, and

$$0 = mW^\alpha(0,t)W_{xx}(0,t) + g_n(t)U_{nx}(0,t) \leqq m\varepsilon_n^m W_{xx}(0,t)$$

since $g_n(t) \leqq 0$. We conclude that

$$W_{xx} \geqq -C \qquad (\text{since } \alpha(\alpha-1)<0 \text{ and } L(-C)\leqq 0).$$

Therefore $v_{xx}^m \geqq -C$ in the sense of distribution. Thus by Proposition 3.2, $v_x^m + Cx$ is monotone increasing w.r.t. x for each t in the common sense, which shows that $\lim_{x\to 0^+} v_x^m(x,t)$ exists everywhere, we write it as $v_x^m(0,t)$.

On the other hand, by Theorem 2.2,

$$\Psi(\lambda(t))x = \int_0^x (x-y)(v(y,t)-\phi(y))dy - \int_0^t v^m(x,s)ds - \int_0^t\int_0^x \lambda'(s)v(y,s)dyds$$

Therefore for every given $x>0$ it follows that

$$\Psi(\lambda(t)) = \int_0^t v_x^m(x,s)ds = \int_0^x (v(y,t)-\phi(y))dy - \int_0^t \lambda'(s)v(x,s)ds. \tag{3.4}$$

Since $v_x^m \in C(Q_{o,T}) \cap L^\infty(\bar{Q}_{o,T})$ and $v \in C^{1/m,1/2m}(\bar{Q}_{o,T})$, we see that $v_x^m(0,t) \in L^1(Q_{0,T}) \cap L^\infty(Q_{0,T})$. Letting $x\to 0^+$ in (3.4), we complete the proof.

<u>Remark 3.1</u> Proposition 3.3 implies that if ϕ^m is convex, then v^m is also convex in x. Moreover the condition "$(\phi^m) \geqq -C$" is not necessary. Indeed, choose $\phi_n \in C^\infty$ and $\phi_n = 1/n$ when $x>R_n$ ($R_n\to\infty$, $n\to\infty$), $0\leqq\phi_n\leqq\|\phi\|_{L^\infty}$, $\phi_n\downarrow$ if $n\uparrow$, $\phi_n\to\phi$ uniformly on any compact subset. Then the standard regularity theory [19] yields that

$$\frac{d^k}{dx^k}(U_n(x,t)-1/n)\to 0, \qquad x\to\infty$$

holds uniformly for $t\in[0,T]$ (k integer). Hence, when ε is small enough, one has

$$(U_n^m)_{xx}(x,\varepsilon) \geqq -K/\varepsilon \qquad \text{on } R_+^1, \tag{3.5}$$

$$(U_n^m)_{xx}(0,t) \geqq 0 \qquad \text{on } [\varepsilon,T] \quad (\text{known}) \tag{3.6}$$

where $K = \|\phi\|_{L^\infty}/(m-1)$. Write $W^* = (U_n^m)_{xx}$, By (3.2) we get $LW^*=0$ where

$$LV=\frac{\partial v}{\partial t}-mW^{\alpha}\frac{\partial^2 v}{\partial x^2}-(2mW_x^{\alpha}+g_n(t))\frac{\partial v}{\partial x}-m\alpha(\alpha-1)W^{\alpha-2}(W_x)^2v-m\alpha W^{\alpha-1}v^2$$

and $\alpha=(m-1)/m$. Also $L(-K/t)\leqq 0$ since $0<\alpha<1$. Therefore it follows by (3.5) and (3.6) that $W^*\geqq -K/t$ $(t>0)$ i.e. $(v^m)_{xx}\geqq -K/t$ $(t>0)$ in the sense of distribution.

Proposition 3.4 If ϕ satifies (H) and $\phi>0$ when $x>0$, then $u>0$ in $P_{\lambda,T}=\{x>\lambda(t), t\in(0,T)\}$.

Proof: Consider the function due to Barenblatt [4]

$$B(x,t)=t^{-1/(m+1)}((1-C_m x^2/t^{2/(m+1)})^+)^{1/(m+1)}$$

where $C_m=(m-1)/2m(m+1)$, which satisfies the porous medium equation, i.e. $B_t=B^m_{xx}$ in $Q=(0,\infty)\times R^1$. Set

$$w(x,t)=(\sigma/R^2)^{1/(m-1)}B(R(x-x_o),\sigma t+\delta) \tag{3.7}$$

where x_o, R, σ, $\delta>0$ are constants. Then w still satisfies the porous medium equation.

By the assumptions, there are constants ε_o, $\delta_o>0$ such that $x_o-\delta_o>0$ and $\phi\geqq\varepsilon_o>0$ on $[x_o-\delta_o,\ x_o+\delta_o]$. We wish to choose a series of w_i in the form of (3.7) such that

$$\text{supp } w_i|_{t=0}\subset[x_o-\delta_o,\ x_o+\delta_o],\ w_i(x,0)\leqq\varepsilon_o,\ \text{meas}(\text{supp } w_i|_{t=0})\leqq 2/i.\quad (i=1,2,\ldots) \tag{3.8}$$

Note the free boundary of w intersects t-axe at

$$t=(C_m x_o^2)^{(m+1)/2}R^{m+1}/\sigma-\delta/\sigma$$

If there is $\bar{t}\in[0,T]$ such that $\lambda(t)=0$ on $[0,\bar{t}]$, we set

$$R_i=i\delta^{1/(m+1)}/C_m^{1/2},\qquad \sigma_i=\varepsilon_o^{m-1}i^2\delta/C_m^{1/2}\quad\text{and}$$

$$t_i=C_m x_o^{m+1}i^{m-1}/\varepsilon_o^{m-1}-C_m/\varepsilon_o^{m-1}i^2$$

and
$$w_i(x,t)=(\sigma_i/R_i)^{1/(m-1)}B(R_i(x-x_o),\ \sigma_i t+\delta).$$

Then w_i satisfies (3.8) and w_i solves the porous medium equation with $w_i(0,t)=0$ on $[0,t_i]$.

On the other hand, in the sense of distribution

$$u_t=u^m_{xx}\ \text{ in } Q_{0,t_i},\quad u(x,0)=\phi(x)\ \text{ on } R^1_+,\quad u(0,t)=0 \text{ on } [0,t_i].$$

Therefore by the comparison principle [7] and [19], it follows that

$$u\geqq w_i\qquad \text{in } Q_{0,t_i}$$

In particular
$$u(x_o,t)\geqq w_i(x_o,t)>0,\quad 0<t<t_i,\quad \text{for any } x_o>0.$$

Noticing $t_i\to\infty(i\to\infty)$, we conclude that $u>0$ in $Q_{0,\bar{t}}$.

If $\lambda<0$ on $(0,T]$, i.e. $\bar{t}=0$, then we take $\delta=\frac{1}{2}$ and

$$R_i=i/C_m^{\frac{1}{2}}2^{1/(m+1)}, \qquad \sigma_i=\varepsilon_0^{m-1}i^2/C_m 2^{(4m-3)/2(m+1)}.$$

Let $\mu_i(t)$ be the left free boundary of w_i. Then

$$\mu_i(t)=x_0-i^{-1}2^{-(3m-4)/2(m+1)^2}(i^2\varepsilon_0^{m-1}C_m^{-1}t + 2^{(2m-5)/2(m+1)})^{1/(m+1)}$$

For any $s\in(0,t)$, choosing $x_0=i^{-1}2^{-(3m-4)/2(m+1)^2}(i^2\varepsilon_0^{m-1}C^{-1}s + C_m')^{1/(m+1)}$

Then $\mu_i(s)=0$. It is clear that $\mu_i(t)=\mu_i(t;s)$ is increasing in s. Set $\mu_s(t)=\mu_1(t)$.

Now if $(x',t')\in Q_{0,T}$, i.e. $x'>0$, since $u(0,t)\geqq 0$, as we proved, we have $u(x',t')>0$. If $(x',t')\in P_{\lambda,T}\setminus Q_{0,T}$, we take $\mu_s(t)$ and the corresponding $w_s(x,t)$. By the choice of R_i and σ_i we see that (3.8) is satisfied. Let $t^*(s)$ be such that $\mu(t^*(s))=\lambda(t^*(s))$(if $\mu(t)=\lambda(t)$ on some interval, we take $t^*(s)$ to be the right end-point of this interval). In the region $Q_s^*=\{(x,t);0<t<t^*(s),\ \mu_s(t)<x<\infty\}$, as we have done, comparing w_s with u we conclude that

$$u>w_s \quad \text{in } Q_s^*$$

In particular $\qquad u(x,t)>w_s>0$ in $A_s=\{(x,s);\ 0<t<t^*(s),\mu_s(t)<x<0\}$.

Since $\lambda(t)$ and $\mu_s(t)$ are continuous, and $\bigcup_{0<s<T}A_s=P_{\lambda,T}\setminus Q_{0,T}$, we have $u>0$ in $P_{\lambda,T}\setminus Q_{0,T}$. To sum up $u>0$ in $P_{\lambda,T}$.

Theorem 3.1 Suppose that (H) is satisfied. Then the problem (2.1)-(2.4) has a unique strong solution (u,λ). And if $\phi>0$ for $x>0$, then $u>0$ in $P_{\lambda,T}$.

Remark 3.2 The hypothesis (H) can be weakened as

(H)': ϕ is measurable and $0\leqq\phi(x)\leqq k_1$, $-M\leqq\psi\leqq-\varepsilon_0<0$.

Indeed, choose a series of C_0^∞ functions ϕ_n such that $\{\phi_n\}$ is a decreasing series, $\phi_n\to\phi$ in $L^1_{loc}[0,\infty)$ and $0\leqq\phi_n\leqq K_1$. Consider

$$(FBP)_n \quad \begin{cases} u_t = u^m_{xx} & 0<t<T,\ \lambda_n(t)<x<\infty,\\ u(\lambda_n(t),t) = 0 & 0<t<T,\\ u(x,0) = \phi_n(x) & 0<x<\infty,\\ u^m_x(\lambda_n(t),t) = \psi(\lambda_n(t))\lambda_m'(t) & 0<t<T. \end{cases}$$

From §2 we know that $(FBP)_n$ has a unique weak solution (u_n,λ_n). And the comparison principle (Corollary 2.1) and Proposition 3.3 yield that

$$\{u_n\} \text{ is decreasing and } 0\leqq u_n\leqq K_1,$$

$$\{\lambda_n\} \text{ is increasing and } \lambda_1 \leqq \lambda_n \leqq 0,$$

$$\lambda_n \in \text{Lip } [0,T] \text{ and each } \lambda_n \text{ is nonincreasing on } [0,T]$$

$$u_x^m(\lambda_n(t),t)=\psi(\lambda_n(t))\lambda_n'(t) \quad \text{a.e on } [0,T]$$

Therefore $\lim_{n\to\infty} u_n$ and $\lim_{n\to\infty} \lambda_n$ exist, we write them as

$$\lambda(t)= \lim_{n\to\infty} \lambda_n \qquad 0\leqq t<T$$

$$u(x,t)= \lim_{n\to\infty} u_n(x,t) \qquad \text{in } P_{\lambda,T}$$

From Definition 2.1, u satisfies the porous medium equation in the sense of distribution. Therefore $u \in C^{\alpha}(P_{\lambda,T,\varepsilon_o})$ for some $\alpha \in (0,1]$, see [8], where $P_{\lambda,T,\varepsilon_o} = \{(x,t);\ \varepsilon_o<t<T,\ \lambda(t)+\varepsilon_o<x<\infty\}$, and $\varepsilon_o>0$ is arbitrary. Hence

$$u(x,t)= \lim_{n\to\infty} u_n(x,t) \tag{3.10}$$

uniformly on any compact subset in $P_{\lambda,T,\varepsilon_o}$.

By Proposition 3.1, there exists a contant C depending only on K_1,ε such that

$$|u_{nx}^{m-1}(1,t)| \leqq C \qquad t\in[\varepsilon,T-\varepsilon] \ (\varepsilon \in (0,T)) \tag{3.11}$$

Thus by Remark 3.1 we have

$$0 \leqq \psi(\lambda_n(t))\lambda_n'(t) = u_{nx}^m(\lambda_n(t),t) \tag{3.12}$$

$$\leqq u_{nx}^m(1,t)+\frac{K}{\varepsilon}(1-\lambda_n(t)) \leqq K_1C+\frac{K}{\varepsilon}(1-\lambda_1(T))\equiv C_1$$

where K is a constant independent of n. Using (3.11), (3.12) and Remark 3.1, as the proof of Lemma 2.2, we can prove that $|u_{nx}^m|\leqq C_2$ in $\{(x,t);\ \lambda_n(t)\leqq x\leqq 1,\ \varepsilon\leqq t\leqq T-\varepsilon\}$ where C_2 is a constant independent of n. Thus $u \in C^{\alpha}(\overline{P_{\lambda,T} \cap \{\varepsilon\leqq t\leqq T-\varepsilon\}})$ for some $\alpha \in (0,1]$. Hence the covergence in (3.10) is uniform on any compact subset in $(\overline{P_{\lambda,T} \cap \{\varepsilon\leqq t\leqq T-\varepsilon\}})$.

Again, from (3.12) and the proof of Theorem 2.3 it follows that there is a function $f_\varepsilon\leqq 0$ which is defined on $[\varepsilon, T-\varepsilon]$ and $f_\varepsilon \in L^\infty[\varepsilon,T-\varepsilon]$ such that

$$\lambda_n'\to f_\varepsilon \text{ in } L^1[\varepsilon, T-\varepsilon] \qquad \text{weakly}$$

for any $\varepsilon \in (0,T)$. Thus there exists a function $f\leqq 0$, defined on $(0,T)$, such that

$$\lambda_n'\to f \text{ in } L^1[\varepsilon, T-\varepsilon] \qquad \text{weakly}$$

for any $\varepsilon \in (0,T)$. Therefore $f \in L^1[0,T]$ and $\lambda_n'\to f$ in $L^1[0,T]$ weakly and

$$\lambda(t) = \int_0^t f(s)ds,$$

Since

$$\lambda(T-\varepsilon) - \lambda(\varepsilon) = \lim_{n\to\infty} (\lambda_n(T-\varepsilon) - \lambda_n(\varepsilon)) =$$

$$= \lim_{n\to\infty} \int_\varepsilon^{T-\varepsilon} \lambda_n'(t)dt = \int_\varepsilon^{T-\varepsilon} f(t)dt$$

$$\int_0^T |f(t)|dt = - \int_0^T f(t)dt = -\lambda(T) < \infty$$

which shows $f \in L^1$ and $\lambda_n' \to f$ in $L^1[0,T]$ weakly, also

$$\lambda(t) - \lambda(\varepsilon) = \int_\varepsilon^t f(s)ds$$

letting $\varepsilon \to 0$, we get

$$\lambda(t) = \int_0^t f(s)ds.$$

We then conclude that $\lambda_n' \to \lambda'$ in $L^1[0,T]$ weakly.

Now set $v_n(x,t)=u_n(x+\lambda_n(t),t)$ and $v(x,t)=u(x+\lambda(t),t)$ in $Q_{0,T}$. From

$$\Psi(\lambda_n(t)) = \int_0^x \frac{x-y}{x}(v_n(y,t) - \phi_n(y))dy - \frac{1}{x}\int_0^t v_n^m(x,s)ds -$$

$$- \frac{1}{x}\int_0^t\int_0^x \lambda_n'(s)v_n(y,s)dyds$$

and

$$\int_0^T\int_0^r (v_n f_t + v_n^m f_{xx} - \lambda_n' v_n f_x)dxdt$$

$$= - \int_0^r \phi_n(x)f(x,0)dx + \int_0^T v_n^m f_x\Big|_0^r dt$$

with f as in Definition 2.1, iii), Letting $n\to\infty$, we obtain

$$\int_0^T\int_0^r (vf_t + v^m f_{xx} - \lambda' vf_x)dxdt = - \int_0^r\phi(x)f(x,0)dx + \int_0^T v^m f_x\Big|_0^r dt$$

$$\Psi(\lambda(t)) = \int_0^x \frac{x-y}{x}(v(y,t) - \phi(y))dy - \frac{1}{x}\int_0^t v^m(x,s)dx +$$

$$+ \frac{1}{x}\int_0^t\int_0^x \lambda'(s)v(y,s)dyds$$

Thus (v,λ) is a solution of the problem (2.5), (2.6), (2.7), (2.8) when ψ, ϕ satisfy (H)'. Obviously Theorem 2.4 and Proposition 3.3 are still true in this case. Hence we obtain the following

Theorem 3.2 Suppose that (H)' holds, i.e.

$$0 \leq \phi \leq C_1$$

$$-M_o \leq \psi \leq -\varepsilon_o \leq 0.$$

Then the problem (2.1), (2.2), (2.3), (2.4) admits a unique solution.

§4. Regularity of Free Boundary

In this section we discuss the properties of the free boundaries. For simplicity we assume that the condition (H) is satisfied through out the section except for another explanation.

Let (u,λ) be the solution of the problem (2.1)-(2.4). If $u>0$, then u is a classical solution. Therefore we have, by Theorem 2.4, that

$$\Psi(\lambda(t)) = \int_{\lambda(t)}^{x} u(y,t)\,dy - \int_{o}^{t} u_x^m(x,s)ds - \int_{o}^{x} \phi(y)dy \qquad (4.1)$$

for given $x>0$. It follows by differentiating (4.1) that

$$\psi(\lambda(t))\lambda'(t) = \int_{\lambda(t)}^{x} u_t(y,t)dy - u_x^m(x,t) \quad (x>0). \qquad (4.2)$$

Hence if we can prove, without the condition $u>0$, that (4.2) is valid for some $x>0$ near the free boundary, then (3.2) and Proposition 3.2 yield that to prove the continuity of λ', it remains to prove that

$\int_{\lambda(t)}^{x} u_t(y,t)dy$ is continuous in t provided that ψ is continuous.

Now suppose

$$u_x^m(\lambda(t),t) \geqq \delta > 0 \quad \text{on } I \qquad (4.3)$$

where I is some nonempty interval. Then we have

<u>Proposition 4.1</u> Write $\eta=(\eta_1,\eta_2)$ and $N(t_o,\eta)=\{(x,t);|x-\lambda(t_o)|<\eta_1,\ |t-t_o|<\eta_2\}$ $\cap P_{\lambda,T}$. Then there are $\eta=(\eta_1,\eta_2)$ with $\eta_1,\eta_2>0$ and positive constants C_1 and C_2 such that

$$C_1 \leqq \frac{u^m(x,t)}{|x-\lambda(t)|} \leqq C_2 \qquad \text{in } N(t_o,\eta) \qquad (4.4)$$

where $t_o \in I$ and (4.3) is satisfied.

<u>Proof</u>: Since u^m is Lipschitz continuous in x on $\bar{P}_{\lambda,T}$, it follows that

$$u^m(x,t) \leqq C_2|x-\lambda(t)| \qquad \text{in } \bar{P}_{\lambda,T}\ ,$$

for some constant $C_2>0$.

Secondly, via Taylor's formula and Proposition 3.3 (or Remark 3.1) we obtain that for every $t \in I$

$$u^m(x,t) \geqq u_x^m(\lambda(t),t)(x-\lambda(t)) - \frac{C}{2}(x-\lambda(t))^2 \geqq (\delta - \frac{C}{2}|x-\lambda(t)|)|x-\lambda(t)|$$

by (4.3). Hence

$$u_x^m(x,t) \geq \tfrac{1}{2}\,\delta|x-\lambda(t)| \quad \text{when} \quad |x-\lambda(t)| \leq \delta/C \text{ and } t \in I.$$

This completes the proof.

Proposition 4.2 If (4.3) is satisfied, then

$$|u_t^m| \leq C, \qquad |u_{tt}^m| \leq C|x-\lambda(t)| \quad \text{in } N(t_o,\eta) \tag{4.5}$$

for some constant $C>0$.

Proof Fix a point $(x_1,t_1) \in N(t_o,\eta)$. Set

$$\nu = |x_1-\lambda(t_1)|/2, \qquad w(x,t)=\nu^{-1}u^m(\nu^{-(m-1)/2m}x,\nu t)$$

Let $N_\nu(t_o,\eta)=\{(\nu^{-(m-1)/2m}x,\nu t),(x,t) \in N(t_o,\eta)\}$. Then w satisfies

$$w_t = mw^\alpha w_{xx} \qquad \text{in } N(t_o,\eta) \tag{4.6}$$

where $\alpha=(m-1)/m$. Denote by (s_1',t_1') the point $(\nu^{-(m-1)/2m}x_1,\nu t_1)$. It follows from Proposition 4.1 that

$$0<mC_1^\alpha \leq mw^\alpha(x_1',t_1') \leq mC_2^\alpha$$

By virtue of the continuity of w and Nash and Schauder's estimates one has

$$|w_t| \leq C^*, \quad |w_{tt}| \leq C^* \text{ (C^* depends only on m,C_1 and C_2)} \tag{4.7}$$

in some neighbourhood N' of (x_1',t_1'). In particular, (4.7) holds at (x_1',t_1'). Returning this to (x,t) and u, the proof is complete.

Remark 4.1 From (4.5) we know that if (4.3) is satisfied, then there exists a neighbourhood N_o of $\{(x,t);\ x=\lambda(t),\ t \in I\}$ such that

$$|u_t(x,t)| \leq \frac{C^*}{m}\,u^{1-m} \qquad \text{in } N_o$$

But this and (4.3) give that

$$|u_t(x,t)| \leq C|x-\lambda(t)|^{1/m-1} \qquad \text{in } N_o \tag{4.8}$$

which shows that $u_t \in L^1(N_o)$.

From Proposition 4.1 and 4.2, we obtain

Theorem 4.1 If (4.7) is satisfied and $\psi(x)$ is continuous, then

$$\lambda(t) \in C^1(I) \text{ and } \psi(\lambda(t))\lambda'(t) = u_x^m(\lambda(t),t) \text{ for all } t \in I. \tag{4.9}$$

Thus $u_x^m(\lambda(t),t) \in C(I)$.

Proof: As we have shown at the beginning of this section, it suffices to prove that

$$\int_{\lambda(t)}^{x} u_t(y,t)dy$$

is of meaning and continuous on I for some $x>0$.

Given t_o, there is $N(t_o,\eta)\ni(\lambda(t_o),t_o)$ such that (4.4) and (4.8) hold. Thus $u\in C^\infty(N(t_o,\eta))$ and $u_t\in L^1(N(t_o,\eta))$. Therefore (4.2) holds if we choose $x>0$ appro-poriately small.

For any t: $(x,t)\in N(t_o,\eta)$ and $\varepsilon>0$, setting

$$J(t,t_o) = \left|\int_{\lambda(t)}^{x} u_t(y,t)dy - \int_{\lambda(t_o)}^{x} u_t(y,t_o)dy\right| \tag{4.10}$$

then

$$J(t,t_o) \leqq \left|\int_{\lambda(t)}^{\lambda(t)+\varepsilon} u_t(y,t)dy\right| + \left|\int_{\lambda(t_o)}^{\lambda(t_o)+\varepsilon} u_t(y,t)dy\right|$$

$$+ \left|\int_{\lambda(t)+\varepsilon}^{x} (u_t(y,t)-u_t(y,t_o)\ dy\right|$$

$$\leqq mC(\varepsilon^{1/m}+|\lambda(t_o)+\varepsilon-\lambda(t)|^{1/m}) \int_{o}^{x}|u_t(y,t)dy-u_t(y,t_o)|dy$$

(by (4.8)). Since (y,t_o), $(y,t)\in N(t_o,\eta)$ (if $\varepsilon>0$ is small enough) and $u\in C^\infty(N(t_o,\eta))$, we get

$$\lim_{t\to t_o} J(t,t_o) \leqq 2C^m\varepsilon^{1/m} \qquad (\varepsilon>0 \text{ is small}).$$

This completes the proof.

To study the continuity of $\lambda'(t)$, we shall construct a class of comparison functions. Consider the free boundary problem

$$\begin{cases} u_t=u^m_{xx} & \lambda(t)<x<\infty,\ 0<t<T \\ u(x,0)=\phi(x) & 0<x<\infty, \\ u(\lambda(t),t)=0 & 0<t<T, \\ u^m_x(\lambda(t),t)=\alpha\lambda'(t) & 0<t<T, \end{cases} \tag{4.11}$$

where $\alpha<0$ is a constant and $\phi\geqq 0$. We assume (4.11) has free boundary $x=\xi_o t$ for some constant $\xi_o<0$ and solution $u(x,t)=v(\xi)$ with $\xi=x-\xi_o t$. Then v satisfies

$$(v^m)'' = -\xi_o v' \qquad 0<\xi<\infty \tag{4.12}$$

$$v(x)=\phi(x) \qquad 0<x<\infty \tag{4.13}$$

$$v(0)=0 \qquad (v^m)'(0)=\alpha\xi_o \tag{4.14}$$

From these it follows that

$$\int_{o}^{v} \frac{ms^{m-1}}{\alpha\xi_o-\xi_o s}\, ds = \xi$$

Since $ms^{m-1}/(\alpha\xi_o-\xi_o s)>0$ $(s>0)$, one gets that (4.12)-(4.14) has a unique solution, thus (4.11) has a solution $u(x,t)=v(x-\xi_o t)$ with the initial data $\phi(x)=v(x)$. Moreover, for any $A>0$ there is a constant $C=C(m,\xi_o,A)$ with the property $C\to 1^+$ as $A\to 0$ such that

$$v^m(\xi) \geqq \alpha\xi_o\xi \qquad \xi>0 \tag{4.15}$$

$$v^m(\xi) \leqq -\alpha C\xi \qquad 0<\xi<A \tag{4.16}$$

See [24].

Now we can prove the following

<u>Proposition 4.3</u> Suppose that ψ is continuous. Then $\lambda'(t+0)$ exists everywhere and

$$\psi(\lambda(t))\lambda'(t+0)=u_x^m(\lambda(t),t) \text{ for all } t \in [0,T]. \tag{4.17}$$

<u>Proof</u>: Without loss of generality we may only prove (4.17) at $t=0$. Since ψ is continuous, for any $\sigma>0$ there is a $\delta_\sigma>0$ such that

$$\psi(0)-\sigma \leqq \psi(x) \leqq \psi(0)+\sigma(<0) \qquad \delta_\sigma<x<0$$

Setting $\alpha_o=u_x^m(\lambda(0),0)$, then either $\alpha_o>0$ or $\alpha_o=0$.

Case I: $\alpha_o>0$. In this case, for $\varepsilon>0$, there exists $\delta_\varepsilon>0$ such that

$$(\alpha_o-\varepsilon)x \leqq u^m(x,0) \leqq (\alpha_o+\varepsilon)x \qquad 0<x<\delta_\varepsilon \tag{4.18}$$

and thus

$$\frac{1}{2}(\alpha_o-\varepsilon)\delta \leqq u^m(\frac{1}{2}\sigma,t) \leqq \frac{1}{2}(\alpha_o+\varepsilon)\delta \qquad 0<t<\delta_1 \text{ for some } \delta_1>0. \tag{4.19}$$

Choosing $\xi_o=(\alpha_o+\varepsilon)/(\psi(0)+\sigma)$ and $\xi_o=(\alpha_o-2\varepsilon)/(\psi(0)-\sigma)$, we obtain two solutions of (4.11) u_1 and u_2 respectively. Taking $\alpha=\psi(0)+\sigma$ and $\alpha=\psi(0)-\sigma$, and A small enough such that $C<\varepsilon/(\alpha_o-2\varepsilon)+1$ where C is in (4.16). Denote by λ_1 and λ_2 the free boundaries of u_1 and u_2. By (4.18), (4.19) and the comparison principal (Corollary 2.1) we conclude that

$$\lambda_1(t) \geqq \lambda(t) \geqq \lambda_2(t) \qquad 0<t<\delta^{**}$$

for some $\delta^{**}>0$, i.e.

$$\frac{\alpha_o+\varepsilon}{\psi(0)+\sigma}t \geqq \lambda(t) \geqq \frac{\alpha_o-2\varepsilon}{\psi(0)-\sigma}t \qquad 0<t<\delta^{**}$$

Taking the upper and lower limits, one has that $\lim_{t\to 0+} \lambda(t)/t$ exists and is equal to $\alpha_o/\psi(0)$, i.e.

$$\psi(\lambda(0))\lambda'(0+0) = \alpha_o = u_x^m(\lambda(0),0).$$

Case II: $\alpha_o=0$. In this case, we only use u_2 to obtain that

$$\lambda(t) \geq -\frac{2\varepsilon}{\psi(0)+\sigma}\, t$$

which implies $\lim_{t\to 0^+} \lambda(t)/t \geq 0$. On the other hand $\lambda(t) \leq 0$, thus

$$\lim_{t\to 0^+} \lambda(t)/t=0 \quad \text{i.e. } \lambda'(0+0)=0.$$

Proposition 4.4 Suppose that ψ is continuous. Then for any s>0 there are two functions $\xi(t)$ and $\eta(t)$ in which ξ is convex and $\eta \in C^{1,1}$ such that $\lambda(t)=\xi(t)+\eta(t)$.

Proof We first assume $-\lambda'(s_o)>0$ and prove a fundemental inequality for our proof. Set

$$U(x,t)=\begin{cases} u(x,t+s_o) & \lambda_1(t)<x<k,\ 0<t<T,\\ u(2k-x,t+s_o) & k<x<2k-\lambda_1(t),\ 0<t<T,\\ \psi(x) & \text{otherwise}\end{cases}$$

where $\lambda_1(t)=\lambda(t+s_o)$ and we define $\lambda_1(t)=0$ for $t>T-s_o$.

Let v_o be the solution in [24] corresponding to $\alpha=-M_o$ there. Setting

$$v(x,t)=(\sigma/R^2)^{1/(m+1)}\ v_o(R(x-x_o),\sigma t+\tau).$$

then v satisfies

$$v_t = v^m_{xx} \qquad \mu(t)<s<\infty,\quad 0<t<T,$$

$$v(x,0) = (\sigma/R^2)^{1/(m+1)}\ v_o(R(x-x_o),\tau) \quad \mu(0)<x<\infty,$$

$$v(\mu(t),t) = 0 \quad \text{and } v^m_x(\mu(t),t) = -M_o\mu'(t) \qquad 0<t<T$$

where $\mu(t)=x_o+\xi_o(\sigma t+\tau)^{\frac{1}{2}}/R$. Set

$$V(x,t)=\begin{cases} v(x,t) & \mu(t)<x<k,\quad 0<t<T\\ v(2k-x,\ t_o+s_o) & k<x<2k-\mu(t),\quad 0<t<T\\ -M_o & \text{otherwise}\end{cases}$$

Since $U^m_x(\lambda(s_o),s_o)=\psi(\lambda(s_o))\lambda'(s_o)>0$, there is a constant $k_o>0$ such that

$$U^m(x,0) \geq \tfrac{1}{2}\lambda'(s_o)\psi(\lambda(s_o))\ (x-\lambda(s_o)) \text{ for } \lambda(s_o)<x<\lambda(s_o)+k_o. \tag{4.20}$$

Note that from [24] we have

$$V^m(x,0)\leq(\sigma/R^2)^{1/(m+1)}(m_oR^2|\xi_o|(x-\lambda(s_o))/2\tau^{\frac{1}{2}} \text{ for } x>0. \tag{4.21}$$

Thus if we choose $x_o=0$, $k=k_o$, $\sigma=\tau=\frac{1}{2}|\psi(\lambda(s_o))\lambda'(s_o)\lambda(s_o)|$

$$R = C_m(|\psi(\lambda(s_o))\lambda'(s_o)\lambda(s_o)|)^{(3m+1)(m-1)/(2m-1)^2}$$

$$\xi_o = C^*_m(|\psi(\lambda(s_o))\lambda'(s_o)\lambda(s_o)|)^{(3m+1)/(2m-1)}\psi(\lambda(s_o))\lambda'(s_o)$$

where $C^*_m<0$ and $C_m>0$ are constants depending only on m, M_o, then $\mu(0)=\lambda_1(0)$ and

$\mu'(0)=\lambda_1'(0)$, and $U(x,0)\geqq V(x,0)$ on R^1. Therefore, as shown in Remark 2.2, we have

$$\iint_{Q_{k_o,T}} (U-V)h\,dx\,dt \geqq 0$$

for any $h \in C_o^\infty(Q_{k_o,T})$, $h\geqq 0$ and h is symmetric about $x=k_o$. Noting that U and V are all symmetric about $x=k_o$, we get $U\geqq V$ and

$$\lambda_1(s)-\lambda_1(0)-\lambda_1'(0)s \leqq \mu(s)-\mu(0)-\mu'(0)s$$

Also
$$\mu''(s) = \mu''(0) + 0(s) = -C\lambda'(s_o) + 0(s)$$

where C depends only on m, M_o. Hence we have

$$\lambda(s+s_o)-\lambda(s_o)-\lambda'(s_o+0)s \leqq -\frac{1}{2}C\lambda'(s_o)s^2+0(s^3)$$

without loss of generality we assume $0(s^3) \geqq 0$.

Obviously, if $\lambda'(s_o+0)=0$, the above inequality is still valid, since $\lambda(t)$ is nonincreasing. Therefore we obtain

$$\lambda(t+s)-\lambda(t)-\lambda'(t)s \leqq -\frac{1}{2}C\lambda'(t)s^2+0(s^3) \quad \text{for } t\geqq s(\ s>0). \tag{4.22}$$

Now as did in [6] and [8], one gets the proposition and

$$\lambda''(t) + C\lambda'(t) = \mu \leqq 0 \tag{4.23}$$

where μ is a measure. Thus $\lambda'(t-0)$ exists everywhere and $\lambda'(t-0)\geqq\lambda'(t+0)$.

From (4.23) we have

Corollary 4.1 For any $t_1\leqq t_2$

$$\lambda'(t_2-0)e^{Ct_2}\leqq\lambda'(t_1+0)e^{Ct_1}$$

$$\lambda'(t_2+0)e^{Ct_2}\leqq\lambda'(t_1+0)e^{Ct_1} .$$

Hence if $\lambda(t)=$ const. in some interval $s_1\leqq t\leqq s_2$, then we must have $\lambda'(t)=0$ in $[0,s_2]$. In particular, when $s_1=s_2$, we get

Proposition 4.5 If there is $t_o \in (0,T)$ such that $\lambda'(t_o+0)=0$, then $\lambda'(t)=0$ on $[0,t_o]$.

To combine Theorem 4.1 with Proposition 4.5 and 4.6, one has

Theorem 4.2 Suppose that ψ is continuous. Then there exists $t^* \in [0,T]$ such that $\lambda(t) \in C^1((0,t^*)U(t^*,T))$ and $\psi(\lambda(t))\lambda'(t)=u_x^m(\lambda(t),t)$ on $(0,t^*)U(t^*,T)$.

It is natural to ask: When does t^* equal zero? Below we give some answers.

Proposition 4.6 Suppose that ψ is continuous and that

$$\exists\ c,x_o>0,\ \beta<1/(m-1) \text{ such that } \phi(x)\geqq cx^{\beta} \text{ for } x\in(0,x_o) \tag{4.23}$$

Then $t^*=0$.

Proof: We may assume $\beta>1/m$ and prove first that

$$\exists\ \delta_o,t_o>0 \text{ such that } u(x,t)>\tfrac{1}{2}C(x-\lambda(t))^{\beta}\ \ \lambda(t)<x<\lambda(t)+\delta_o,\ 0<t<t_o \tag{4.24}$$

Let (v,λ) be the solution of (2.5)——(2.8) and let v_n be the solution of $(IVP)_n$ where we assume that $0<\varepsilon_n<1$ and $\phi_n(x)\geqq cx^{\beta}+\varepsilon_n$, $x\in(0,x_o)$ and $\phi_n(x)\geqq\varepsilon_n$. Define the nonlinear operator L as follows:

$$Lv=\frac{\partial v}{\partial t}-\frac{\partial^2}{\partial x^2}v^m-g_n(t)\frac{\partial v}{\partial x}$$

Then $Lv_n=0$ in $Q_{o,n,T}$, see §2. Put $G_n(x,t)=x^{\beta}(\frac{1}{2}c+\varepsilon_n x)$. Then

$$\begin{cases} v_n(x,0)\geqq G_n(x,0) & 0<x<\min(x_o,1)\\ v_n(0,t)\geqq G_n(0,t) & 0<t<T\end{cases} \tag{4.25}$$

Choose $\delta_o\in(0,\min(x_o,1))$ such that $\delta_o<\frac{1}{2}c$, then

$$v_n(\delta_o,0)\geqq G_n(\delta_o,0)$$

Therefore there exists $t_o>0$ independent of n such that

$$v_n(\delta_o,t)\geqq G_n(\delta_o,t)=G_n(\delta_o,0)\qquad 0<t<t_o, \tag{4.26}$$

since v_n converges uniformly on any compact subset and $\{G_n\}$ is a equicontinuous family. By a complicated computation we get

$$-LG_n\geqq x^{m-2}g_n^*(x)$$

where

$$g_n^*(x)=m(\tfrac{1}{2}c+\varepsilon_n x)^{m-2}((m-1)(\tfrac{1}{2}c\beta+\varepsilon_n(\beta+1)x)^2+(\tfrac{1}{2}c+\varepsilon_n x)(\tfrac{1}{2}c\beta(\beta-1)+\varepsilon_n(\beta+1)x)$$

$$-K(\tfrac{1}{2}c\beta+\varepsilon_n(\beta+1)x)x^{1-(m-1)\beta}.$$

Since $1/m<\beta<1/(m-1)$ and $c>0$, each g_n^* is continuous and g_n^* is equicontinuous. Note $g_n^*(0)=m(\frac{1}{2}c)^m\beta(m\beta-1)>0$. There is $\delta_1>0$ such that

$$g_n^*(x)>0 \qquad 0<x<\delta_1.$$

Therefore

$$-LG_n\geqq 0 \qquad 0<x<\delta_o'=\min(\delta_o,\delta_1)$$

From (4.25), (4.26) and the comparison principal [20] we get

$$v_n\geqq G_n \qquad 0<t<t_o,\ \ 0<x<\delta_o',$$

i.e.,

$$v(x,t)\geqq\tfrac{1}{2}cx^{\beta} \qquad 0<t<t_o,\ \ 0<x<\delta_o'.$$

Returning to u, we have (4.24)

Now if $t^*>0$, then

$$v_t = v^m_{xx} \text{ in } Q_{o,t^*} \quad \text{in the sense of distribution}$$

$$v(x,0)=\phi(x) \quad \text{on } R^1_+,\ v(0,t)=0 \text{ on } (0,t^*)$$

and $v^m_x(0,t)=0$ on $(0,t^*)$. Define $\phi_1(x)=\phi(x)\,(x>0)$ and $=0$ $(x<0)$, $v_1(x,t)=v(x,t)$ in Q_{o,t^*} and $=0$ otherwise. Then

$$v_{1t} = v^m_{1xx} \quad \text{in } Q_{t^*}\text{, in the sense of distribution}$$

$$v_1(x,0)=\phi_1(x) \qquad \text{on } R^1$$

where $Q_{t^*}=R^1 x(0,t^*)$. By the known results $v_1 \in C^\alpha(Q_{t^*})$ and $v^m_1 \in C(Q_{t^*})$, see [8], it is easy to prove that v_1 is a weak solution of the porous medium equation in Q_{t^*} with the initial data ϕ_1. Therefore the well-known theory on the Cauchy problem of the porous medium equation (see [1],[2],[3],[6],[8],[17]) yields that

$$v_1 \in C^{1/(m-1),1/2(m-1)}(R^1 x(s,t')) \qquad (s>0).$$

In Particular, $|v_1(x,t)-v_1(0,t)| \leqq C_s|x|^{1/(m-1)}$ $\qquad t>s$

thus $\qquad v(x,t) \leqq C_s x^{1/(m-1)} \qquad x>0,\ t>s.$

Fix $x \in (0,t^*)$ and (4.24) gives that $C_s x^{1/(m-1)} \geqq \frac{1}{2}cx^\beta$ $\quad 0<x<\delta'_o$ but $\beta<1/(m-1)$, a contradiction. Thus $t^*=0$.

Remark 4.2 The proof of Proposition 4.7 implies that if $\psi=0$, then the problem (2.1)-(2.4) is in fact the Cauchy problem for the porous medium equation. At the same time we get the estimate for u near the vertical free boundary:

$$u(x,t) \leqq C_s\ x^{1/(m-1)} \tag{4.27}$$

Thus if $t^*>0$, then

$$u(x,t^*) \leqq C_s\ x^{1/(m-1)} \qquad x>0 \text{ for some } s \in (0,t^*),$$

which implies $u^m_x(0,t^*)=0$. Therefore, $\lambda'(t^*+0)=0$. Obvously $\lambda'(t^*-0)=0$. Thus we have obtained

Theorem 4.3 Suppose that ψ is continuous. Then $\lambda \in C^1$ and (2.4) holds everywhere.

We continue the discussion on t^*.

Proposition 4.7 If $(\phi^m)''\geqq 0$ in the sense of distribution and $\psi\neq 0$ on a set of positive measure, then $t^*=0$.

Proof: If $t^*>0$, then u satisfies

$$u_t = u^m_{xx} \text{ in } Q_{o,t*}, \text{ in the sense of distribution}$$

$$u(0,t) = u^m_x(0,t) = 0 \qquad \text{on } (0,t*).$$

Remark 3.2 and the proof of Proposition 3.3 give that

$$u^m_{xx} \geqq 0 \text{ in } \mathcal{D}'(Q_{o,t*}),$$

i.e.,

$$u_t \geqq 0 \text{ in } \mathcal{D}'(Q_{o,t*}).$$

By Proposition 3.2 we get that for any $x>0$

$$\int_0^x (x-y)(u^{m+1}(y,t_2)-u^{m+1}(y,t_1))dy \leqq \frac{1}{2}(m+1)\int_{t_1}^{t_2} u^{2m}(x,t)dt$$

$t_1 \leqq t_2 \in (0,t*)$. Note that u is nondecreasing in t. We conclude that for any given $X \in (0,x)$, one has

$$\int_0^x \frac{x-y}{x}(u^{m+1}(y,t_2)-u^{m+1}(y,t_1))dy \leqq (m+1)TK*/x$$

where $K* = (\sup_{Q_{o,T}} u)^{2m} \leqq K_1^{2m}$. Letting $x \to \infty$, the Lebesgue domination convergence theorem gives that

$$\int_0^X (u^{m+1}(y,t_2)-u^{m+1}(y,t_1))dy = 0$$

for any $X>0$ and any $t_1 \leqq t_2 \in (0,t*)$. This shows that u is independent of t. Therefore $u(0,t) = u^m_x(0,t) = 0$ implies $u=0$ in Q, a contradiction. Thus $t*=0$.

Proposition 4.8 If $\phi(x) \geqq \varepsilon_o > 0$, then $t*=0$.

Proof: Use the comparison functions in [24].

Proposition 4.9 If there exist constants c, $x_o > 0$ such that

$$\phi(x) \leqq cx^\alpha, \; 0<x<x_o \quad \text{and } \alpha > 2/(m-1),$$

then $t*>0$.

Proof: Use the comparison functions $(u*,\lambda*) = ((\frac{Ax^2}{T_1-t})^{1/(m-1)},0)$.

Proposition 4.10 If $\phi \in L^1 \cap L^\infty$, then for every t, $u(x,t) \in L^1(R^1_{\lambda(t)})$ where $R^1_{\lambda(t)} = (\lambda(t),\infty)$ and

$$\psi(\lambda(t)) = \int_{\lambda(t)}^\infty u(y,t)dy - \int_0^\infty \phi(y)dy$$

Thus

$$\int_0^\infty u(y,t)dy = \int_0^\infty \phi(y)dy \text{ for } 0<t\leqq t*, \quad \int_{\lambda(t)}^\infty u(y,t)dy < \int_0^\infty \Phi(y)dy \text{ for } t*<t<T.$$

Remark 4.4 Proposition 4.10 shows that the energy of the motion decreases as the time increases. Hence the free boundary condition is exactly the mathematical description of the consumption of energy in motion. The reader may compare these with those in the case of the porous medium equation [8].

From Theorem 2.3 we see that, if we assume that ψ is continuous a.e., then all the results we have obtained in this section are still valid. Below we show that this condition is also necessary for $\lambda \in C^1$.

We see, from the proof of Proposition 4.5, that Proposition 4.5 and Corollary 4.1 are still true even if ψ only satisfies (H) and no matter whether ψ is continuous a.e.. Therefore if $\lambda(t)$ is strictly decreasing for $t>t^*$ and $t^*<\infty$, then for any given $\varepsilon>0$ there is a constant $C(\varepsilon)>0$ such that

$$\lambda'(t) \leqq -C(\varepsilon) \text{ for } t>t^*+\varepsilon$$

Hence, by Proposition 3.3 we have

$$u_x^m(\lambda(t),t) \geqq \varepsilon_0 C(\varepsilon) \equiv C_1 \quad \text{a.e. for } t>t^*+\varepsilon$$

As we did in the proof of Proposition 4.1, we get

$$u^m(x,t) \geqq \frac{1}{4}C|x-\lambda(t)| \text{ if } |x-\lambda(t)| \leqq C_1/2C$$

for almost all $t>t^*+\varepsilon$, and then for all $t>t^*+\varepsilon$ since u and λ are continuous. This shows that Proposition 4.1 holds in this case and therefore (4.1) is valid,

$$\Psi(\lambda(t)) = \int_{\lambda(t)}^{x} u(y,t)dy - \int_0^t u_x^m(x,s)ds - \int_0^x \phi(y)dy \equiv r(x,t)$$

for $t>t^*$.

According to the proof of Theorem 4.1, we know that $r(x,t)$ is continuously differentiable in t for $x>\lambda(t)$. Fix $x=1$ and set $r(t)=r(1,t)$. We have

$$\psi(\lambda(t))\lambda'(t) = r'(t) \quad \text{a.e.} \quad \text{for } t>t^*$$

i.e.,

$$\psi(\lambda(t)) = r'(t)/\lambda'(t) \quad \text{a.e.} \quad \text{for } t>t^*$$

If $\lambda'(t)$ is continuous, noting that λ is strictly decreasing for $t>t^*$, we conclude that $\lambda^{-1}(z)$ exists and is continuous for $z>0$, and

$$\psi(z) = r'(\lambda^{-1}(z))/\lambda'(\lambda^{-1}(z)) \quad \text{a.e.} \quad \text{for } z>0$$

which implies that ψ is continuous a.e..

With this and Proposition 3.4, Remark 3.1 and 3.2, we get

Theorem 4.3 Suppose that (H)' is satisfied. Then $\lambda \in C^1$ if and only if ψ is continuous a.e. unless $\lambda \equiv 0$.

Remark 4.5 All of the results in $Q_{o,T}$ in the paper can be extented to $Q=R_+^1 \times (0,\infty)$ without any difficulty.

References

1. Aronson D.G., Regularity properties of flows through porous media; SIAM J. Appl. Math. Vol. 17, No.2(1969) 461-467.

2. ___________ Regularity properties of flow through porous media; The interface, Arch. Rat. Math. Anal. Vol.37, No.1(1970) 1-10.

3. ___________ Regularity properties of flow through porous media; A counterexample, SIAM J. Appl. Math. Vol.19, No.2(1970) 299-307.

4. Barenblatt G.I., On some unsteady motions of a liquid or a gas in a porous medium; Prikl. Mat. Mech. 16(1952) 67-78.

5. Brezis H. & Crandall M.G., Uniqueness of solutions of the initial value problem for $u_t - \phi(u)=0$; J. Math. Pure Appl. 58(1979) 153-163.

6. Caffarelli L.A. & Friedman A., Regularity of the free boundary for the one-dimensional flow of gas in a porous medium; Amer. J. Math. Vol.101, No.6(1979) 1193-1218.

7. Friedman A., Partial Differential Equations of Parabolic Type; Printice-Hall Englewood, Cliffs. N. J. 1964.

8. ___________ Variational Principles and Free-Boundary Problems; John Wiley & Sons. N.Y. 1982.

9. ___________ Analytisity of the free boundary for the Stefan problem; Arch. Rat Mech. Anal. 61, 97-125(1976).

10. Gilbarg D. & Trudinger N.S., Elliptic Partial Differential Equations of Second Order; Springer-Verlag 2nd Edi. 1984.

11. Gilding B.H., Holder continuity of solutions of parabolic equations; J. London Math. Soc. 12(1976).

12. ___________ & Peletier L.A., The Cauchy problem for an equation in the theory of infiltration; Arch. Mech. Anal. Vol.61, No.2(1976) 127-140.

13. Jensen R., The smoothness of the free boundary in the Stefan problem with supercooled water; Ill.J. Math. 22, 623-629(1978).

14. Kalashinikov A.S., The propogation of disturbance in problems of nonlinear heat condition with absorption; Zh. Vychisl. Mat. mat. Fiz. 144(1971)(890-907) 70-85.

15. Kamennomostskaja S.L., On Stefan problem; Mat. Sb.53(95) 485-514(1965).

16. Kindelerer D., The smoothness of the free boundary in the one-phase Stefan problem; Comm. Pure Appl. Math. 31, 257-282(1978).

17. Knerr B.F., The porous medium equation in one dimension; Trans. Amer. Math. Soc. Vol.234, No.2(1977) 381-415.

18. Kruzhkov S.N., Results concerning the nature of the continuity of solutions of parabolic equations and some of their applications;Matematicheskic, Zam. Vol.6, No.1-2(1969)(97-108) 517-523.

19. Ladyzenskaja O.A. el., Linear and Quailinear Equations of Parabolic Type; Amer. Math. Soc. Transl. R.J. 1968.

20. Oleinik O.A. el., The Cauchy's problem and boundary problems for equations of the type of nonstationary filtration; Izv. Akad. Mauk SSSR. Ser. Mat. 22(1958) 667-704.

21. Peletier L.A., Lecture Notes in Mathematics 415; Springer-Verlag 1974, 412-416.

22. ____________ A necessary and sufficient condition for the existence of an interface in flows thought porous medium; Arch. Rat. Mech. Math. Anal. Vol.56, No.2(1974) 183-190-

23. Vol'pert A.I. & Hudjaev S.I., Canchy's problem for degenerate second order quasilinear parabolic equations; Mat. Sb. 78(1969) 365-387.

24. Wu Zhuoqun, A free boundary problem for degenerate quasilinear parabolic equations; MRC. Tsch. Sum. Rep. #2656, 1983.

GLOBAL PERTURBATION OF THE RIEMANN PROBLEM FOR THE SYSTEM OF ONE-DIMENSIONAL ISENTROPIC FLOW

Li Ta-tsien & Zhao Yan-chun
Fudan University
Shanghai, China

I. Introduction and main results

In this paper we consider the discontinuous initial value problem for the system of one-dimensional isentropic flow

$$\begin{aligned} &\frac{\partial \tau}{\partial t} - \frac{\partial u}{\partial x} = 0 , \\ &\frac{\partial u}{\partial t} + \frac{\partial p(\tau)}{\partial x} = 0 , \end{aligned} \tag{1.1}$$

where $\tau>0$ is the specific volume, u the velocity and $p=p(\tau)$ the pressure. For polytropic gases,

$$p(\tau) = A\tau^{-\gamma} , \tag{1.2}$$

A is a positive constant and $\gamma>1$, the adiabatic exponent. Introducing Riemann invariants

$$\begin{aligned} r &= \frac{1}{2} \left(u-\int_\tau^\infty \sqrt{-p'(\eta)}\,d\eta\right) , \\ s &= \frac{1}{2} \left(u+\int_\tau^\infty \sqrt{-p'(\eta)}\,d\eta\right) \end{aligned} \tag{1.3}$$

as new unknown functions, system (1.1) can be rewritten as

$$\begin{aligned} &\frac{\partial r}{\partial t} + \lambda(r,s)\frac{\partial r}{\partial x} = 0 , \\ &\frac{\partial s}{\partial t} + \mu(r,s)\frac{\partial s}{\partial x} = 0 , \end{aligned} \tag{1.4}$$

where

$$-\lambda(r,s)=\mu(r,s)=\sqrt{-p'(\tau(s-r))}= a(s-r)^{\frac{\gamma+1}{\gamma-1}} \quad (a>0, \text{ constant}) . \tag{1.5}$$

We consider the discontinuous initial value problem for system (1.4) with the following discontinuous initial data

$$t=0: \ (r,s) = \begin{cases} (r_0^-(x),s_0^-(x)), & x\leqq 0, \\ (r_0^+(x),s_0^+(x)), & x\geqq 0, \end{cases} \tag{1.6}$$

where $(r_0^-(x),s_0^-(x))$ and $(r_0^+(x),s_0^+(x))$ are smooth functions on $x\leqq 0$ and $x\geqq 0$

respectively with a discontinuity at the origin:

$$(r_o^-(0),\ s_o^-(0)) \neq (r_o^+(0),\ s_o^+(0)). \tag{1.7}$$

problem (1.4)-(1.6) can be regarded as a perturbation of the corresponding Riemann problem for system (1.4) with the following piecewise constant initial data

$$t=0:\ (r,s)= \begin{matrix} (r_-,s_-), & x\leqq 0, \\ (r_+,s_+), & x\geqq 0, \end{matrix} \tag{1.8}$$

where

$$(r_\pm,s_\pm)=(r_o^\pm(0),\ s_o^\pm(0))\ . \tag{1.9}$$

It is well-known (cf. R.Courant and K.O.Friedrichs [1]) that there are four different situations for the solution to Riemann problem (1.4),(1.8), that is, the solution is composed of (see figure 1)

(a). a backward centered rarefaction wave and a forward centered rarefaction wave; or

(b). a backward centered rarefaction wave and a forward typical shock; or

(c). a backward typical shock and a forward centered rarefaction wave; or

(d). a backward typical shock and a forward typical shock.

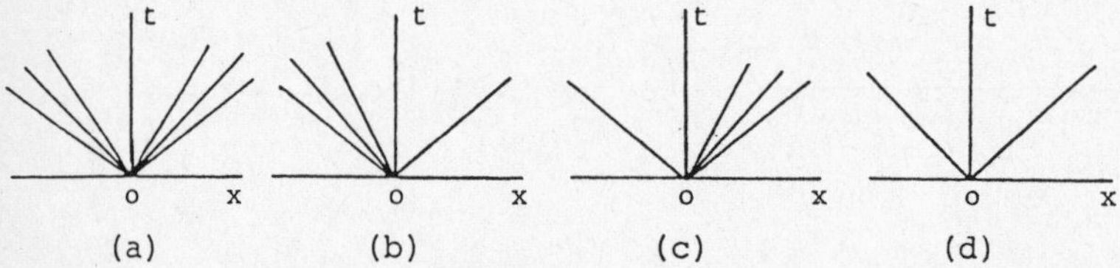

figure 1

It is also well-known (cf. Gu Chao-hao, Li Ta-tsien and Hou Zong-Yi [2], Li Ta-tsien and Yu Wen-ci [3]) that the discontinuous initial value problem (1.4)-(1.6) admits a unique solution locally in time in a class of piecewise continuous and piecewise smooth functions, and this solution has a structure similar to the solution of the corresponding Riemann problem (1.4),(1.8), namely, in a local domain $R(\delta)$ $=\{(t,x)\,|\,0\leqq t\leqq\delta,\ -\infty<x<+\infty\}$, the solution is composed of (see figure 2)

(a). a backward centered wave and a forward centered wave (In this case the initial discontinuity disappears immediately and the solution becomes continuous for $0<t\leqq\delta$); or

(b). a backward centered wave and a forward shock; or

(c). a backward shock and a forward centered wave; or

(d). a backward shock and a forward shock respectively.

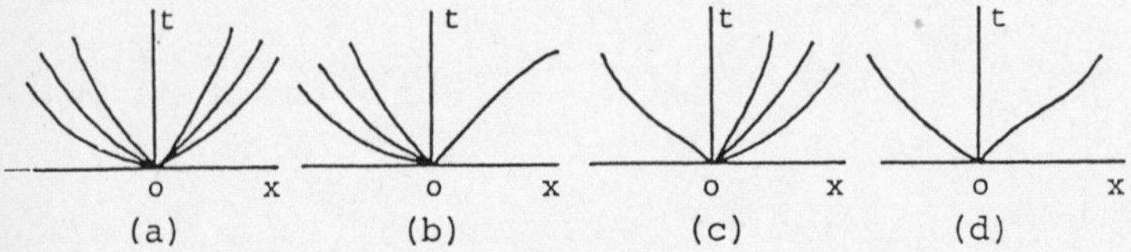

figure 2

Generally speaking, these kinds of structure of the solution can not be kept for a longer time and new singularities (for instance, new shocks) may occur in a finite time. We shall point out, however, that under certain reasonable assumptions on the initial data (1.6), (1.4)-(1.6) admits a unique global solution in a class of piecewise continuous and piecewise smooth functions and the solution has a global structure similar to that of the corresponding Riemann problem (1.4),(1.8). That is, the solution contains no shock(in case (a)), only one shock(in case (b) or (c)), or only two shocks (in case (d)) on $t \geqq 0$ respectively.

Case (a). In this case, we have

$$r_- < r_+ , \quad s_- < s_+ , \tag{1.10}$$

and to avoid the vacuum state, we suppose

$$s_- - r_+ > 0 . \tag{1.11}$$

According to the well-known local existence theorem for this problem and the global existence theorem for the initial value problem with smooth initial data (cf. P.D. Lax [4]) we can easily obtain the following (also see Lin Long-wei [5])

Theorem 1: Suppose that $(r_o^-(x), s_o^-(x))$ and $(r_o^+(x), s_o^+(x))$ are bounded, C^1 functions on $x \leqq 0$ and $x \geqq 0$ respectively and

$$\begin{aligned} &r_o^{-\prime}(x) \geqq 0, \quad s_o^{-\prime}(x) \geqq 0, \quad x \leqq 0, \\ &r_o^{+\prime}(x) \geqq 0, \quad s_o^{+\prime}(x) \geqq 0, \quad x \geqq 0. \end{aligned} \tag{1.12}$$

Suppose furthermore that there is no vacuum state on both $x \leqq 0$ and $x \geqq 0$ at the initial time, i.e.,

$$\begin{aligned} &s_o^-(x) - r_o^-(x) > 0, \quad x \leqq 0, \\ &s_o^+(x) - r_o^+(x) > 0, \quad x \geqq 0. \end{aligned} \tag{1.13}$$

Then problem (1.4)-(1.6) admits a unique global solution $(r(t,x), s(t,x))$ on $t \geqq 0$ in a class of piecewise continuous and piecewise smooth functions. This solution is composed of a backward centered wave and a forward centered wave with the origin as their center and then continuous for $t > 0$. Moreover, there never exists any vacuum state on $t \geqq 0$.

Case (b). Case (b) and Case (c) are similar, we only consider Case(b) here. In this case, instead of (1.6), we only consider the following initial data:

$$t=0: (r,s) = \begin{cases} (r_o^-(x), s_-), & x \leqq 0, \\ (r_o^+(x), s_+), & x \geqq 0. \end{cases} \tag{1.14}$$

We have the following

Theorem 2: Suppose that $r_o^-(x)$ and $r_o^+(x)$ are bounded, C^1 functions on $x \leqq 0$ and on $x \geqq 0$ respectively with

$$\begin{aligned} &r_o^{-\prime}(x) \geqq 0, \quad x \leqq 0, \\ &r_o^{+\prime}(x) \geqq 0, \quad x \geqq 0 \end{aligned} \tag{1,15}$$

and

$$s_- - \bar{r}_\circ(x) > 0, \quad x \leqq 0,$$
$$s_+ - \overset{+}{r}_\circ(x) > 0, \quad x \geqq 0. \tag{1.16}$$

Then Problem (1.4),(1.14) admits a unique global discontinuous solution $(r(t,x),s(t,x))$ on $t\geqq 0$ in a class of piecewise continuous and piecewise smooth functions. This solution contains only a backward centered rarefaction wave with the origin as its center and a forward shock passing through the origin. Moreover, there is no vacuum state on $t\geqq 0$.

Case (d).

In this case, there exists $(r_\circ,s_\circ)$ such that the solution to the corresponding Riemann problem (1.4),(1.8) is composed of a backward typical shock

$$(r.s) = \begin{matrix} (r_-,s_-), & x\leqq Ut \\ (r_\circ,s_\circ), & x\geqq Ut \end{matrix} \tag{1.17}$$

and a forward typical shock

$$(r.s) = \begin{matrix} (r_\circ,s_\circ), & x\leqq Vt, \\ (r_+,s_+), & x\geqq Vt, \end{matrix} \tag{1.18}$$

where U and V are the corresponding shock speeds. Moreover, according to the entropy condition and the Rankine-Hugoniot condition, we have

$$\begin{aligned} &U<\lambda(r_-,s_-)<\mu(r_-,s_-), \\ &\lambda(r_+,s_+)<\mu(r_+,s_+)<V, \\ &\lambda(r_\circ,s_\circ)<U<V<\mu(r_\circ,s_\circ), \end{aligned} \tag{1.19}$$

$$\begin{aligned} &s_\circ-r_\circ>s_--r_- , \\ &s_\circ-r_\circ>s_+-r_+ , \end{aligned} \tag{1.20}$$

and

$$s_\circ=f(r_-,s_-,r_\circ), \tag{1.21}$$

$$r_\circ=g(r_+,s_+,s_\circ), \tag{1.22}$$

$$U=F(r_-,s_-,r_\circ,s_\circ)\overset{\Delta}{=}-\sqrt{-\frac{p(\tau(s_\circ-r_\circ))-p(\tau(s_--r_-))}{\tau(s_\circ-r_\circ)-\tau(s_--r_-)}}, \tag{1.23}$$

$$V=G(r_+,s_+,r_\circ,s_\circ)\overset{\Delta}{=}\sqrt{-\frac{p(\tau(s_\circ-r_\circ))-p(\tau(s_+-r_+))}{\tau(s_\circ-r_\circ)-\tau(s_+-r_+)}}, \tag{1.24}$$

where (1.21) and (1.22) are determined from

$$(r_\circ+s_\circ)-(r_-+s_-)=-\sqrt{-(p(\tau(s_\circ-r_\circ))-p(\tau(s_--r_-)))(\tau(s_\circ-r_\circ)-\tau(s_--r_-))} \qquad (1.25)$$

and

$$(r_\circ+s_\circ)-(r_++s_+)=\sqrt{-(p(\tau(s_\circ-r_\circ))-p(\tau(s_+-r_+)))(\tau(s_\circ-r_\circ)-\tau(s_+-r_+))} \qquad (1.26)$$

respectively. Moreover, we have

$$0\leqq\frac{\partial f}{\partial r_\circ}(r_-,s_-,r), \qquad \frac{\partial g}{\partial s_\circ}(r_+,s_+,s)<1. \qquad (1.27)$$

we have the following

<u>THEOREM 3</u>: Suppose that $(\overset{-}{r_\circ}(x),\overset{-}{s_\circ}(x))$ and $(\overset{+}{r_\circ}(x),\overset{+}{s_\circ}(x))$ are C^1 functions on $x\leqq 0$ and $x\geqq 0$ respectively with

$$s_--r_->0, \qquad s_+-r_+>0. \qquad (1.28)$$

Then there exist constants $\varepsilon>0$ and $\eta>0$ small enough such that if

$$\begin{aligned} &|\overset{-}{r_\circ}(x)-r_-|,|\overset{-}{s_\circ}(x)-s_-|\leqq\varepsilon, && x\leqq 0,\\ &|\overset{+}{r_\circ}(x)-r_+|,|\overset{+}{s_\circ}(x)-s_+|\leqq\varepsilon, && x\geqq 0 \end{aligned} \qquad (1.29)$$

and

$$\begin{aligned} &|\overset{-}{r_\circ}{}'(x)|,|\overset{-}{s_\circ}{}'(x)|\leqq\frac{\eta}{|x|}, && x<0,\\ &|\overset{+}{r_\circ}{}'(x)|,|\overset{+}{s_\circ}{}'(x)|\leqq\frac{\eta}{|x|} && x>0, \end{aligned} \qquad (1.30)$$

then Problem (1.4)-(1.6) admits a unique global discontinuous solution $(r(t,x),s(t,x))$ on $t\geqq 0$ in a class of piecewise continuous and piecewise smooth functions. This solution contains only a backward shock $x=x_1(t)$ and a forward shock $x=x_2(t)$ passing through the origin, and there exists no vacuum state on $t\geqq 0$. Moreover, the solution has the following properties:

$$|r(t,x)-r_-|,|s(t,x)-s_-|\leqq\varepsilon \qquad \text{for } x\leqq x_1(t), \qquad (1.31)$$

$$|r(t,x)-r_+|,|s(t,x)-s_+|\leqq\varepsilon \qquad \text{for } x\geqq x_2(t), \qquad (1.32)$$

and on the angular domain between two shocks:

$$R=\{(t,x)\,|\,t\geqq 0,\ x_1(t)\leqq x\leqq x_2(t)\}, \qquad (1.33)$$

$(r(t,x),s(t,x))\ \varepsilon\ C^1$ and

$$|r(t,x)-r_\circ|,|s(t,x)-s_\circ|\leqq K_\circ\varepsilon, \qquad (1.34)$$

$$\left|\frac{\partial r}{\partial x}(t,x)\right|,\left|\frac{\partial s}{\partial x}(t,x)\right|,\left|\frac{\partial r}{\partial t}(t,x)\right|,\left|\frac{\partial s}{\partial t}(t,x)\right|\leqq\frac{K_1\eta}{t},\ t>0. \qquad (1.35)$$

Moreover, we have

$$|x_1'(t)-U|,|x_2'(t)-V|\leqq K_2\varepsilon,\quad t\geqq 0, \qquad (1.36)$$

$$|x_1''(t)|,|x_2''(t)|\leqq\frac{K_3\eta}{t},\quad t>0, \qquad (1.37)$$

where $K_i\,(i=0,1,2,3)$ are suitable constants.

II. The proof of main results

The proof of Theorems 1 and 2 can be found in Li Ta-tsien and Zhao Yan-chun [6]-[8], we only give the sketch of the proof of Theorem 3 here. To do this, we need some results on the global existence of classical solutions (on an angular domain) for the typical free boundary problem for first order quasilinear hyperbolic systems as follows.

For the first order quasilinear strictly hyperbolic system

$$\begin{aligned} &\frac{\partial r}{\partial t}+\lambda(r.s)\frac{\partial r}{\partial x} = 0, \\ & \qquad\qquad\qquad\qquad (\lambda(r,s)<\mu(r,s)), \\ &\frac{\partial s}{\partial t}+\mu(r,s)\frac{\partial s}{\partial x} = 0, \end{aligned} \tag{2.1}$$

we consider the following typical free boundary problem on an angular domain

$$R=\{(t,x)\,|\,t\geqq 0,\ x_1(t)\leqq x\leqq x_2(t)\}. \tag{2.2}$$

On the free boundary $x=x_1(t)(x_1(0)=0)$:

$$s=f(a(t,x),r), \tag{2.3}$$

$$x_1'(t)=F(\alpha(t,x),r,s),\quad x_1(0)=0\ ; \tag{2.4}$$

On the free boundary $x=x_2(t)$ $(x_2(0)=0)$

$$r=g(b(t,x),s), \tag{2.5}$$

$$x_2'(t)=G(\beta(t,x),r,s),\quad x_2(0)=0. \tag{2.6}$$

We give the following hypotheses:

(H1). $\lambda,\mu,f,g,a,b,F,G,\alpha,\beta,\varepsilon C^1$.

(H2). There exists a unique state $(r_\circ,s_\circ)$ such that

$$s_\circ=f(a(0,0),r_\circ),\quad r_\circ=g(b(0,0),s_\circ). \tag{2.7}$$

(H3). There is no characteristic passing through the origin and entering into the angular domain R, i.e.,

$$\lambda(r_\circ,s_\circ)<F(\alpha(0,0),r_\circ,s_\circ)<G(\beta(0,0),r_\circ,s_\circ)<\mu(r_\circ,s_\circ). \tag{2.8}$$

(H4). The minimal characterizing number of this problem (cf. Li Ta-tsien and Yu Wen-ci [3]) is less than 1, namely,

$$\left|\frac{\partial f}{\partial r}(a(0,0),r_\circ)\frac{\partial g}{\partial s}(b(0,0),s_\circ)\right|<1. \tag{2.9}$$

Then we have the following (cf. Li Ta-tsien and Zhao Yan-chun [9])

<u>THEOREM 4</u>: Under Hypotheses (H1)-(H4), there exist suitably small numbers $\varepsilon>0$ and $\eta>0$ such that if

$$|a(t,x)-a(0,0)|,|\alpha(t,x)-\alpha(0,0)|\leqq\varepsilon,\quad x=x_1(t),\ t\geqq 0, \tag{2.10}$$

$$|b(t,x)-b(0,0)|,|\beta(t,x)-\beta(0,0)|\leqq\varepsilon,\quad x=x_2(t),\ t\geqq 0, \tag{2.11}$$

$$\left|\frac{\partial a}{\partial t}(t,x)\right|,\left|\frac{\partial a}{\partial x}(t,x)\right| \leq \frac{\eta}{t}, \qquad x=x_1(t),\ t>0, \tag{2.12}$$

$$\left|\frac{\partial b}{\partial t}(t,x)\right|,\left|\frac{\partial b}{\partial x}(t,x)\right| \leq \frac{\eta}{t}, \qquad x=x_2(t),\ t>0, \tag{2.13}$$

then on the angular domain R, Problem (2.1)-(2.6) admits a unique global classical solution: $(r(t,x),s(t,x))\varepsilon C^1$ and $x_1(t),x_2(t)\varepsilon C^2$. Moreover, we have

$$|r(t,x)-r_\circ|,|s(t,x)-s_\circ| \leq K_\circ\varepsilon, \qquad (t,x)\varepsilon R, \tag{2.14}$$

$$\left|\frac{\partial r}{\partial x}(t,x)\right|,\left|\frac{\partial s}{\partial x}(t,x)\right|,\left|\frac{\partial r}{\partial t}(t,x)\right|,\left|\frac{\partial s}{\partial t}(t,x)\right| \leq \frac{K_1\eta}{t}, \qquad (t,x)\varepsilon R,\ t>0 \tag{2.15}$$

and

$$|x_1'(t)-x_1'(0)|,|x_2'(t)-x_2'(0)| \leq K_2\varepsilon, \qquad t\geqq 0, \tag{2.16}$$

where $K_\circ,K_1,K_2$ are positive numbers and

$$x_1'(0)=F(\alpha(0,0),r_\circ,s_\circ),\ x_2'(0)=G(\beta(0,0),r_\circ,s_\circ). \tag{2.17}$$

We now use Theorem 4 to prove Theorem 3.

First of all, by means of the initial data on $x\geqq 0$, we solve the corresponding Cauchy problem for System (1.4). By (1.19), there exist

$\xi_+=\frac{V+\mu(r_+,s_+)}{2}$ and $\delta>0$ such that

$$V=\xi_++2\delta,\quad \mu(r_+,s_+)=\xi_+-2\delta\ . \tag{2.18}$$

Setting

$$\hat{R}_+=\{(t,x)\,|\,t\geqq 0,\ x\geqq\xi_+t\}, \tag{2.19}$$

we have

Lemma 1: Under the assumptions of Theorem 3, the Cauchy problem for System (1.4) with the initial data $(r_\circ^+(x),s_\circ^+(x))$ on $x\geqq 0$ admits a unique global C^1 solution $(r_+(t,x),s_+(t,x))$ on the domain $\hat{R}_+$, provided that $\varepsilon>0$ and $\eta>0$ are suitably small. Moreover, we have

$$s_+(t,x)-r_+(t,x)>0, \qquad (t,x)\varepsilon\hat{R}_+, \tag{2.20}$$

$$|r_+(t,x)-r_+|,|s_+(t,x)-s_+| \leq \varepsilon, \qquad (t,x)\varepsilon\hat{R}_+, \tag{2.21}$$

$$\left|\frac{\partial r_+}{\partial x}(t,x)\right|,\left|\frac{\partial s_+}{\partial x}(t,x)\right|,\left|\frac{\partial r_+}{\partial t}(t,x)\right|,\left|\frac{\partial s_+}{\partial t}(t,x)\right| \leq \frac{M\eta}{t}, \qquad (t,x)\varepsilon\hat{R}_+,\ t>0, \tag{2.22}$$

where M is a positive constant.

Proof of Lemma 1: According to (1.28) and (2.18), we can choose $\varepsilon>0$ small enough such that if

$$|r-r_+|\leqq\varepsilon,\quad |s-s_+|\leqq\varepsilon\ , \tag{2.23}$$

then it holds that

$$s-r>0 \tag{2.24}$$

and

$$\mu(r,s)\leqq\xi_{+}-\delta \ . \tag{2.25}$$

Hence, there always exists a unique local C^1 solution on $\hat{R}_{+}$. In order to get the global existence of C^1 solution$(r_{+}(t,x),s_{+}(t,x))$ on R_{+}, we only need to prove the estimates (2.20)-(2.22).

(2.20)-(2.21) come directly from System (1.4) and (1.28)-(1.29), provided that $\varepsilon>0$ is small enough. Therefore, by (2.25), any characteristic curve passing through the point $(t,x)=(0,\beta)(\beta\geqq0)$ must intersect the boundary $\{x=\xi_{+}t,t\geqq0\}$ of $\hat{R}_{+}$ in a finite time. Let $x=x_1(t,\beta)$ be the backward characteristic curve passing through the point $(t,x)=(0,\beta)$ $(\beta\geqq0)$ and $(T,\xi_{+}T)$ the intersection point of $x=x_1(t,\beta)$ with the boundary $\{x=\xi_{+}t,\ t\geqq0\}$. From (2.25), it is easy to see that

$$x_1(t,\beta)\leqq(\xi_{+}-\delta)t+\beta, \tag{2.26}$$

then

$$\xi_{+}T=x_1(T,\beta)\leqq(\xi_{+}-\delta)T+\beta, \tag{2.27}$$

i.e.

$$T\leqq\frac{\beta}{\delta} \ . \tag{2.28}$$

Therefore, for any t such that the point (t,x) on $x=x_1(t,\beta)$ belongs to $\hat{R}_{+}$, we have

$$t\leqq T\leqq\frac{\beta}{\delta} \ . \tag{2.29}$$

By means of the transformation of P.D.Lax, it is easy to conclude that

$$\begin{aligned}\frac{\partial r_{+}}{\partial x}(t,x_1(t,\beta))=&\,\overset{+}{r}_{o}{}'(\beta)e^{h(\overset{+}{r}_{o}(\beta),\overset{+}{s}_{o}(\beta))-h(\overset{+}{r}_{o}(\beta),s_{+}(t,x_1(t,\beta)))}\\ &\cdot\Big[1+\int_{o}^{t}\overset{+}{r}_{o}{}'(\beta)e^{h(\overset{+}{r}_{o}(\beta),\overset{+}{s}_{o}(\beta))-h(\overset{+}{r}_{o}(\beta),s_{+}(\tau,x_1(\tau,\beta)))}\\ &\cdot\frac{\partial\lambda}{\partial r}(\overset{+}{r}_{o}(\beta),s_{+}(\tau,x_1(\tau,\beta)))d\tau\Big]^{-1},\end{aligned} \tag{2.30}$$

where $h=h(r,s)$ is defined by

$$\frac{\partial h}{\partial s}=\frac{\partial\lambda}{\partial s}/\lambda-\mu \ . \tag{2.31}$$

By (2.21),we have

$$e^{2|h|}\leqq M_1, \quad \left|\frac{\partial\lambda}{\partial r}\right|e^{2|h|}\leqq M_2 \ , \tag{2.32}$$

where M_1 and M_2 are positive constants, and then we can choose $\eta>0$ small enough such that

$$\frac{M_2\eta}{\delta}\leqq\frac{1}{2} \ . \tag{2.33}$$

Hence, the combination of (2.30) and (2.29) gives

$$\begin{aligned}\left|\frac{\partial r_{+}}{\partial x}(t,x_1(t\ ,\ \beta))\right| &\leqq M_1\frac{\eta}{\beta}(1-M_2\frac{\eta}{\beta}T)^{-1}\\ &\leqq M_1\frac{\eta}{\beta}(1-\frac{M_2\eta}{\delta})^{-1}\leqq 2M_1\frac{\eta}{\beta}\end{aligned} \tag{2.34}$$

$$\leq \frac{2M_1}{\delta}\frac{\eta}{t}, \qquad t>0,\ (t,x_1(t,\beta))\in\hat{R}_+ ,$$

which is nothing else but the first estimate in (2.22). Then the third estimate in (2.22) can be obtained immediately from System (1.4). The rest two in (2.22) can be proved in a similar way. The proof of Lemma 1 is completed.

On the other hand, we solve the corresponding Cauchy problem for System (1.4) by means of the initial data on $x\leq 0$. Setting $\xi_-=\frac{U+\lambda(r_-,s_-)}{2}$, it follows from (1.19) that

$$U<\xi_-<\lambda(r_-,s_-). \tag{2.35}$$

In a similar way, we can also obtain a global C^1solution $(r_-(t,x),s_-(t,x))$ on the domain

$$\hat{R}_-=\{(t,x)|\,t\geq 0,\ x\leq \xi_-t\}\ . \tag{2.36}$$

Moreover, we have the following estimates

$$s_-(t,x)-r_-(t,x)>0, \qquad (t,x)\in\hat{R}_- , \tag{2.37}$$

$$|r_-(t,x)-r_-|,\ |s_-(t,x)-s_-|\leq\varepsilon, \qquad (t,x)\in\hat{R}_- , \tag{2.38}$$

$$\left|\frac{\partial r_-}{\partial x}(t,x)\right|,\left|\frac{\partial s_-}{\partial x}(t,x)\right|,\left|\frac{\partial r_-}{\partial t}(t,x)\right|,\left|\frac{\partial s_-}{\partial t}(t,x)\right|\leq\frac{M\eta}{t}, \tag{2.39}$$
$$(t,x)\in\hat{R}_-,\ t>0.$$

Now we return to the proof of Theorem 3. According to the local existence theorem (cf. Gu Chao-hao, Li Ta-tsien and Hou Zong-Yi [2], Li Ta-tsien and Yu Wen-ci [3]), this discontinuous initial value problem (1.4)-(1.6) admits a unique discontinuous solution at least on a local domain $R(\delta)=\{(t,x)|\ 0\leq t\leq\delta,\ -\infty<x<\infty\}$ in a class of piecewise continuous and piecewise smooth functions and this solution contains only a backward shock $x=x_1(t)$ and a forward shock $x=x_2(t)$ passing through the origin. Noticing (1.19), $x=x_1(t)$ must be located on the left side of $x=\xi_-t$, $x=x_2(t)$ on the right side of both $x=\xi_+t$ and $x=x_1(t)$. Therefore, the solution on the left side of $x=x_1(t)$ and on the right side of $x=x_2(t)$ will be furnished by $(r_-(t,x),$ $s_-(t,x))$ and $(r_+(t,x),\ s_+(t,x))$ respectively, and then the construction of a global discontinuous solution only containing two shocks asks us to solve the following typical free boundary problem on the angular domain (1.33) for System (1.4):

On the free boundary $x=x_1(t)\,(x_1(0)=0)$,

$$s=f(r_-(t,x),s_-(t,x),r), \tag{2.40}$$

$$\frac{dx_1}{dt}=F(r_-(t,x),s_-(t,x),r,s); \tag{2.41}$$

On the free boundary $x=x_2(t)\,(x_2(0)=0)$,

$$r=g(r_+(t,x),s_+(t,x),s), \tag{2.42}$$

$$\frac{dx_2}{dt}=G(r_+(t,x),s_+(t,x),r,s), \tag{2.43}$$

where f,g,F and G are defined by (1.21)-(1.24). Moreover, according to the entropy condition, the solution has to satisfy the following properties:

On $x=x_1(t)$,

$$s-r>s_-(t,x)-r_-(t,x)>0, \tag{2.44}$$

and on $x=x_2(t)$,

$$s-r>s_+(t,x)-r_+(t,x)>0. \tag{2.45}$$

Noticing (1.27), we have

$$\left|\frac{\partial f}{\partial r}(r_-,s_-,r_\circ)\frac{\partial g}{\partial s}(r_+,s_+,s_\circ)\right|<1. \tag{2.46}$$

Using Lemma 1, it comes immediately from Theorem 4 that if $\varepsilon>0$ is suitably small, then this typical free boundary problem (1.4),(2.40)-(2.43) (together with (2.44)-(2.45)) admits a unique global C^1 solution $(r(t,x), s(t,x))$ or the angular domain R and all the estimates (1.34)-(1.37) hold. This completes the proof of Theorem 3.

References

[1] R. Courant and K.O.Friedrichs, Supersonic flow and shock waves, New York, 1948.

[2] Gu Chao-hao, Li Ta-tsien and Hou Zong-Yi, The Cauchy problem of quasilinear hyperbolic systems with discontinuous initial values I, II and III (in Chinese), Acta Math. Sinica 4 (1961), 314-323, 324-327, and 2 (1962), 132-143.

[3] Li Ta-tsien and Yu Wen-ci, Boundary value problems for quasilinear hyperbolic systems, Duke University Mathematics Series V, 1985.

[4] P.D.Lax, Development of singularities of solutions of nonlinear hyperbolic partial differential equations, J. Math. Phys. 5 (1964), 611-613.

[5] Lin Long-wei, On the vacuum state for the equations of isentropic gas dynamics (in chinese), Journal of Huachiao Univ. 2 (1984), 1-4.

[6] Li Ta-tsien and Zhao Yan-chun, Vacuum problems for the system of one dimensional isentropic flow (in chinese), Chin. Quart. J. of Math. 1 (1986) 41-45.

[7] Li Ta-tsien and Zhao Yan-chun, Globally defined classical solutions to free boundary problems with characteristic boundary for quasilinear hyperbolic sys tems, to appear in Chin. Ann. of Math.

[8] Li Ta-tsien and Zhao Yan-chun, Global discontinuous solutions to a calss of discontinuous initial value problems for the system of isentropic flow and applications, to appear in Chin. Ann. of Math.

[9] Li Ta-tsien and Zhao Yan-chun, Globally defined classical solutions to typical free boundary problems for quasilinear hyperbolic systems, to appear.

ANALYSIS OF C^∞-SINGULARITIES FOR A CLASS OF OPERATORS WITH VARYING MULTIPLE CHARACTERISTICS

Qiu Qing-jiu & Qian Si-xin
Dept. of Math. Nanjing Univ.,
Nanjing, China

Introduction.

Since P. Lax and L. Nirenberg raised the problem of the reflection of singularities, many mathematicians have studied this topic and got lots of results. In particular, R.B. Melrose proved a series of important results for operators with simple characteristics, which are included in the new book by L. Hörmander ([1] chap. 14). J. Chazarain ([2]) extended the theorem of Lax and Nirenberg ([3]) to operators with characteristics of constant multiplicity. But there are very few results for operators with varying multiple case (e.g.[4]).

Roughly, the operators with characteristics of varying multiplicity are divided into two parts: effective and non-effective. In this paper, as a beginning of studying this problem for operators with varying multiple case, we concern with a class of non-effective operators, and study the main step of this problem - analysis of C^∞-Singularities for Cauchy problem. We will explore a unusual phenomenon which disappears to the operators with simple or constant multiple characteristics, that is, some singularities do not propagate along some bicharacteristics. In the final part of this paper, we take an example to illustrate this phenomenon. In another paper we will deal with this problem for effective operators.

The operators discussed here are the typical non-effective ones which were studied by many mathematicians (such as O. Oleinik, A. Menikoff etc.) and whose Cauchy problems, we know, are C^∞-wellposed and possessed of parametrices under certain hypothese. Therefore we can make use of them to get a type of "reflection" theorem in this paper and then may extend this result to obtainning the theorem of Lax-Nirenberg type with the above phenomenon for operators $P=H_1QH_2$, where Q elliptic, H_1 and H_2 of the type of those above operators.

1. A class of operators and their parametrices.

Let $t\in R$, $x=(x_1,\cdots,x_n)\in R^n$, $D_x=(D_{x_1},\cdots,D_{x_n})$, $D_{x_j}=-i\frac{\partial}{\partial x_j}$, $D_t=-i\frac{\partial}{\partial t}$. We consider the following linear partial differential operator P:

$$P(t,x,D_t,D_x)=P_m+P_{m-1}+\cdots+P_o, \tag{1}$$

project supported by the Science Fund of the Chinese Academy Sciences.

where $P_{m-j} = \sum_{k=o}^{m-j} P_{kj}(t,x,D_x) D_t^{m-j-k}$, $P_{kj}(t,x,\xi) \in C^\infty([0,T] \times R^n \times R_n)$, $0 \leq j \leq m$, $0 \leq k \leq m-j$, are homogeneous polynomials of degree k in ξ.

We make the following assumptions.

(I): The principal symbol P_m is smoothly factorizable in the form:

$$P_m(t,x,\tau,\xi) = \prod_{j=1}^{m} (\tau - t^{\ell}\lambda_j(t,x,\xi)),$$

where $\ell>0$ is an integer, $\lambda_j(t,x,\xi) \in C^\infty([0,T] \times R^n \times R_n \setminus \{0\})$ are real valued, distinct and $\lambda_j(0,x,\xi) \neq 0$ $(j=1,\cdots, m)$.

(II): $P_{kj}(t,x,\xi) = t^{k\ell-j}\tilde{P}_{kj}(t,x,\xi)$, $k\ell-j \geq 0$, $m-j-k \geq 0$, where $\tilde{P}_{kj}(t,x,\xi) \in C^\infty([0,T] \times R^n \times R_n)$.

By [5], the Cauchy problem of (1) is C^∞ - wellposed, and by [6], the Cauchy problem

$$\begin{aligned} &p\, u(t,x) = 0 \\ &D_t^h u\Big|_{t=0} = u_h(x) \in \mathcal{E}'(R^n), \quad 0 \leq h \leq m-1 \end{aligned} \tag{2}$$

has the parametrix

$$u(t,x) = (2\pi)^{-n} \sum_{h=1}^{m} \sum_{j=1}^{m} \int e^{i\phi_j(t,x,\xi)} [a_{hj}(t,x,\xi) + \tilde{a}_{hj}(t,x,\xi)] \hat{u}_{h-1}(\xi) d\xi$$
$$\in C^\infty([0,T]; \mathcal{D}'(R^n)), \tag{3}$$

where $\phi_j(t,x,\xi)$ is the phase function corresponding to $\tau - t^{\ell}\lambda_j$, that is,

$$\frac{\partial \phi_j}{\partial t} = t^{\ell}\lambda_j(t,x,\nabla_x \phi_j), \quad \phi_j\Big|_{t=0} = \langle x,\xi \rangle. \tag{4}$$

The amplitudes a and $\tilde{a}$ are as follows:

$$a_{hj}(t,x,\xi) \in \bigcap_{\varepsilon>o} S^{(\ell+1)^{-1}(m_j-h+1+\xi),\, m_j+\xi}$$

$$\tilde{a}_{hj}(t,x,\xi) \in \bigcap_{\varepsilon>o} S^{(\ell+1)^{-1}(m_j-h+1+\xi),\, \infty}.$$

(For the definition of $S^{q,r}$ see [6]). $\tilde{a}$ are flat at t=0, the number m_j is the supremun for $(x,\xi) \in R^n \times (R_n \setminus \{0\})$ of $\mathrm{Re}\{-H_j(x,\xi)/G_j(x,\xi)\}$ where

$$G_j(x,\xi) = \sum_{k=o}^{m-1} (m-k)\lambda_j(0,x,\xi)^{m-k-1}\tilde{P}_{ko}(0,x,\xi)$$

$$H_j(x,\xi) = \frac{\ell}{2} \sum_{k=o}^{m-2} (m-k)(m-k-1)\lambda_j(0,x,\xi)^{m-k-1}\tilde{P}_{ko}(0,x,\xi) + $$
$$+ i \sum_{k=1}^{m-1} \lambda_j(0,x,\xi)^{m-k-1}\tilde{P}_{k1}(0,x,\xi).$$

Set $$L_o = \sum_{\substack{k\ell \geq j \\ m-k-j \geq o}} \tilde{P}_{kj}(0,x,\xi) t^{k\ell-j} D_t^{m-k-j}$$

and $$C_j^*(t,x,\xi) = \exp[i(\ell+1)^{-1} t^{\ell+1} \lambda_j(0,x,\xi)].$$

Let $C_j^*(t,x,\xi)V_j(t,x,\xi)$, $1\leq j\leq m$, be a fundamental system of solutions of L_0 in $t>0$ with the following property (see [8])

$$V_j(t,x,\xi)\sim t^{\mu_j(x,\xi)} \sum_{r=0}^{\infty} e_{jr}(x,\xi)t^{-r}, \quad \text{as } t\to\infty ,$$

where $e_{jo}(x,\xi)=1$ and the symbol "$\sim$" denotes an asymptotic expansion uniform with respect to the parameters $x,\xi,(|\xi|=1)$. Therefore, by the stretching transformation $t\to|\xi|^{(\ell+1)^{-1}}t$, we may assume that this asymptotic expansion holds in $t\in[0,T]$ and ξ large. Thus the solution of Cauchy problem

$$L_o U_h = 0, \qquad D_t^k U_h\Big|_{t=0} = \delta_{k,h-1}, \quad 0\leq k\leq m-1 \tag{5}$$

can be expressed as follows:

$$U_h(t,x,\xi) = \sum_{j=1}^{m} T^{(h,j)}(x,\xi)C_j^*(t,x,\xi)V_j(t,x,\xi), \quad 1\leq h\leq m,$$

here $T^{(h,j)}$ are called the central connection coefficients, and we have

$$a_{hj}(t,x,\xi) - T^{(h,j)}(x,\xi)V_j(t,x,\xi)\in \bigcap_{\varepsilon>0} S^{(\ell+1)^{-1}(m_j-h+1+\varepsilon),m_j+1+\varepsilon},$$

hence the "principal" part of $a_{hj}+\tilde{a}_{hj}$ in (3) is $T^{(h,j)}(x,\xi)V_j(t,x,\xi)$.

2. Analysis of C^∞-Singularities

We denote the bicharacteristics corresponding to $\tau-t^{\ell}\lambda_j(t,x,\xi)$ issued from $(0,x_o,0,\xi_o)$ by $\gamma_j(x_o,\xi_o)$, $j=1,\cdots,m$. It is interesting that they are transversal to $t=0$ at the multiple characteristic point $(0,x_o,0,\xi_o)$ though $H_{P_m}^{(N)}t=0$ for every natural N, i.e., $(0,x_o,0,\xi_o)\in G^\infty$, ([1]).

We first prove the following microlocal regularity lemma of the Cauchy problem of (1). For the definition of WF_b see [1].

LEMMA 1 Suppose that $Pu=f$, $(x_o,\xi_o)\notin WF_b(f)$, $(x_o,\xi_o)\notin WF(u_h)$, $h=0,\cdots,m-1$, $u\in C^\infty(\bar{R}_+,\mathcal{D}'(R^n))$. Then there exists $\delta>0$ such that

$$WF_b(u)\cap\gamma(x_o,\xi_o)\cap\{0\leq t\leq\delta\} = \phi \tag{6}$$

where $\gamma(x_o,\xi_o)$ denotes any bicharacteristic of the family $\{\gamma_j(x_o,\xi_o)\}$ from (x_o,ξ_o)

PROOF. Taking a pseudodifferential operator $Q(x,D_x)$ supported in a conical neighbourhood $U_1\times W_1$ of (x_o,ξ_o) and elliptic at this point, we have, by (3)

$$Q(x,D_x)u(t,x) = (2\pi)^{-n}\sum_{h=1}^{m}\sum_{j=1}^{m}\int e^{i\phi_j(t,x,\xi)}b_{hj}(t,x,\xi)\hat{u}_{h-1}(\xi)d\xi ,$$

where $b_{hj}(t,x,\xi)=\exp(-i\phi_j)Q(x,D_x)((a_{hj}+\tilde{a}_{hj})e^{-i\phi_j})$ which has the asymptotic expansion (see [7])

$$\sum_{\alpha} \frac{1}{\alpha!}\partial_\xi^\alpha Q(x,\partial_x\phi_j)D_y^\alpha((a_{hj}+\tilde{a}_{hj})e^{-i\rho_j})\Big|_{y=x},$$

where $\rho_j(t,x,y,\xi)=\phi_j(t,x,\xi)-\phi_j(t,y,\xi)-\langle\partial_x\phi_j(t,x,\xi),x-y\rangle$.

Since $(x_o,\xi_o)\notin WF(u_{h-1})$, there exists such a conical neighbourhood $U\times W$ of (x_o,ξ_o) that $(U\times W)\cap WF(u_{h-1})=\phi$, and then by (4), we can take a conical neighbourhood $U_1\times W_1$ of $(x_o,\xi_o)$and $\delta_1>0$ such that $(x,\xi)\notin U\times W$ implies $(x,\partial_x\phi_j(t,x,\xi))\notin U_1\times W_1$ for every $t\in[0,\delta_1]$. Therefore, it is easy to know that $Q(x,D_x)u\in C^\infty([0,\delta_1]\times R^n)$ by the choice of Q and the application of the asymptotic expansion of b_{hj}. Thus by using Duhamel's principle it turns out $(x_o,\xi_o)\notin WF_b(u)$. This means that there exist $\delta_2>0$ and $U_2\times W_2$ such that $WF(u)\cap B(\delta_2,U_2,W_2)=\phi$, where

$$B(\delta_2,U_2,W_2) = \{(t,x,\tau,\xi);t\in[0,\delta_2],(x,\xi)\in U_2\times W_2,\ \tau\in R\}.$$

In particular $WF_b(u)\cap B(\delta_2,U_2,W_2)\cap\{0<t\leqq\delta_2\}=\phi$ and then

$$WF_b(u)\cap\gamma(x_o,\xi_o)\cap\{0<t\leqq\delta_2\} = \phi.$$

So Lemma 1 holds.

We rewrite (3) as follows:

$$u(t,x) = \sum_{j=1}^m\Big(\sum_{h=1}^m \bar{\psi}_j\circ A_j^h u_{h-1}\Big) \equiv \sum_{j=1}^m I_j \tag{7}$$

where $\bar{\psi}_j$ is the Fourier integral operator with phase ϕ_j and amplitude 1, A_j^h is the pseudodifferential operator with amplitude $a_{hj}+\tilde{a}_{hj}$ whose principal part is $T^{(h,j)}V_j$.

We establish the following lemma concerning the analysis of C^∞-singularities.

<u>LEMMA 2</u> If $\gamma_{j_o}(x_o,\xi_o)\cap WF(u)=\phi$, $t>0$, for some $j_o\in\{1,\cdots,m\}$, then

$$(x_o,\xi_o)\notin WF\Big[\Big(\sum_{h=1}^m A_{j_o}^h u_{h-1}\Big)(t)\Big],\quad t>0\ . \tag{8}$$

<u>PROOF</u>. Since $(t,x(t),\tau(t),\xi(t))\notin WF(u)$ for $(t,x(t),\tau(t),\xi(t))\in\gamma_{j_o}(x_o,\xi_o)$ near $(0,x_o,0,\xi_o)$, $t>0$, and (1) is strictly hyperbolic for $t>0$, we have $(t,x(t),\tau(t),\xi(t))\notin\gamma_k(x_o,\xi_o)$ for $\forall\ k$, $k\neq j_o$, and then $(t,x(t),\tau(t),\xi(t))\notin WF(I_k)$. Noting that $I_{j_o}=u-\sum_{k\neq j_o} I_k$ and $\gamma_{j_o}(x_o,\xi_o)\cap WF(u)=\phi$, $t>0$, we obtain $(t,x(t),\tau(t),\xi(t))\notin WF(I_{j_o})$. This means that $(x(t),\xi(t))\notin WF(I_{j_o}|_t)$, (cf.[3] remark 9.1), where $I_{j_o}|_t$ is the restriction

of I_{j_o} at t and is also the acting of the elliptic Fourier integral operator $\bar{\psi}_{j_o}\big|_t$ on $g_{j_o}\big|_t=(\sum_{h=1}^{m} A_{j_o}^h u_{h-1})\big|_t$.

Since $WF(\bar{\psi}_{j_o}(g)\big|_t)=\Lambda'_{\phi_{j_o}|t}\circ WF(g\big|_t)$, here $\Lambda'_{\phi_{j_o}|t}=\{(x(t),\xi(t),x_o,\xi_o);(x(t),\xi(t))\in\gamma_{j_o}(x_o,\xi_o)\big|_t\}$ is the Lagrange manifold corresponding to $\phi_{j_o}\big|_t$, it turns out that $(x_o,\xi_o)\notin WF(g_{j_o}\big|_t)$ for every $t>0$, which is exactly the result of Lemma 2.

Using the above two lemmas, we can prove our main results.

<u>THEOREM 1</u> Assume that $(x_o,\xi_o)\notin WF_b(f)$, and that $\gamma_{j_o}(x_o,\xi_o)\cap WF(u)=\phi$, $t>0$, $(x_o,\xi_o)\notin WF(u_{h-1})$, $h\neq k_o$, for some j_o, $k_o\in\{1,\cdots,m\}$. Then if $T^{(k_o,j_o)}(x_o,\xi_o)\neq 0$, we have $(x_o,\xi_o)\notin WF(u_{h-1})$, $h=1,\cdots,m$, and $\gamma_j(x_o,\xi_o)\cap WF_b(u)=\phi$, $j=],\cdots,m$.

<u>PROOF</u>. By Lemma 2, $(x_o,\xi_o)\notin WF[(\sum_{h=1}^{m} A_{j_o}^h u_{h-1}(t)]$, $t>0$, and then by the pseudolocal property of $A_{j_o}^h$ and the assumption $(x_o,\xi_o)\notin WF(u_{h-1})$, it turns out that $(x_o,\xi_o)\notin WF[(A_{j_o}^h u_{h-1})(t)]$, $h\neq k_o$. So we have $(x_o,\xi_o)\notin WF[(A_{j_o}^{K_o} u_{k_o-1})(t)]$.

On the other hand, $T^{(K_o,j_o)}(x_o,\xi_o)\neq 0$ implies that (x_o,ξ_o) is an elliptic point of $A_{j_o}^{K_o}\big|_t$. Therefore, we conclude that $(x_o,\xi_o)\notin WF(u_{k_o-1})$, hence $(x_o,\xi_o)\notin WF(u_{h-1})$, $h=1,\ldots,m$.

The remaining part of this theorem follows from Lemma 1.

A similar proof gives the following theorem.

<u>THEOREM 2</u> Assume that $(x_o,\xi_o)\notin WF_b(f)$, $\gamma_j(x_o,\xi_o)\cap WF(u)=\phi$, $t>0$, $j=1,\ldots,k$, $k<m$. and that $(x_o,\xi_o)\notin WF(u_{h-1})$, $1\leq h\leq m-k$. Then if the matrix:

$$\begin{pmatrix} T^{(m-k+1,1)} & \cdots\cdots & T^{(m,1)} \\ \cdots & \cdots & \cdots \\ T^{(m-k+1,k)} & \cdots\cdots & T^{(m,k)} \end{pmatrix}_{(x_o,\xi_o)}$$

is non-singular, we have (x_o,ξ_o) $WF(u_{h-1})$, $h=1,\ldots,m$, and $\gamma_j(x_o,\xi_o)\cap WF_b(u)=\phi$, $j=1,\ldots,m$.

REMARK. Theorem 1 implies that if $T^{(K_o,j_o)}(x_o,\xi_o)\neq 0$ for some k_o and j_o, then $\gamma_{j_o}(x_o,\xi_o)\subset WF(u)$ because $(x_o,\xi_o)\in\{WF_b(u)\setminus(WF_b(Pu)\cup WF(u_{h-1}))\}$, $h\neq k_o$, $h=1,\dots,m$. In other words, the singularities of u at $(0,x_o,0,\xi_o)$ propagate into $t>0$ along the bicharacteristic $\gamma_{j_o}(x_o,\xi_o)$. On the other hand, if u has the singularities along some bicharacteristic $\gamma(x_o,\xi_o)$ in a conical neighbourhood of $(0,x_o,0,\xi_o)$ in $t>0$, then $(x_o,\xi_o)\in WF_b(u)$ since $WF_b(u)$ is closed. Therefore, there is the following possibility: the singularities of u at $(0,x_o,0,\xi_o)$ may not propagate into $t>0$ along $\gamma_{j_o}(x_o,\xi_o)$ if $T^{(k_o,j_o)}(x_o,\xi_o)=0$ although $(x_o,\xi_o)\in\{WF_b(u)\setminus(WF_b(Pu)\cup WF(u_{h-1}))\}$, $h\neq k_o$. In this case we have to examine the lower order parts of $a_{hj}+\tilde{a}_{hj}$, which are also connected with the operator L_o.

There is a similar case for Theorem 2.

3. An example

To clarify the above inferences, we discuss the case for m=2. Set

$$\chi_j = (-1)^{j+1}(\ell+1)^{-1}[\lambda_1(0,x,\tfrac{\xi}{|\xi|})-\lambda_2(0,x,\tfrac{\xi}{|\xi|})]^{-1}[\ell\lambda_j(0,x,\tfrac{\xi}{|\xi|})+i\tilde{P}_{11}(0,x,\tfrac{\xi}{|\xi|})],$$

When $\chi_j,\chi_j+(\ell+1)^{-1}\notin Z$, $j=1,2$, we can use [8] (theorem 3 in [8]) to compute $T^{(h,j)}$, and at the exceptional values χ_j or $\chi_j+(\ell+1)^{-1}\in Z$, we can also compute $T^{(h,j)}$ directly.

Now we give an example to verify the possibility of non-propagating case mentioned in the remark above.

Consider

$$P = \partial_t^2 - t^{2\ell}\sum_{j=1}^{n}\partial_{x_j}^2 - iat^{\ell-1}\Lambda \tag{9}$$

where Λ is the pseudodifferential operator with symbol $|\xi|$, $\ell>1$ integer, a is constant number.

Its Cauchy problem is

$$Pu = 0,\quad u|_{t=0}=u_o(x),\quad u_t|_{t=0}=u_1(x). \tag{10}$$

Taking the partial Fourier transformation in x and using the theory of Fourier integral operator, it is not difficult to show that, by [9], the solution of (10) is

$$u(t,x) \equiv I_+ + I_-,$$

where
$$I_\pm = \int e^{i\phi_\pm}[W_\pm^{(0)}(t,\xi)\hat{u}_o(\xi) + W_\pm^{(1)}(t,\xi)\hat{u}_1(\xi)]d\xi \tag{11}$$

$$\phi_\pm = \langle x,\xi\rangle \pm\omega t^{\ell+1}|\xi|,\quad \omega=(\ell+1)^{-1},\quad \alpha=\tfrac{\omega}{2}(\ell+a),\ \beta=\ell\omega,$$

$$z = 2i\omega t^{\ell+1}|\xi|,$$

and
$$W_+^{(0)}(t,\xi)\sim\frac{\Gamma(\beta)}{\Gamma(\alpha)}z^{\alpha-\beta}\,U_+(t,\xi),\quad W_+^{(1)}\sim\frac{\Gamma(2-\beta)}{\Gamma(1+\alpha-\beta)}(2i\omega|\xi|)^{-\omega}z^{\alpha-\beta}U_+,$$

$$W_-^{(0)}\sim\frac{\Gamma(\beta)}{\Gamma(\beta-\alpha)}(e^{-\pi i}z)^{-\alpha}U_-(t,\xi),\quad W_-^{(1)}\sim\frac{\Gamma(2-\beta)}{\Gamma(1-\alpha)}e^{\pi i\omega}(e^{-\pi i}z)^{-\alpha}(2i\omega|\xi|)^{-\omega}U_-$$

$$U_+(t,\xi) = 1+\sum_{k=1}^{\infty}\frac{(\beta-\alpha)_k(1-\alpha)_k}{k!}z^{-k},$$

$$U_-(t,\xi) = 1+\sum_{k=1}^{\infty}(-1)^k\frac{(\alpha)_k(1+\alpha-\beta)_k}{k!}z^{-k}$$

here $(\alpha)_k=\alpha(\alpha+1)\dots(\alpha+k-1)$, Γ is the Γ- function, and the symbol "$\sim$" denotes the asymptotic expansion as $t^{\ell+1}|\xi|\to\infty$.

The solution of (10) consists of four Fourier integral operators, and we notice that every symbol has a factor of the type $\frac{\Gamma(A)}{\Gamma(B)}$, which vanishes when B is an integer $\leqq 0$. Hence Theorem 1 gives the following results, where m denotes an arbitray integer $\geqq 0$.

1. Suppose $(x_o,\xi_o)\notin WF(u_o)$:

(1) When $(1+\alpha-\beta)\neq -m$, then $\gamma_+(x_o,\xi_o)\notin WF(u)$ implies $(x_o,\xi_o)\notin WF_b(u)$;

when $(1+\alpha-\beta)=-m$ or $a=-\{2m(\ell+1)+\ell+2\}$, then $\gamma_+(x_o,\xi_o)$ does not propagate singularities though $(x_o,\xi_o)\in WF_b(u)$.

(2) When $(1-\alpha)\neq -m$, then $\gamma_-(x_o,\xi_o)\notin WF(u)$ implies $(x_o,\xi_o)\notin WF_b(u)$;

when $(1-\alpha)=-m$ or $a=2m(\ell+1)+\ell+2$, then $\gamma_-(x_o,\xi_o)$ does not propagate singularities though $(x_o,\xi_o)\in WF_b(u)$.

2. Suppose $(x_o,\xi_o)\notin WF(u_1)$:

The discussion is similar to the above one.

On the other hand, the operator L_o corresponding to (9) is

$$L_o = D_t^2-t^{2\ell}|\xi|^2 + iat^{\ell-1}|\xi|$$

and $\chi_1=\alpha$, $\chi_2=\alpha-\beta$. Since ℓ is an integer >1, then χ_2 is not an exceptional value if χ_1 is, and vice versa. Moreover, the poles of the Γ- functions in the denominators of the factors in $W_\pm^{(0)}$ and $W_\pm^{(1)}$ are exactly those exceptional values. By comutation, it is easy to know that $T^{(h,j)}$ is just the factor in $U_\pm$.

REFERENCES

[1] L. Hörmander. The Analysis of Linear Partial Differential Operators III, 1985, Springer-Verlag.

[2] J. Chazarain. Reflection of C^∞ Singularities for a Class of Operators with Multiple Characteristics, Publ. RIMS. Kyoto Univ., 12 Suppl., 1977, 39-52.

[3] L. Nirenberg. Lectures on Linear Partial Differential Equations, 1973, Amer. Math. Society.

[4] R.B. Melrose & G.A. Uhlman. Microlocal Structure of Involutive Conical Reflection, Duke Math. J., 46:3, 1979, 571-582.

[5] H. Uryu. The Cauchy Problem for Weakly Hyperbolic Equations, C.P.D.E., 5(1), 1980, 23-40.

[6] G. Nakamura & H. Uryu. Parametrix of Certain Weakly Hyperbolic Operators, C.P.D.E., 5(8), 1980, 837-896.

[7] Qiu Qing-jiu, Chen Shuxing et al. The Theory of Fourier Integral Operators and its Applications, 1985, Science Press (in Chinese).

[8] K. Amano & G. Nakamura. Branching of singularities for Degenerate Hyperbolic Operators, Publ. RIMS. Kyoto Univ., 20, 1984, 225-275.

[9] K. Taniguchi & Y. Tozaki. A Hyperbolic Equation with Double Characteristics which has a Solution with Branching Singularities, Math. Jaonica, 25:3, 1980, 279-300.

AN INVERSE PROBLEM FOR NONLOCAL ELLIPTIC BVP AND RESISTIVITY IDENTIFICATION

Tan Yongji

Department of Mathematics
Fudan University
Shanghai, China

Oil geophysical resistivity logs and some other remote sensitive problems can be formulated to inverse problems for quasiharmonic equation under nonlocal boundary conditions. We first discuss the existence, uniqueness and stability of the solution for such a problem and then apply it to resistivity identification.

I. Motivation

In oil prospeciting, well log is usually an important technique to detect the resistivity of the layers. After a well has been drilled, we put a log tool into the well. The log tool is an insulation rod whose lateral surface is covered by metal memberence as electrodes. While it works, the electrodes discharge a current of fixed intensity, then the potential on the electrode is measured. The goal of the well log is to determine the resistivity of the objective layer by the potential data on the electrode.

Figure 1 shows the configuration used in modelling, provided that the layers are symetric about the well axis and the middle plane, where Ω_1 is the wellbore filled with mud of resistivity R_m, Ω_2 is the surrounding rock of resistivity R_s, Ω_3 and Ω_4 are two parts of the objective layer and the shaded part the area occupied by the log tool (see [1]). Usually the objective layer is sandy rock which is porous material. The mud filter fluid penetrates into the porosity and changes the resistivity of the domain Ω_3. Therefore, Ω_3 is called invaded area and we denote the resistivity in this domain by R_{xo}. The resistivity R_t in Ω_4, the part of the objective layer which is not invaded by the conductive fluid, will be detected.

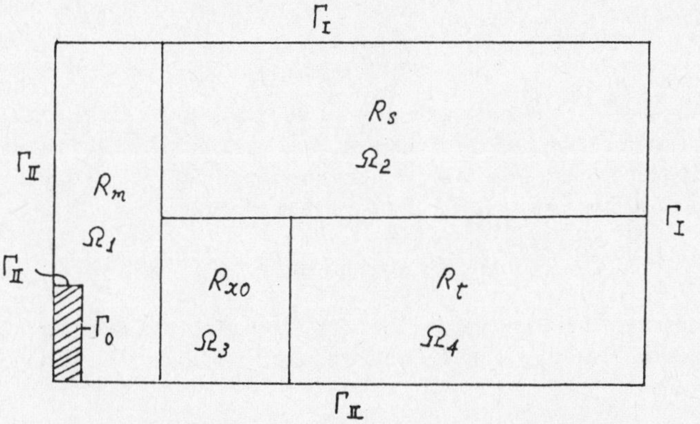

Figure 1

It is known that the potential function $u(x)$ of the field in the layers is governed by a quasiharmonic equation:

$$\frac{\partial}{\partial x_1}\left(\frac{1}{R}\frac{\partial u}{\partial x_1}\right) + \frac{\partial}{\partial x_2}\left(\frac{1}{R}\frac{\partial u}{\partial x_2}\right) + \frac{\partial}{\partial x_3}\left(\frac{1}{R}\frac{\partial u}{\partial x_3}\right) = 0 \tag{1.1}$$

where R is piecewisely constant, and

$$R = \begin{cases} R_m & x \in \Omega_1 \\ R_s & x \in \Omega_2 \\ R_{xo} & x \in \Omega_3 \\ R_t & x \in \Omega_4 \end{cases}$$

On the interface in two different subdomains with different resistivities, $u(x)$ satisfies the interface conditions as follows:

$$u^- = u^+ \tag{1.2}$$

$$\left(\frac{1}{R}\frac{\partial u}{\partial n}\right)^- = \left(\frac{1}{R}\frac{\partial u}{\partial n}\right)^+ \tag{1.3}$$

where the symbols "+" and "-" represent the limits of the function to which the function tends when the points tend to the point on the interface from left and right sides respectively, and n is the unit normal vector of the interface which is oriented.

On the surface of the earth (the upper boundary in Fig.1), the potential vanishes. And the potential on the right boundary, which is usually very far from the well axis and is called infinitely distant boundary, can be neglected. We denote these two parts of the boundary by Γ_I. So we have:

$$u|_{\Gamma_I} = 0 \tag{1.4}$$

On the symetric axis (the left boundary in Fig. 1) and the symetric plane (the bottom boundary in Fig. 1) the normal derivative of potential u must vanish, and on the rubber top of the log tool the insulation condition $\frac{\partial u}{\partial n} = 0$ is satisfied. We denote these three parts of the boundary by Γ_{II}. We have:

$$\frac{\partial u}{\partial n}|_{\Gamma_{II}} = 0 \tag{1.5}$$

Since in an electric field the surface of a metal body is a surface with equal potential, the potential keeps constant on the surface of the electrode. Without lossing generality, we consider the situition of simpified tool where there is only one electrode, i.e., the lateral surface of the tool is entirely an electrode. We denote it by Γ_0, hence we have:

$$u|_{\Gamma_0} = \text{constant} \tag{1.6}$$

But this constant has not been known yet. Besides, the intensity of the current discharged by the electrode is known. Denoting it by I_0, we have another boundary condition on Γ_0:

$$\int_{\Gamma_0} \frac{1}{R}\frac{\partial u}{\partial n} dx = I_0 \tag{1.7}$$

The boundary conditions (1.6) and (1.7) are known as equally valued surface boundary condition. It is obvious that they are nonlocal.

If the domain and the coefficient were known, (1.1)–(1.7) would be a nonlocal boundary value problem for quasiharmonic equation which has been well studied in [2]. Nevertheless, the practical purpose of the well log is to determine the resistivity of the objective layer, i.e., the coefficient of the equation in Ω_4 by the value of the potential, which is the solution of the boundary value problem, on the boundary Γ_0. It means that we are going to find some information of the coefficient of the PDE by some information of its solution. This problem is known as an inverse problem. So we have to solve an inverse problem for a nonlocal boundary value problem which has not been studied yet.

We introduce a function set

$$V = \{v(x) \in H^1(\Omega) \quad v|_{\Gamma_I} = 0, \quad v|_{\Gamma_0} = \text{constant}\} \tag{1.8}$$

and a functional defined on it as follows:

$$J(v) = \frac{1}{2}\int_{\Omega}\frac{1}{R}|\nabla v|^2 dx - I_0 \cdot v|_{\Gamma_0} \tag{1.9}$$

where $\Omega = \sum_{i=1}^{4}\Omega_i$, $H^1(\Omega)$ is a Sobolev space as usual, $v|_{\Gamma_I}$ or $v|_{\Gamma_0}$ are defined in the sense of trace (see [3], [4]).

It is known that finding the H^1 generalized solution $u(x)$ for boundary value problem (1.1)-(1.7) is equivalent to finding the minimizer of functional (1.9) in V i.e., finding $u \in V$ such that

$$J(u) = \inf_{v \in V} J(v) \tag{1.10}$$

Therefore, in the sense of generalized solution, the inverse problem can be posed like this: To find R_t, the coefficient in Ω_4, such that the trace on Γ_0 of the minimizer u of functional (1.9) in V is just equal to a given value.

II. Existence, uniqueness and stability

We investigate a slightly extended problem. Let Ω be a domain in n-dimentional space of variable $x = (x_1, \cdots, x_n)$, $n = 2, 3$ (cf. Fig. 2). Assume that:

(1) $\Omega = \bigcup_{i=1}^{L}\Omega_i$ where Ω_i $(i = 1, \cdots, L)$ are subdomains of $\Omega, \Omega_i \bigcap \Omega_j = \phi$ $(i \neq j,\ i, j = 1, \cdots, L)$.

(2) Denoting the boundarys of Ω and Ω_i by Γ and Γ_i respectively, we know Γ and Γ_i $(i = 1, \cdots, L)$ are all Lipschtz boundarys which are regular almost everywhere (see [3]).

(3) $\Gamma = \Gamma_0 \bigcup \Gamma_I \bigcup \Gamma_{II}$, where Γ_i $(i = 0, I, II)$ are relative open sets in $\Gamma, \Gamma_i \bigcap \Gamma_j = \phi$ $(i \neq j, i, j = 0, I, II)$, and Γ_I and Γ_0 are disjoint.

(4) $\Gamma_0 \bigcap \Gamma_L = \phi$, $\Gamma_L \bigcap \Gamma_I \neq \phi$, $(\Gamma \setminus \Gamma_L) \bigcap \Gamma_I \neq \phi$.

Let

$$V = \{v(x) \in H^1(\Omega) | v|_{\Gamma_I} = 0, \quad v|_{\Gamma_0} = \text{constant}\} \tag{2.1}$$

and

$$J_k(v) = \frac{1}{2}\sum_{i=1}^{L-1}\int_{\Omega_i} k_i|\nabla v|^2 dx + \frac{1}{2}\int_{\Omega_L} k|\nabla v|^2 dx - I_0 \cdot v|_{\Gamma_0} \tag{2.2}$$

The inverse problem is that, for a given domain Ω and fixed positive numbers k_i, to find a positive number k such that the trace on Γ_0 of the mimizer for functional (2.2) in V is just equal to a given positive number c.

Lemma 1. Let Ω be a domain fulfilling the hypotheses (1)-(4) then there exists a unique $u_k \in V$ such that

$$J_k(u_k) = \min_{v \in V} J_k(v)$$

and the minimizer u_k is characterized by

$$\sum_{i=1}^{L-1}\int_{\Omega_i} k_i \nabla u_k \cdot \nabla\varphi \, dx + \int_{\Omega_L} k \nabla u \cdot \nabla\varphi \, dx = I_0 \cdot \varphi|_{\Gamma_0} \quad \forall\varphi \in V \tag{2.3}$$

Proof. Due to the trace theorem (see [3]), $f(\varphi) = I_0 \cdot \varphi|_{\Gamma_0}$ is a bounded linear functional in V, and it is easy to see that V is also a Hilbert space. By using Fridrichs inequality, it shows that

$$\sum_{i=1}^{L-1}\int_{\Omega_i} k_i |\nabla\varphi|^2 dx + \int_{\Omega_L} k|\nabla\varphi|^2 dx$$

is an equivalent norm in $H^1(\Omega)$. By use of Riesz representation theorem, the lemma is not difficult to prove.

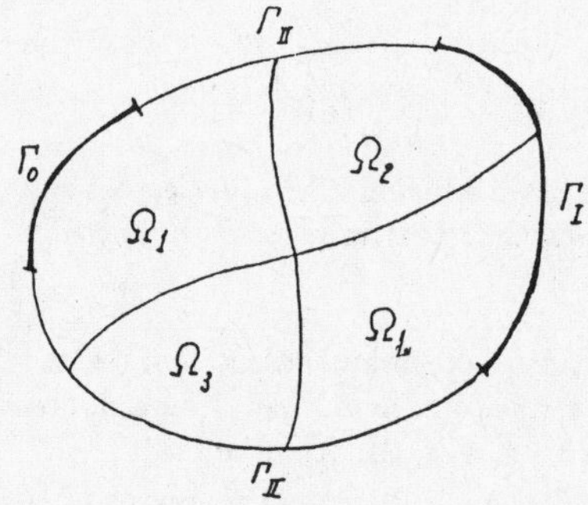

Figure 2

Applying Lemma 1, for given domain Ω, function space V and positive numbers $k_1, \cdots, k_{L-1}$, I_0, $k \mapsto u|_{\Gamma_0}$ defines a function in $(0, +\infty)$ through minimizing functional (2.2). We denote this function by

$$c = u|_{\Gamma_0} = g(k) \tag{2.4}$$

By use of Green's formula (see [5]) to (2.3), it is easy to show that u_k satisfies equation

$$\sum_{i=1}^{n} \frac{\partial}{\partial x_i}\left(\mu \frac{\partial u_k}{\partial x_i}\right) = 0$$

in the sense of distribution, $u|_{\Gamma_I} = 0$, $\frac{\partial u}{\partial n}|_{\Gamma_{II}} = 0$ in the sense of trace and

$$\left\langle v, \frac{\partial u_k}{\partial n} \right\rangle_{H^{\frac{1}{2}}(\Gamma_0), H^{-\frac{1}{2}}(\Gamma_0)} = I_0 \cdot v|_{\Gamma_0}, \quad \forall v \in V$$

where

$$\mu = \begin{cases} k_i & x \in \Omega_i \quad (i = 1, \cdots, L-1) \\ k & x \in \Omega_L \end{cases}$$

and

$$<\cdot,\cdot>_{H^{\frac{1}{2}}(\Gamma_0),H^{-\frac{1}{2}}(\Gamma_0)}$$

is the duality between $H^{\frac{1}{2}}(\Gamma_0)$ and $H^{-\frac{1}{2}}(\Gamma_0)$.

Let $\varphi = u_k$ in (2.3). By Fridrichs inequality, we get

Corollary 2. Let $c = g(k)$ be a function defined above, then

$$g(k) \geq 0 \quad \forall k \in (0,+\infty) \tag{2.5}$$

and the equality is valid only if $u = 0$.

Theorem 3. Suppose that Ω is a domain fulfilling hypotheses (1)-(4) and $k_1, k_2, \cdots, k_{L-1}, I_0$ are given positive constants, then the function defined (2.4) is a strictly montone decreasing function in $(0, +\infty)$.

Proof. Assume that $u, u' \in V$ satisfy

$$J_k(u) = \inf_{v \in V} J_k(v) \tag{2.6}$$

and

$$J_{k'}(u') = \inf_{v \in V} J_{k'}(v) \tag{2.7}$$

respectively.

By Lemma 1, we have

$$\begin{aligned}
&I_0 \cdot u|_{\Gamma_0} \\
&= - J_k(u) \\
&= - (\sum_{i=1}^{L-1} \int_{\Omega_i} k_i | \nabla u|^2 dx + k \int_{\Omega_L} | \nabla u|^2 dx - 2 \cdot u|_{\Gamma_0} \cdot I_0) \\
&= - (\sum_{i=1}^{L-1} \int_{\Omega_i} k_i | \nabla u|^2 dx + k' \int_{\Omega_L} | \nabla u|^2 dx - 2 \cdot u|_{\Gamma_0} \cdot I_0) + (k' - k) \cdot \int_{\Omega_L} | \nabla u|^2 dx \\
&= - J_{k'}(u) + (k' - k) \cdot \int_{\Omega_L} | \nabla u|^2 dx
\end{aligned}$$

Noticing (2.7) and using Lemma 1, we have

$$\begin{aligned}
u|_{\Gamma_0} \cdot I_0 &\leq - J_{k'}(u') + (k' - k) \int_{\Omega_L} | \nabla u|^2 dx \\
&= u'|_{\Gamma_0} \cdot I_0 + (k' - k) \int_{\Omega_L} | \nabla u|^2 dx
\end{aligned}$$

Consequently

$$u|_{\Gamma_0} - u'|_{\Gamma_0} \leq \frac{k' - k}{I_0} \int_{\Omega_L} | \nabla u|^2 dx$$

i.e.,

$$u|_{\Gamma_0} \leq u'|_{\Gamma_0}$$

when $k > k'$ and the equality is valid only if

$$\int_{\Omega_L} | \nabla u|^2 dx = 0 \tag{2.8}$$

But that is impossible. Assume the contrary; then by noticing, $\Gamma_L \bigcap \Gamma_I \neq \phi$, we know that (2.8) implies $u = 0$ a.e. in Ω_L. Without loss of generality, we assume that Ω_j adjoints Ω_L. By the trace theorem, both $u = 0$ and $\frac{\partial u}{\partial n} = 0$ are valid on $\Gamma_j \bigcap \Gamma_L$. Since u is regular inside Ω_j and in a neighborhood of the regular point on $\Gamma_j \bigcap \Gamma_L$ (see [6]). According to [7], it follows from $u = 0$ and $\frac{\partial u}{\partial n} = 0$ on $\Gamma_j \bigcap \Gamma_L$ that u vanishes in the neighborhood of any ragular point on $\Gamma_j \bigcap \Gamma_L$, hence u vanishes in Ω_j. Furthermore, since Ω is connected, u vanishes in the whole Ω, hence $\frac{\partial u}{\partial n}|_{\Gamma_0} = 0$ and $\langle V, \frac{\partial u}{\partial n} \rangle_{H^{-\frac{1}{2}}(\Gamma_0), H^{-\frac{1}{2}}(\Gamma_0)} = 0$, $\forall v \in V$. This is contradictory since we should have $\langle V, \frac{\partial u}{\partial n} \rangle_{H^{\frac{1}{2}}(\Gamma_0), H^{-\frac{1}{2}}(\Gamma_0)} = I_0 \cdot v|_{\Gamma_0}$ by Green's formula. This shows that $\int_{\Omega_L} |\nabla u|^2 dx \neq 0$ and

$$u|_{\Gamma_0} < u'|_{\Gamma_0}$$

i.e.,

$$g(k) < g(k')$$

<u>Theorem 4</u>. Under the hypotheses of Theorem 3, the function $c = g(k)$ is continuous in $(0, +\infty)$.

Proof. Assume that u and u' satisfy (2.6) and (2.7) respectively. By Lemma 1, for any $\varphi \in V$ we have

$$\sum_{i=1}^{L-1} \int_{\Omega_i} k_i \nabla u \cdot \nabla \varphi \, dx + \int_{\Omega_L} k \nabla u \cdot \nabla \varphi \, dx = \varphi|_{\Gamma_0} \cdot I_0 \tag{2.9}$$

and

$$\sum_{i=1}^{L-1} \int_{\Omega_i} k_i \nabla u' \cdot \nabla \varphi \, dx + \int_{\Omega_L} k' \nabla u' \cdot \nabla \varphi \, dx = \varphi|_{\Gamma_0} \cdot I_0 \tag{2.10}$$

In (2.9) and (2.10), replacing φ by u' and substracting (2.10) from (2.9), we have

$$\sum_{i=1}^{L-1} \int_{\Omega_i} k_i \nabla u' \cdot \nabla (u - u') dx + k \int_{\Omega_L} \nabla u' \cdot \nabla (u - u') dx \tag{2.11}$$

$$+ (k - k') \int_{\Omega_L} |\nabla u'|^2 dx = 0$$

Replacing φ by u in (2.9) and (2.10) then substracting each other we get

$$\sum_{i=1}^{L-1} \int_{\Omega_i} k_i \nabla u \cdot \nabla (u - u') dx + k \int_{\Omega_L} \nabla u \cdot \nabla (u - u') dx \tag{2.12}$$

$$+ (k - k') \int_{\Omega_L} \nabla u' \cdot \nabla u \, dx = 0$$

Substracting (2.11) from (2.12) we show

$$\sum_{i=1}^{L-1} \int_{\Omega_i} k_i |\nabla (u - u')|^2 dx + k \int_{\Omega_L} |\nabla (u - u')|^2 dx$$

$$= (k' - k) \int_{\Omega_L} \nabla u' \cdot \nabla (u' - u) dx$$

By noticing that Γ_I is not empty, it follows from Fridrichs inequality and Schwartz inequality that

$$\|u - u'\|_{H^1(\Omega)} \leq m \cdot |k' - k| \{\int_{\Omega_L} |\nabla u'|^2 dx\}^{\frac{1}{2}} \tag{2.13}$$

where m is a positive constant independent of k'.

As k' tends to k, we may assume $k' \in (\frac{k}{2}, \frac{3}{2}k)$. If $\tilde{u}$ satisfies

$$J_{\frac{k}{2}}(\tilde{u}) = \inf_{v \in V} J_{\frac{k}{2}}(v)$$

by Theorem 3,

$$u'|_{\Gamma_0} < \tilde{u}|_{\Gamma_0}$$

holds. Therefore,

$$\int_{\Omega_L} |\nabla u'|^2 dx \leq \frac{1}{k'} [\sum_{i=1}^{L-1} \int_{\Omega_i} |\nabla u'|^2 dx + k' \cdot \int_{\Omega_L} |\nabla u'|^2 dx]$$
$$= \frac{1}{k'} u'|_{\Gamma_0} \cdot I_0 < \frac{2}{k} \tilde{u}|_{\Gamma_0} \cdot I_0$$

Then (2.13) can be rewritten into

$$\|u - u'\|_{H^1(\Omega)} \leq m' \cdot |k' - k| \tag{2.14}$$

where m' is a positive constant independent of k'. It follows from (2.14) that as $k' \to k$.

$$u' \overset{H^1(\omega)}{\longrightarrow} u$$

Applying trace theorem, we show

$$u'|_{\Gamma_0} \longrightarrow u|_{\Gamma_0}$$

This proves the theorem.

According to Theorem 3 and Theorem 4, the inverse function $k = g^{-1}(c)$ of $c = g(k)$ exists and is continuous. Therefore, for any given c in the region of $g(k)$, we are able to determine a unique k which depends on c continuously.

Remark 5. In the case of $k_1, \cdots, k_{L-1}$ being positive functions of x, the conclusions of Theorem 3 and Theorem 4 are still true.

III. Limiting behaviour of $g(k)$

Let

$$\tilde{\Omega} = \Omega \setminus \Omega_L \tag{3.1}$$

we denote the minimizer of the functional

$$\tilde{J}(v) = \frac{1}{2} \sum_{i=1}^{L-1} \int_{\Omega_i} k_i |\nabla v|^2 dx - v|_{\Gamma_0} \cdot I_0 \tag{3.2}$$

in

$$V_0 = \{v(x) \in H^1(\tilde{\Omega}) | v|_{\Gamma_I \cap \partial\tilde{\Omega}} = 0, v|_{\Gamma_0} = \text{constant}\} \tag{3.3}$$

and

$$V_\infty = \{v(x) \in H^1(\tilde{\Omega}) | v|_{(\Gamma_I \bigcup \Gamma_L) \bigcap \partial\tilde{\Omega}} = 0, v|_{\Gamma_0} = \text{constant}\} \tag{3.4}$$

by u_0 and u_∞ respectively.

Lemma 6. Under the hypotheses of Theorem 3, the function $g(k)$ defined by (2.4) is bounded and

$$g(k) \le u_0|_{\Gamma_0} \qquad \forall k \in (0, +\infty)$$

holds.

Proof. By the definition of u_0,

$$\sum_{i=1}^{L-1} \int_{\Omega_i} k_i \nabla u_0 \cdot \nabla\varphi \, dx = \varphi|_{\Gamma_0} \cdot I_0 \qquad \forall \varphi \in V_0 \tag{3.5}$$

holds. So we have

$$\tilde{J}(u_0) = -u_0|_{\Gamma_0} \cdot I_0 \tag{3.6}$$

Suppose that $u \in V$ such that

$$J_k(u) = \inf_{v \in V} J(v)$$

we have

$$J_k(u) = -u|_{\Gamma_0} \cdot I_0 \tag{3.7}$$

Noticing that the restriction $u|_{\tilde{\Omega}}$ of u on $\tilde{\Omega}$ belongs to V_0, we get

$$\begin{aligned}(u_0|_{\Gamma_0} - u|_{\Gamma_0}) \cdot I_0 &= J_k(u) - \tilde{J}(u_0)\\ &\ge \tilde{J}(u|_{\tilde{\Omega}}) - \tilde{J}(u_0) \ge 0\end{aligned}$$

Theorem 7. Let the hypotheses of Theorem 3 be fulfilled, and u_k be the minimizer of $J_k(v)$ in V, then

$$u_k|_{\Gamma_0} \longrightarrow u_0|_{\Gamma_0}, \qquad \text{as } k \to 0 \tag{3.8}$$

$$u_k|_{\Gamma_0} \longrightarrow u_\infty|_{\Gamma_0}, \qquad \text{as } k \to +\infty \tag{3.9}$$

Proof. (1). We first prove (3.8). We extend u_0 to the whole Ω, and denote it by $\bar{u}_0$, such that $\bar{u}_0 \in H^1(\Omega)$ and

$$\bar{u}_0|_{\Gamma_I \bigcap \Gamma_L} = 0$$

It is obvious that such extension exists and that $\bar{u}_0 \in V$.

We know that u_k satisfies

$$\sum_{i=1}^{L-1} \int_{\Omega_i} k_i \nabla u_k \cdot \nabla\varphi \, dx + k \int_{\Omega_L} \nabla u_k \cdot \nabla\varphi \, dx = \varphi|_{\Gamma_0} \cdot I_0, \quad \forall \varphi \in V \tag{3.10}$$

Replacing φ by u_0 and $\bar{u}_0$ in (3.5) and (3.10) respectively, then substracting them, we obtain

$$\sum_{i=1}^{L-1}\int_{\Omega_i} k_i \nabla u_0 \cdot \nabla(u_k - u_0)dx + k\int_{\Omega_L} \nabla u_k \cdot \nabla \bar{u}_0 dx = 0 \tag{3.11}$$

We denote the restriction of u_k on $\tilde{\Omega}$ by $u_k|_{\tilde{\Omega}}$. Replacing φ by u_k and $u_k|_{\tilde{\Omega}}$ in (3.10) and (3.5) respectively, then substracting each other, we obtain

$$\sum_{i=1}^{L-1}\int_{\Omega_i} k_i \nabla u_k \cdot \nabla(u_k - u_0)dx + k\int_{\Omega_L} |\nabla u_k|^2 dx = 0 \tag{3.12}$$

Substracting equalities (3.11) from (3.12), we have

$$\sum_{k=1}^{L-1}\int_{\Omega_i} k_i|\nabla (u_k - u_0)|^2 dx + k\int_{\Omega_L} |\nabla u_k|^2 dx = k\int_{\Omega_L} \nabla u_k \cdot \nabla \bar{u}_0 dx$$

It shows that

$$\sum_{i=1}^{L-1}\int_{\Omega_i} |\nabla (u_k - u_0)|^2 dx \le \frac{k}{2}\int_{\Omega_L} |\nabla \bar{u}_0|^2 dx \tag{3.13}$$

by applying schwartz inequality.

Since $\partial\tilde{\Omega} \bigcap \Gamma_I$ is not empty, using Fridrichs inequality, we obtain

$$\|u_k|_{\tilde{\Omega}} - u_0\|^2_{H^1(\tilde{\Omega})} \le M \cdot k \tag{3.14}$$

where M is a positive constant independent of k. It follows from the trace theorem that

$$u_k|_{\Gamma_0} \longrightarrow u_0|_{\Gamma_0}$$

as $k \to 0$.

(2). Now we prove (3.9). It is easy to see that

$$g(k) = u_k|_{\Gamma_0} \longrightarrow A, \quad \text{as} \quad k \longrightarrow +\infty \tag{3.15}$$

since $g(k)$ is strictly monotone decreasing and is greater than zero, and where A is a positive constant. What we should prove is that $A = u_\infty|_{\Gamma_0}$.

Lemma 6 follows

$$\begin{aligned}&\sum_{i=1}^{L-1}\int_{\Omega_i} k_i|\nabla u_k|^2 dx + k\int_{\Omega_L} |\nabla u_k|^2 dx \\ &= u_k|_{\Gamma_0} \cdot I_0 \le u_0|_{\Gamma_0} \cdot I_0\end{aligned} \tag{3.16}$$

If $k > K$, where K is a fixed positive constant, we have

$$\sum_{i=1}^{L-1}\int_{\Omega_i} k_i|\nabla u_k|^2 dx + K\int_{\Omega_L} |\nabla u_k|^2 dx \le u_0|_{\Gamma_0} \cdot I_0 \tag{3.17}$$

Owing to Fridrichs inequality, (3.17) shows that $\{u_k\}$ is a bounded set in $H^1(\Omega)$. It follows from the weak compactness of Hilbert space that there exists a subsequence $\{u_{k_l}\}$ of $\{u_k\}$ such that

$$u_{k_l} \overset{H^1(\omega)}{\longrightarrow} \bar{u} \in V, \quad \text{as } l \to +\infty \tag{3.18}$$

But (3.17) shows that

$$k \int_{\Omega_L} |\nabla u_k|^2 dx \leq m_1$$

where m_1 is a positive constant, hence

$$\int_{\Omega_L} |\nabla u_{k_l}|^2 dx \longrightarrow 0, \quad \text{as } 1 \longrightarrow +\infty \tag{3.19}$$

Using (3.19) and Fridrichs inequality, we show that

$$\|u_{k_l}\|_{H^1(\Omega_L)} \longrightarrow 0, \quad \text{as } 1 \longrightarrow +\infty \tag{3.20}$$

It follows from (3.18) and (3.20) that the restriction of $\bar{u}$ in Ω_L is zero almost everywhere. Therefore,

$$\bar{u}|_{\Gamma_L} = 0 \tag{3.21}$$

We extend u_∞ to the whole Ω by zero and denote it by $\bar{u}_\infty$, i.e.,

$$\bar{u}_\infty = \begin{cases} u_\infty & x \in \tilde{\Omega} \\ 0 & x \in \Omega \setminus \tilde{\Omega} \end{cases} \tag{3.22}$$

It is easy to see that $\bar{u}_\infty \in V$.

In the equation (3.10) which is satisfied by any element of $\{u_{k_l}\}$, replacing φ by $\bar{u}_\infty$ we obtain

$$\sum_{i=1}^{L-1} \int_{\Omega_i} k_i \nabla u_{k_l} \cdot \nabla u_\infty dx = u_\infty|_{\Gamma_0} \cdot I_0$$

Let $1 \to +\infty$. Noticing the weak convergence of $\{u_{k_l}\}$, we pass the limit and show

$$\sum_{i=1}^{L-1} \int_{\Omega_i} k_i \nabla \bar{u} \cdot \nabla u_\infty dx = u_\infty|_{\Gamma_0} \cdot I_0 \tag{3.23}$$

Nevertheless, u_∞ satisfies

$$\sum_{i=1}^{L-1} \int_{\Omega_i} k_i \nabla u_\infty \cdot \nabla \varphi \, dx = \varphi|_{\Gamma_0} \cdot I_0 \quad \forall \varphi \in V_\infty$$

It follows from (3.21) that the restriction of $\bar{u}$ on $\tilde{\Omega}$ belongs to V_∞. Therefroe,

$$\sum_{i=1}^{L-1} \int_{\Omega_i} k_i \nabla u_\infty \cdot \nabla \bar{u} \, dx = \bar{u}|_{\Gamma_0} \cdot I_0 \tag{3.24}$$

It follows from (3.23) and (3.24) that

$$\bar{u}|_{\Gamma_0} = u_\infty|_{\Gamma_0}$$

Using the fact that u_{k_l} is weakly convergent to $\bar{u}$, we find that

$$\lim_{l\to\infty} u_{k_l}|_{\Gamma_0} = \bar{u}|_{\Gamma_0} = u_\infty|_{\Gamma_0} \tag{3.25}$$

It shows that there exists a subsequence of $\{u_k\}$, which is convergent to $u_\infty|_{\Gamma_0}$. This implies that the limit of $u_k|_{\Gamma_0}$ must be $u_\infty|_{\Gamma_0}$, i.e. $A = u_{\infty|\Gamma_0}$.

Summrizing Theorem 3, Theorem 4 and Theorem 7, denoting $u_0|_{\Gamma_0}$, $u_\infty|_{\Gamma_0}$ by U_0, U_∞ respectively, we have

Theorem 8. Let the hypotheses on Ω of Theorem 3 be fulfilled, then for any $c \in (U_\infty, U_0)$, there exists a unique k such that the trace on Γ_0 of the minimizer u_k of functional (2.2) in the function space V defined by (2.1) is just equal to c and k depends on c continuously.

IV. Applications

The results can be directly applied to the well logs. For example, in the cases of the simplified electric well log mentioned in Section I or the three electrodes lateral well log, the resistivity in the objective layer can be determined uniquely by the potential measured on the electrode, and the error of the resistivity caused by the measuring error of the potential is small, provided that the measuring error is small. Therefore, the resistivity of the objective layer is identifiable by means of the well logs.

Theorem 3 implies that, when the current discharged by the eletrode is invariant, the bigger the resistivity of the objective layer R_t goes, the bigger the electrode potential will be. This coincides with Ohm's law.

Theorem 7 shows that, as the resistivity of the objective layer tends to infinity, the electrode potential tneds to the theoretical value of the electrode potential when the objective layer becomes insulated. And as the resistivity of the objective layer tends to zero, the electrode potential tends to the theoretical value of the potential when the potential on both interfaces between the objective layer and the surrounding rock or the objective layer and the invaded area is equal to zero. This result coincides with the physical phynomina that as the resistivity decreases to zero, the whole objective layer Ω_4 turns to an equalpotential body with zero potential.

Paper [8] offers an effective method to solve the positive problem, i.e., the BVP with nonlocal boundary concition for quasiharmonic equation. According to Theorem 8, we can apply the results for positive problem in solving the inverse problem numerically, and code the artificial intelligent interpreting software for well logs.

The author gratefully acknowledges Prof. Li Daqian for his helpful suggestions.

Reference

[1]Li Daqian, Zheng Songmu, Tan Yongji, et al. (1980) Applications of FEM in Electric Logs, Petroleum Industry Press, Beijin.

[2] Li Daqian, Zheng Songmu, Tan Yongji, et al. (1976) On boundary value problem of the self-adjoint elliptic equation with equal value boundary conditions (I), Fudan Journal, No. 1, 61-71.

[3] Necas (1967) Les Methods Directes en Theorie des Equations Elliptiques, Masson, Paris.

[4] Grisvard, P. (1985) Ellipti Problem in Nonsmooth Domain, Pitman, Boston.

[5] Lions, J., Maganes, E. (1968) Problem aux Limites Non Homogenes et Applications, Dunod, Paris.

[6] Nirenberg, L., On elliptic partial differential equation, Annali S.N.S.Pisa, 13, 115-162.

[7] Homander, L. (1976) Linear Partial Differential Operators, Springer-Verlag.

[8] Li Daqian, Zheng Songmu, Tan Yongji, et al. (1976) On boundary value problems of the self-adjoint elliptic equation with equal value boundary conditions (II), Fudan Journal, No. 3- 4, 137-145.

LOCAL ELLIPTICITY OF F AND REGULARITY OF F MINIMIZING CURRENTS[1]

Jean E. Taylor
Mathematics Department, Rutgers University
New Brunswick, NJ 08903

Abstract. Local convex ellipticity of a parametric integrand F at a given point in space and given tangent plane direction is defined. Using this definition, the same type of local regularity is proved for $\mathbf{F}$-minimizing currents near points whose tangent planes are directions at which F is convexly elliptic, as holds for currents minimizing a globally elliptic integrand. Various notions of ellipticity are discussed. Finally, some examples are given of physically interesting parametric integrands that are only locally elliptic, and a conjecture is made as to the overall structure of surfaces minimizing the integrals of such integrands.

The surface energy of an interface between a crystalline material and another phase is often represented as an integral over that interface of a continuous surface energy density which depends on tangent plane (or, equivalently, exterior normal) directions. Such energy density functions are naturally parametric integrands as defined below. Regularity almost everywhere for sets minimizing an elliptic parametric integral was proved nearly twenty years ago in [A1] and the methods were carried over to integral currents in [F 5.3.14]. However, the energy density functions characteristic of crystalline or otherwise ordered materials need not be convex or elliptic, except perhaps in (possibly small) open subsets of directions. For example, a statistical mechanics computation of surface energy [RW] predicts that in certain cases the equilibrium shape of a single crystal would be a distorted sphere with six flat facets (see figure 1 in section 6). No elliptic surface energy density function could have such a shape for its equilibrium crystal shape, as will be proved in section 6 below. But there are elliptic integrands which would agree, on a rather large open set of tangent plane directions, with an integrand producing that shape.

We define here for the first time a notion of such local ellipticity and show (as a special case of a more general theorem) that local ellipticity implies corresponding local smoothness of the interface, thus answering the question posed by the author in [B]. Very little about the local geometric structure of the surface is assumed *a priori* (in particular, it is not *assumed* locally to be the graph of a function nor the image of a piece of $\mathbf{R}^2$ under a continuous map).

By a *parametric integrand* we mean a continuous function $F : \mathbf{R}^n \times \wedge_k \mathbf{R}^n \to \mathbf{R}$, $2 \le k < n$, which is positively homogeneous in its second variable, i.e. $F(X, \lambda \boldsymbol{\xi}) = \lambda F(X, \boldsymbol{\xi})$ for every $\lambda \ge 0$, $X \in \mathbf{R}^n$, and $\boldsymbol{\xi} \in \wedge_k \mathbf{R}^n$. The *integral* $\mathbf{F}(\mathrm{T})$ of F over a rectifiable current $T = t(A, \theta, \boldsymbol{\tau})$ is defined as follows:

$$\mathbf{F}(T) = \int_{X \in A} F(X, \boldsymbol{\tau}(X)) \theta(X) \, d\mathcal{H}^k X.$$

Here A is a rectifiable set (a set which coincides with k-dimensional C^1 submanifolds except on sets of arbitrarily small area), θ is positive-integer-valued summable "multiplicity" function defined on A, and

[1] This research was partially supported by an NSF grant

$\boldsymbol{\tau}$ is a measurable simple-unit-k-vector-valued orientation function for the tangent spaces to A, defined at every $X \in A$ such that the k-dimensional density of $(A \sim M) \cup (M \sim A)$ at X is zero for some approximating C^1 manifold M. Removing a set of measure 0 from A if necessary, we can and do assume that each point of A is a Lebesgue point of both θ and $\boldsymbol{\tau}$ (with respect to k-dimensional area measure).

We say that F is *locally convexly elliptic at* $(X_0,\boldsymbol{\xi}_0)$ if F is Lipschitz in its first variable near X_0, $C^{2,\alpha}$ in its second variable near $\boldsymbol{\xi}_0$, and there is a parametric integrand G which is elliptic in the sense of [SS] (which means that formula (5*) below holds for *all* $(X_0,\boldsymbol{\xi}_0)$) such that $F \geq G$ everywhere and $F = G$ in some neighborhood of $(X_0,\boldsymbol{\xi}_0)$. An alternative characterization is that for some $c > 0$ and some $\zeta > 0$,

$$(5^*) \qquad Bc(d^2/dt^2)F(X,\boldsymbol{\xi}+t\boldsymbol{\eta})|_{t=0} \geq |\boldsymbol{\xi}|^2 - (\boldsymbol{\xi}\bullet\boldsymbol{\eta})^2, \qquad |\boldsymbol{\eta}| = 1, |\boldsymbol{\xi}-\boldsymbol{\xi}_0|, |X - X_0| < \zeta,$$

and $F(\boldsymbol{\xi}_1) + F(\boldsymbol{\xi}_2) \geq F(\boldsymbol{\xi})$ whenever $\boldsymbol{\xi}_1 + \boldsymbol{\xi}_2 = \boldsymbol{\xi}$ and $|\boldsymbol{\xi} - \boldsymbol{\xi}_0| < \zeta$. (Other alternative characterizations appear in section 5 below.) Finally, we say that T is $\mathbf{F}$-minimizing if $\mathbf{F}(T) \leq \mathbf{F}(S)$ for any rectifiable current with $\partial S = \partial T$ (the boundaries being defined by Stokes's Theorem). One result proved here is:

(Codimension 1 Regularity). *If* $n = k+1$, *F is a positive and bounded parametric integrand,* $T = t(A,\theta,\boldsymbol{\tau})$ *is* $\mathbf{F}$*-minimizing,* $X_0 \in A$ *satisfies* $dist(X_0, spt\partial T)$ *and F is locally convexly elliptic at* $(X_0,\boldsymbol{\tau}(X_0))$, *then* $spt(T)$ *is an embedded* $\mathbf{F}$*-minimizing* $C^{1,1/2}$ *manifold in a neighborhood of* X_0.

This result is a corollary to a theorem that holds for all codimensions (the Regularity Theorem below); however, in general codimensions it is not known (even assuming global ellipticity) that a required lower density bound holds for even almost every point with a tangent plane (this is the enormous "bubbling" problem which has recently been solved for the area integrand in [A3]).

The major part of the proof of this theorem consists in carefully going through the proof of regularity in the case that F is globally elliptic given in [SS], in order to verify that the global ellipticity assumed there need not in fact be used. The proof in [SS] was used, rather than the one in [F] or [A2], because it invokes a Lipschitz approximation theorem, thereby enabling one to use local convex ellipticity only for directions near the original fixed direction. (It might have been better to modify the proof in [A2], since the context of "$(\mathbf{F},\epsilon,\delta)$ minimal sets" is a broader and more natural context for applications than integral currents are, but that proof uses ellipticity in a way that looks very difficult to localize.) The reader does not need to have read [SS] nor to have a copy of [SS] in hand in order to read this paper; however, the parts of the proof in [SS] that do not need to be changed are merely summarized here.

The figures in this paper were drawn by F. J. Almgren, who also provided many helpful conversations.

1. Notation and statment of Regularity Theorem and Corollary.

We follow the notation of [SS]. In particular, equation numbers without asterisks will denote equations from [SS]. The numbering of equations here is thus neither consecutive nor monotone, since many equations there are either omitted or used in a different order. Equation numbers with asterisks (such as (4*)) are modifications of the corresponding equations in [SS] and replace those equations. Equation numbers such as (52A*) indicate equations needed in addition to the corresponding equation in [SS].

$\mathcal{L}^k$ will denote ordinary k-dimensional Lebesgue measure in $\mathbf{R}^k$; $\mathbf{R}^k$ will often be identified with $\mathbf{R}^k \times \{0\} \subset \mathbf{R}^k \times \mathbf{R}^{n-k} = \mathbf{R}^n$.

$\mathbf{p}$ will denote the orthogonal projection of $\mathbf{R}^n$ onto $\mathbf{R}^k$.

$\mathbf{q}$ will denote the orthogonal projection of $\mathbf{R}^n$ onto $\mathbf{R}^{n-k}$.

$B(y,\rho)$ will denote the *open* ball, with center y and radius ρ, in $\mathbf{R}^k$. The corresponding *closed* ball is denoted by $\bar{B}(y,\rho)$. The volume of such a ball with unit radius is denoted ω_k.

$C(y,\rho) = B(y,\rho) \times \mathbf{R}^{n-k} \subset \mathbf{R}^n$.

$e_1,\ldots,e_n$ denote the standard orthonormal basis for $\mathbf{R}^n$. $\wedge_k\mathbf{R}^n$ will denote the vector space of k-vectors of $\mathbf{R}^n$; $\mathbf{e}^k$ will denote the simple k-vector $e_1 \wedge \cdots \wedge e_k$, and for $\alpha = k+1,\ldots,n$ and $i = 1,\ldots,k$ we will let $\mathbf{e}_i^\alpha$ denote the simple k-vector $(-1)^{i-1}e_\alpha \wedge e_1 \wedge \cdots \wedge e_{i-1} \wedge e_{i+1} \wedge \cdots \wedge e_k$.

We will be considering k-dimensional rectifiable currents T in $\mathbf{R}^n$; for a discussion the reader is referred to [F, 4.1.1–4.1.9, 4.1.28, 4.1.30] or to more recent treatments such as [S]. Also there is a discussion of currents in section A.4 of [ABL] in this volume. As in [SS] and [F], the set of all rectifiable currents will be denoted by $\mathcal{R}_k(\mathbf{R}^n)$ and $\|T\|$ will denote the usual Borel weight measure associated with T, $\Theta^k(\|T\|,\cdot)$ the associated k-dimensional multiplicity function, and $\vec{T}$ the orienting simple k-vectorfield associated with T. In the newer notation used above, $T = t(A,\theta,\boldsymbol{\tau})$, where A is as in the introduction (and equals, up to set of measure 0, the set of points where $\Theta^k(\|T\|,\cdot)$ is positive), so that spt(T) (the support of the measure $\|T\|$) is the closure of A, $\theta(\cdot) = \Theta^k(\|T\|,\cdot)$, $\|T\| = \theta(\cdot)\mathcal{H}^k \llcorner A$, and $\boldsymbol{\tau} = \vec{T}$ $\|T\|$-almost everywhere.

We note that if spt$\partial T \subset \mathbf{R}^n \sim C(y,\rho)$, then by the constancy theorem [F 4.1.7]

$$\mathbf{p}_\#(T \llcorner C(y,\rho)) = m\mathbf{E}^k \llcorner B(y,\rho), \tag{1}$$

where m is an integer and $\mathbf{E}^k$ is the standard k-current obtained by integration of k-forms over $\mathbf{R}^k$. We also note that $DF_X(\boldsymbol{\xi})$ (F_X as defined below) is a linear function from $\wedge_k\mathbf{R}^n$ to $\mathbf{R}$ and thus is naturally isomorphic to a k-covector. We will use this isomorphism implicitly and do not denote it by * as is done in [SS], since * has a common alternative meaning.

$F : \mathbf{R}^n \times \wedge_k\mathbf{R}^n \to \mathbf{R}$ will denote a k-dimensional parametric integrand as defined in the introduction. We need to impose several conditions on F. The most general way to phrase the first of these conditions, the *positivity and boundedness condition,* would be to say that there exists a closed differential k-form Υ on $\mathbf{R}^n$, a constant $0 < c < \infty$, and a constant μ such that

$$|\boldsymbol{\xi}| \le cF_\Upsilon(X,\boldsymbol{\xi}) \le \mu|\boldsymbol{\xi}|, \qquad (X,\boldsymbol{\xi}) \in \mathbf{R}^n \times \wedge_k(\mathbf{R}^n) \tag{2*}$$

where the integrand F_Υ is defined by

$$F_\Upsilon(X,\boldsymbol{\xi}) = F(X,\boldsymbol{\xi}) + \langle \Upsilon(X), \boldsymbol{\xi}\rangle$$

However, since Υ is a closed differential k-form, Stokes' theorem says that $\mathbf{F}_\Upsilon(S) - \mathbf{F}(S)$ depends only on the boundary of S. Thus F_Υ and F have the same minimal surfaces. Therefore we can without loss of generality assume throughout the rest of this paper that F itself, rather than some F_Υ, satisfies (2). Similarly, we can and will assume that $c = 1$. In this case (2*) above reduces to

$$|\boldsymbol{\xi}| \le F(X,\boldsymbol{\xi}) \le \mu|\boldsymbol{\xi}|, \qquad (X,\boldsymbol{\xi}) \in \mathbf{R}^n \times \wedge_k(\mathbf{R}^n). \tag{2}$$

There are a variety of definitions (shown to be equivalent in section 5) that we could take for local convex ellipticity of F. Two are given in the introduction above. The most straightforward definition for this proof, however, is to define F to be *locally convexly elliptic at* $(X_0, \boldsymbol{\xi}_0)$ if and only if for some $\zeta > 0$, F is Lipschitz in its first variable in a ζ neighborhood of X_0, $C^{2,\alpha}$ in its second variable in a ζ neighborhood of $\boldsymbol{\xi}_0$ and

$$(4^*) \qquad (1/2)|\boldsymbol{\xi} - \boldsymbol{\eta}|^2 \le F_X(\boldsymbol{\eta}) - \langle DF_X(\boldsymbol{\xi}), \boldsymbol{\eta}\rangle, \qquad |\boldsymbol{\xi}| = |\boldsymbol{\eta}| = 1 \text{ and } |\boldsymbol{\xi} - \boldsymbol{\xi}_0|, |X - X_0| < \zeta.$$

Here, for any given $X \in \mathbf{R}^n$, we let $F_X : \wedge_k(\mathbf{R}^n) \to \mathbf{R}$ be the "constant coefficient integrand" defined by $F_X(\boldsymbol{\xi}) = F(X, \boldsymbol{\xi})$, and $DF_X(\boldsymbol{\xi})$ denotes the linear map $\wedge_k(\mathbf{R}^n) \to \mathbf{R}$ induced by F_X at $\boldsymbol{\xi}$. (Again, a more general statement of (4*) would have the left side multiplied by some positive constant c, but we can without loss of generality replace F by the integrand $c^{-1}F$ and thus assume that $c = 1$.) Note that if $\zeta = 2$ then for X within distance 2 of X_0, F_X is locally convexly elliptic at all $\boldsymbol{\xi}$ and hence elliptic in the sense of [SS]. Thus we may without any loss assume that $\zeta \le 2$.

Condition (4*) implies (see section 5) that

$$(5^*) \qquad (d^2/dt^2)F_X(\boldsymbol{\xi} + t\boldsymbol{\eta})|_{t=0} \ge |\boldsymbol{\xi}|^2 - (\boldsymbol{\xi} \bullet \boldsymbol{\eta})^2, \qquad |\boldsymbol{\eta}| = 1, |\boldsymbol{\xi} - \boldsymbol{\xi}_0|, |X - X_0| < \zeta,.$$

Equation (50) in [SS] (which can also be found in the text below) shows how to compute $\vec{T^\gamma}(X)$ when T^γ is the current corresponding to the graph of a function with Lipschitz constant γ. A consequence of this is that in this situation there is a ς such that

$$(52^{**}) \qquad |\vec{T^\varsigma} - \mathbf{e}^k| \le \zeta$$

We fix ζ and its associated ς for the rest of this paper.

We make the quantitative assumptions

$$(7^*) \qquad \sup_{|X - X_0| \le \zeta} \|D^2 F_X(\boldsymbol{\xi})\| \le \mu, \qquad |\boldsymbol{\xi} - \boldsymbol{\xi}_0| < \zeta, |\boldsymbol{\xi}| = 1$$

$$(8^*) \qquad \sup_{|X - X_0| \le \zeta} \|D^2 F_X(\boldsymbol{\xi}) - D^2 F_X(\boldsymbol{\eta})\| \le \mu |\boldsymbol{\xi} - \boldsymbol{\eta}|^\beta, \qquad |\boldsymbol{\xi}| = |\boldsymbol{\eta}| = 1, \ |\boldsymbol{\xi} - \boldsymbol{\xi}_0|, |\eta - \boldsymbol{\xi}_0| < \zeta$$

$$(9^*) \qquad \sup_{\boldsymbol{\xi} \in \wedge_k \mathbf{R}^n, |\boldsymbol{\xi}| = 1} |F(X, \boldsymbol{\xi}) - F(Y, \boldsymbol{\xi})| \le \mu_1 |X - Y|, \qquad X, Y \in \mathbf{R}^n, \ |X - X_0|, |Y - X_0| < \zeta$$

($\mu_1 \ge 0$ is a constant). These assumptions differ from those in [SS] only by being localized.

Note that if j is any linear isometry of $\mathbf{R}^n$, F is a parametric integrand, and T minimizes $\mathbf{F}$, then $j_\# T$ minimizes the integral $(j^{-1})_\# \mathbf{F}$, defined by $(j^{-1})_\# \mathbf{F}(S) = \mathbf{F}(j_\#^{-1} S)$ for $S \in \mathcal{R}_k(\mathbf{R}^n)$. One readily checks that if F satisfies (2),(4*), and (7*)-(9*), then $(j^{-1})^\# \mathbf{F}$ satisfies (2),(4*) and (7*)-(9*) with the same constants β, μ, μ_1 but with $(X_0, \boldsymbol{\xi}_0)$ replaced by $(j(X_0), \wedge_k j(\boldsymbol{\xi}_0))$. We can thus without loss of generality assume that $(X_0, \boldsymbol{\xi}_0) = (O, \mathbf{e}^k)$.

The fundamental hypotheses (H*) are the same here as in [SS], except that we explicitly include the hypotheses on F and add a hypotheses on ρ.

$$(\mathrm{H}^*) \qquad \begin{cases} F \text{ satisfies } (2), (4*) - (5*) \text{ and } (7*) - (9*) \\ \rho < \zeta \\ T \text{ minimizes } \mathbf{F} \\ \operatorname{spt}\partial T \subset \mathbf{R}^n \sim C(y, \rho) \\ \Theta^k(\|T\|, X) \ge m \text{ for } \|T\| - a.e. X \in C(y, \rho) \text{ (same } m \text{ as in (1))} \\ E(T, y, \rho) < \epsilon \end{cases}$$

The quantity $E(T,y,\rho)$ appearing here is the *cylindrical excess* of T, defined by

$$(12) \qquad E(T,y,\rho) = (1/2)\rho^{-k} \int_{C(y,\rho)} |\vec{T} - \mathbf{e}^k|^2 \, d\|T\|.$$

As in all proofs of regularity, it plays a major role. Given the fourth and fifth hypotheses of (H*), $E(T,y,\rho)$ is equal to its other definition

$$(13) \qquad E(T,y,\rho) = \rho^{-k}(\|T\|(C(y,\rho)) - \|\mathbf{p}_\# T\|(B(y,\rho))) = \rho^{-k}(\|T\|(C(y,\rho)) - m\omega_k \rho^k).$$

The only difference in the regularity theorem here and in [SS] is that the bound ϵ depends on the size ζ of the neighborhood in which F is elliptic:

REGULARITY THEOREM. *There is a constant $\epsilon = \epsilon(\zeta,\beta,\mu,m,n,k) > 0$ such that the hypotheses (H), together with the hypothesis $\mu_1\rho < \epsilon$, imply that spt$T \cap C(y,\rho/4) =$ graph u, for some $C^{1,1/2}(\bar{B}(y,\rho/4))$ function $u = (u^{k+1},\dots,u^n)$, with*

$$(15) \qquad \sup_{B(y,\rho/4)} \|Du\| + \rho^{1/2} \sup_{x,z\in B(y,\rho/4), x\neq z} \|Du(x) - Du(z)\| \le c_1(E(T,y,\rho) + \mu_1\rho)^{1/2},$$

where $c_1 = c_1(\zeta,\beta,\mu,m,k,n)$.

COROLLARY (CODIMENSION 1 REGULARITY). *If $n = k+1$, T is $\mathbf{F}$-minimizing, $\boldsymbol{\tau}(X_0)$ is defined as in the introduction, dist$(X_0, \text{spt}\partial T) > 0$, and F is a locally convexly elliptic parametric integrand at $(X_0, \boldsymbol{\tau}(X_0))$ which also satisfies (2*), then spt(T) is an embedded $\mathbf{F}$-minimizing $C^{1,1/2}$ manifold in a neighborhood of X_0.*

PROOF OF COROLLARY: As mentioned above, by translating and rotating we may assume that $(X_0, \boldsymbol{\xi}_0) = (O, \mathbf{e}^k)$ and we may without loss of generality assume that F satisfies (2) rather than (2*). (5*) holds since (4*) does. We may decrease μ if necessary so that (7*)-(8*) also hold. The definition of $\boldsymbol{\tau}(X_0)$, together with a corollary in [F 5.1.6] to the lower density ratio bound (also in [F 5.1.6])

$$(11) \qquad c_0^{-1} \le \rho^{-k}\|T\|\{X \in \mathbf{R}^n : |X - Y| < \rho\}$$

(which holds because of property (2) and $\mathbf{F}$-minimality, having nothing to do with ellipticity), implies that spt$T$ has a tangent plane at X_0 and $\boldsymbol{\tau}(X_0)$ is an orientation for it. $\mathbf{F}$-minimality now further implies that the cylindrical excess approaches zero as ρ approaches zero, and thus that the excess is less than ϵ for small enough ρ. We are left only with the lower bound on the density to verify. The fact that $n = k+1$ enables one to "peel off" sheets of spt(T) from highest to lowest in the cylinder. Now each such sheet must be individually $\mathbf{F}$-minimizing (if it were not, T would not be) and have its own excess bounded by ϵ and its boundary on the cylinder of radius ρ. Furthermore, the density is constantly 1 on each sheet. Therefore we may apply the Regularity Theorem to each sheet separately. Finally, we use the maximum principle [F 5.2.19, 5.3.19] to conclude that all sheets are identical, thereby obtaining that spt$T$ is the graph of a $C^{1,1/2}$ function.

2. Lemmas.

The heart of the proof of the Regularity Theorem theorem is the following lemma:

LEMMA 1. *There exist constants* $\theta, \epsilon_* \in (0,1/8)$ ($\epsilon_* \le (\theta/4)^{2k}$), *depending only on* $\zeta, \mu, \beta, m, n, k$, *so that if the hypotheses* (H^*) *hold with* $y = O, \epsilon = \epsilon_*$, *and if the additional hypotheses* $O \in \operatorname{spt}T, \theta^{-2k}\mu_1\rho < \epsilon_*$ *hold, then, taking* $T_0 = T \llcorner C(O, \rho/2)$,

$$\sup_{X \in \operatorname{spt}T_0} |\mathbf{q}(X)| \le \zeta\rho/8 \tag{18}$$

and

$$E(j_\# T_0, O, \theta\rho) \le \theta \max\{E(T, O, \rho), \theta^{-2k}\mu_1\rho\}$$

for some linear isometry j *of* $\mathbf{R}^n$ *with*

$$\|j - 1_{\mathbf{R}^n}\|^2 \le \theta^{-2k} \max\{E(T, O, \rho), \theta^{-2k}\mu_1\rho\} \le \varsigma^2/64.$$

Two other lemmas in [SS] lay the groundwork for the proof of this lemma:

LEMMA 2 (HEIGHT BOUND). *There is* $\epsilon_0 = \epsilon_0(\mu, m, n, k) \in (0, 1/8)$ *so that the hypotheses* (H^*) *with* $\epsilon = \epsilon_0$ *imply*

$$\sup_{X^{(1)}, X^{(2)} \in \operatorname{spt}T \cap C(y, \rho/2)} |\mathbf{q}(X^{(1)}) - \mathbf{q}(X^{(2)})| \le c_6 E^{1/2k}(T, y, \rho)\rho.$$

In the proof of this lemma (originally in [A1] but repeated in [SS]), ϵ_0 is shown to depend on μ only through the dependence of c_0 on μ, where c_0 is the reciprocal of the lower bound of (11) above.

LEMMA 3 (LIPSCHITZ APPROXIMATION LEMMA). *If the hypotheses* (H^*) *hold with* $\epsilon = \epsilon_0$ (ϵ_0 *as in Lemma 2), and* $\gamma \in (0, 1/8)$ *is given, then there is a Lipschitz function* $g = (g^{k+1}, \dots, g^n)$ *from* $B(y, \rho/4)$ *into* $\mathbf{R}^{n-k}$ *with*

$$Lip(g) \le \gamma, \qquad \sup_x |g(x) - g(y)| < c_7 E^{1/2k}(T, y, \rho)\rho \tag{25}$$

$$\rho^{-k}\mathcal{L}^k(B(y, \rho/4) \sim \{x \in B(y, \rho/4) : \mathbf{p}^{-1}(x) \cap \operatorname{spt}T = \{(x, g(x))\}\}) \le c_8\gamma^{-2k}E(T, y, \rho) \tag{26}$$

$$\rho^{-k}\|T - T^g\| C(y, \rho/4) \le c_9\gamma^{-2k}E(t, y, \rho), \tag{27}$$

where T^g *is the rectifiable current corresponding to the graph of* g *with multiplicity* m.

The proof of Lemma 3 in no way uses any ellipticity properties of F; it is rather based on the Besicovitch Covering Theorem (to estimate the measure of the set

$$B = \{z \in B(y, \rho) : E(T, z, \sigma) > \eta \text{ for some } \sigma \in (0, \rho/2)\},$$

where $\eta = (\gamma/2c_6)^{2k}$), Lemma 2 (to show that for $x^{(1)}, x^{(2)} \in A \equiv B(y, \rho/2) \sim B$ and $X^{(1)}, X^{(2)} \in \operatorname{spt}T \cap \mathbf{p}^{-1}(x^{(i)}), i = 1, 2$, we have

$$|\mathbf{q}(X^{(1)}) - \mathbf{q}(X^{(2)})| \le 2c_6\eta^{1/2k}|x^{(1)} - x^{(2)}| = \gamma|x^{(1)} - x^{(2)}|,$$

thereby demonstrating that $\operatorname{spt}T \llcorner \mathbf{p}^{-1}A$ is the graph of a Lipschitz function over A), and Kirzbraun's extension theorem (to extend the Lipschitz function on A to a function on all of $B(y, \rho/4)$). (There is a minor mistake in the proof of Lemma 3 in [SS], in that σ must approximate $2|x^{(1)} - x^{(2)}|$ rather than $|x^{(1)} - x^{(2)}|$, but it has no consequences.)

3. Proof of Lemma 1

We procede to outline the proof of Lemma 1 as given in [SS], indicating where changes must be made.

For some $\sigma \in (1/16, 1/8)$ [SS] uses the values of a mollified version of such a Lipschitz approximation g on $\partial B(y, \sigma\rho)$ as the prescribed boundary values for a solution u of the linearized Euler-Lagrange equation for F. More precisely, the scale can be and is changed so that $\rho = 1$ (so that μ_1 from here on is really $\rho\mu_1$ for the original ρ). We abbreviate $E(T, O, 1)$ by E. We let $\delta = (9k^2)^{-1}$ and denote by $g_\delta : B(0, 1/4) \to \mathbf{R}^{n-k}$ a Lipschitz function as in Lemma 3 corresponding to the choice $\gamma = E^{2\delta}$.

We define $\bar{S}$ to be the current corresponding to the graph of a mollification $\bar{g}_\delta$ of g_δ, with multiplicity m.

We let σ be one of the infinitely many numbers between 1/16 and 1/8 for which (35)-(38) of [SS] hold, and define $T_\sigma = T \llcorner C(O, \sigma)$, $\bar{S}_\sigma = \bar{S} \llcorner C(O, \sigma)$.

Let $V = H_\#([\![0,1]\!] \times \partial T_\sigma)$, where H is the homotopy $H(t, X) = tP(X) + (1-t)X$ and P is the "vertical retraction" $P(X) = (\mathbf{p}(X), \bar{g}_\delta(\mathbf{p}(X)))$, so that $\partial V = \partial T_\sigma - \partial \bar{S}_\sigma$. [SS] goes through a sequence of computations, based on the estimates on g_δ from Lemma 3, to show

(38)
$$\|V\|(\mathbf{R}^n) \leq c_{19} E^{1+\delta}.$$

Let u be the $C^{1,\delta}(\bar{B}(O, \sigma))$ solution of the problem

(39)
$$\begin{aligned} D_i(a_{ij}^{\alpha\beta} D_j u^\beta) &= 0, \quad \alpha = k+1, \ldots, n \quad \text{on } B(O, \sigma) \\ u &= \bar{g}_\delta \quad \text{on } \partial B(O, \sigma), \end{aligned}$$

where the coefficients $a_{ij}^{\alpha\beta}$ are the constants given by

$$a_{ij}^{\alpha\beta} = \frac{\partial^2}{\partial s \partial t} F_O(\mathbf{e}^k + t\mathbf{e}_i^\alpha + s\mathbf{e}_j^\beta)|_{s=t=0}$$

(condition (5*) implies that the above system is strongly elliptic). As in [SS], we have the estimate from linear elliptic PDE theory that

(40)
$$\sup_{x,z \in B(O,\sigma)} |x - z|^{-\delta} \|Du(x) - Du(z)\| + \sup_{B(O,\sigma)} \|Du\| \leq c_{20} E^\delta.$$

Let S be the current associated to the graph of u, with multiplicity m. Since $\bar{g}_\delta$ and u have the same boundary values, $\partial S = \partial \bar{S}$, and so

(43)
$$\partial(V + S - T_\sigma) = 0$$

and thus

(44)
$$(V + S - T_\sigma)(\omega_0) = 0$$

for any constant k-form ω_0.

As in [SS], a direct computation shows that if H is the rectifiable current corresponding to the graph of a Lipschitz function $h : B(O,\sigma) \to \mathbf{R}^{n-k}$, then

$$\vec{H}(x) = (1 + \sum_{j=1}^{k} \sum_{\beta=k+1}^{n} (D_j h^\beta)^2 + |R|^2)^{-1/2}(\mathbf{e}^k + D_i h \alpha \mathbf{e}_i^\alpha) + \bar{R}, \tag{50}$$

with $\bar{R}$ orthogonal to $\mathbf{e}^k$ and $\mathbf{e}_i^\alpha$, and

$$|R|, |\bar{R}| \le c_{24}\|Dh\|^2(1 + \|Dh\|)^{n-k-1}. \tag{51}$$

Applying this with $h = u$ and recalling that $\|Du\| \le E^\delta$ (by (40)) we see immediately that

$$|\vec{S} - \mathbf{e}^k| \le c_{25}^* E^\delta \tag{52A*}$$

and therefore

$$|DF_0(\vec{S}) - DF_0(\mathbf{e}^k)| \le c_{25} E^\delta \tag{52}$$

As mentioned at the beginning, we also see that there is a ς such that the current T_σ^ς corresponding to the graph (with multiplicity m) of g_ς (g_ς is as in Lemma 3 with ς as its Lipschitz constant) has its tangent planes satisfying

$$|\vec{T}_\sigma^\varsigma - \mathbf{e}^k| \le \zeta. \tag{52B*}$$

[SS] now embarks upon the proof that

$$(1/2) \int |\vec{T} - \vec{S}|^2 \, d\|T_\sigma\| \le c_{34} E^{1+\beta\delta} + 2c_{32}\mu_1, \tag{57}$$

provided

$$c_{30} E^{\beta\delta} \le 1/8. \tag{58}$$

(Here β is the Holder exponent from 8∗.) Since the tangent planes of T_σ may be outside the region of ellipticity, this part of the proof must be examined carefully. All the tangent planes of S, however, are within ζ of $\mathbf{e}^k$, by (52*), provided $c_{25}^* E^\delta < \zeta$, and therefore (4*) will hold for each of them if E is that small. Thus, provided we at minimum add this condition to (58) to obtain

$$c_{30} E^{\beta\delta} \le 1/8, \; c_{25}^* E^\delta < \zeta, \tag{58*}$$

we can procede as in [SS] with hope of success. In the following sequence of computations we extend $\vec{S}$ to a function on all of $C(y,\rho)$ by setting $\vec{S}(X) = \vec{S}(\mathbf{p}(X), u(\mathbf{p}(X))), X \in C(O,\sigma)$.

$$\mathbf{F}_O(T_\sigma) - \mathbf{F}_O(S) = \int F_O(\vec{T}_\sigma) d\|T_\sigma\| - \int F_O(\vec{S}_\sigma) d\|S\|$$

$$= \int (F_O(\vec{T}) - \langle DF_0(\vec{S}), \vec{T}\rangle) d\|T_\sigma\| + \int \langle DF_0(\vec{S}), \vec{T}\rangle d\|T_\sigma\| - \int \langle DF_0(\vec{S}), \vec{S}\rangle d\|S\|$$

(by adding and subtracting and using homogeneity, which implies $\langle v, DF_0(v)\rangle = F_0(v)$)

(45) $$\geq (1/2)\int |\vec{T}-\vec{S}|^2\, d\|T_\sigma\| + (T_\sigma - S)(DF_O(\vec{S})).$$

(by (4*) – this is the only place, apart from the ellipticity of the linear PDE above, that ellipticity is invoked)

(46) $$\geq (1/2)\int |\vec{T}-\vec{S}|^2\, d\|T_\sigma\| + (T_\sigma - S)(DF_O(\vec{S}) - DF_O(\mathbf{e}^k)) + V(DF_0(\mathbf{e}^k))$$

(by (44), since $DF_0(\mathbf{e}^k)$ can be regarded as a constant k-form).

The proof of (57), subject to (58*), now proceeds exactly as in [SS], except that T_σ^0 there is here replaced by T_σ^ς as defined above.

None of the rest of the proof of Lemma 1 uses ellipticity, but the final choice of ϵ_* does have to be slightly modified so as to get the ς where needed in the lemma. Thus we outline the remainder of the proof, giving this modification.

One shows that for arbitary $\theta \in (0, \sigma/4)$,

$$\int_{(C(0,2\theta)} |\vec{T}-\vec{S}(O)|^2\, d\|T\| \leq 2(1+2c_{36})\int |\vec{T}-\vec{S}|^2\, d\|T_\sigma\| + 8c_{36}\theta^{k+2}E$$

subject to $E \leq \theta^k$. When this is combined with (57) one obtains

(61) $$(1/2)\theta^{-k}\int_{C(O,2\theta)} |\vec{T}-\vec{S}|^2 d\|T\| \leq (1+4c_{36})\theta^{-2}E + c_{37}\theta^{-k}\mu_1$$

provided

(62*) $$c_{30}E^{\beta\delta} \leq 1/8,\ c_{25}^*e^\delta < \zeta,\ E \leq \theta^k, \text{ and } 2(1+2c_{36})c_{34}E^{\beta\delta} \leq \theta^{2+k}$$

(Note that there is a minor mistake in [SS] here in that the last inequality there mistakenly reads $(1+2c_{36})c_{34}E^{\beta\delta} \leq \theta^2$). One can further add (61) to the definition of E to obtain

(64) $$|\vec{S}(O) - \mathbf{e}^k| \leq c_{38}\theta^{-k}(E+\mu_1),$$

provided (62) holds.

Now (64) implies that there is a linear isometry j of $\mathbf{R}^n$ such that

(65) $$\wedge_k j(\vec{S}(O)) = \mathbf{e}^k \text{ and } \|j - 1_{\mathbf{R}^n}\|^2 \leq c_{39}\theta^{-k}(E+\mu_1).$$

Furthermore, provided $c_{39}\theta^{-k}(E+\mu_1)$ is sufficiently small and (62) holds,

(67) $$\operatorname{spt}\partial j_\# T \subset \mathbf{R}^n \sim C(O,1/4),\ \operatorname{spt} T \cap j^{-1}C(O,\theta) \subset C(O,2\theta)$$

Therefore if we now choose $\theta = (2\max\{c_{37}, 1+4c_{36}\})^{-1}$ and then choose ϵ_* so that

$$c_{30}\epsilon_*^{\beta\delta} \leq 1/8,\quad 2(1+2c_{36})c_{34}\epsilon_*^{\beta\delta} \leq \theta^{2+k},\quad (1+c_{39})(\zeta\theta)^{-2k}\epsilon_* \leq (8(1+c_9))^{-2k}$$

then (62*) holds, the above quantity *is* small enough, and all the conclusions of Lemma 1 hold.

4. Conclusion of the proof of the Regularity Theorem.

Now that it has been shown that the proof of Lemma 1 remains valid under the assumption of just local ellipticity, we examine the rest of the proof of the Regularity Theorem. In fact, what is used is the following Corollary to Lemma 1.

COROLLARY TO LEMMA 1. *Suppose θ, ϵ_* are as in Lemma 1, suppose that the hypotheses (H*) hold with $y = 0$ and $\epsilon = \epsilon_*$, and suppose also that $O \in sptT, \theta^{-2k}\mu_1\rho < \epsilon_*$. Then*

$$E(T,0,r) \le c_2 \max\{E(T,0,\rho), \theta^{-2k}\mu_1\rho\}, \qquad 0 < r \le \rho, \tag{19}$$

where $c_2 = c_2(\beta,\mu,m,k,n)$. Furthermore if $T_0 = T \llcorner C(O,\rho/2)$ then there is a linear isometry j of $\mathbf{R}^n$ such that

$$spt\partial_{\#}T_0 \cap C(O,\rho/4) = \emptyset, \qquad \|j-1\|^2 \le 4\theta^{-2k}\max\{E(T,O,\rho),\theta^{-2k}\mu_1\rho\} \le 1/16$$

and

$$E(j_{\#}T_0,O,r) \le c_3(r/\rho)\max\{E(T,O,\rho),\theta^{-2k}\mu_1\rho\}, \qquad 0 < r \le \rho/4$$

where $c_3 = c_3(\zeta,\beta,\mu,m,k,n)$.

The corollary is proved by applying Lemma 1 iteratively. First one proves (inductively) that, with $T_0 = T \llcorner C(O,\rho/2)$, $j_0 = 1_{\mathbf{R}^n}$, and $\mathcal{E}(S,\sigma) = \max\{E(S,O,\sigma),\theta^{-2k}\mu_1\sigma\}$ for any rectifiable k-current S, there are linear isometries j_q of $\mathbf{R}^n$ for $q = 1,2,\ldots$ so that, with $T_q = j_{q\#}T_0$, the following three statements hold:

$$\sup_{X \in sptT_{q-1}\cap C(O,\theta^{q-1}\rho/r)} |\mathbf{q}(X)| \le \theta^{q-1}\varsigma\rho/2, \tag{21*}$$

$$\mathcal{E}(T_q,\theta^q\rho) \le \theta\mathcal{E}(T_{q-1},\theta^{q-1}\rho) \le \theta^q\mathcal{E}(T_0,\rho), \tag{22}$$

$$\|j_q - j_{q-1}\| \le \theta^{-k}\theta^{(q-1)/2}\mathcal{E}(T_0,\rho), \qquad \|j_q - 1_{\mathbf{R}^n}\| \le \varsigma/8. \tag{23*}$$

(These statements differ from those in [SS] only by the presence of the ς.) One then notes that for each $Q, q \ge 0$,

$$\|j_Q - j_q\| \le \sum_{s=q}^{Q-1}\|j_{s+1}-j_s\| \le \theta^{-k}\sum_{s=q}^{\infty}\theta^{s/2}\mathcal{E}(T_0,\rho)^{1/2} \le 2\theta^{(q/2)-k}\mathcal{E}(T_0,\rho)^{1/2}.$$

Setting $j = \lim_{Q\to\infty} j_Q$, we thus have

$$\|j - j_q\|^2 \le 4\theta^{q-2k}\mathcal{E}(T_0,\rho) \le \varsigma^2/16, \qquad q \ge 0. \tag{24*}$$

The corollary now follows as in [SS] by various combinations of (18) and (21*)-(24*). We thus see that ellipticity versus local ellipticity does not enter directly into this proof, except through keeping the slopes small enough that F remains elliptic and Lemma 1 can indeed be applied iteratively.

To finish the proof of the Regularity Theorem, we let $\epsilon = \min\{\theta^{2k}\epsilon_*, \varsigma^{2k}2^{-k}(2c_6)^{-2k}c_2^{-1}, \epsilon_0\}$ instead of $\epsilon = \min\{\theta^{2k}\epsilon_*, 2^{-k}(2c_6)^{-2k}c_2^{-1}, \epsilon_0\}$ as in [SS], where θ and ϵ_* are as in Lemma 1, ϵ_0 and c_6 are as in Lemma 2, c_2 is as in the Corollary to Lemma 1, and ς is the Lipschitz constant needed to make the tangent planes within ζ of $\mathbf{e}^k$.

Then as in [SS], we note that $E(T,y,\rho) < \epsilon_1$ implies $E(T,z,\rho/2) < 2^k\epsilon_1 \le \epsilon_*$, and the corollary to Lemma 1 (with $\rho/2$ in place of ρ) and the definition of ϵ_1 imply that $E(T,z,r) \le (\varsigma/c_{13})^{2k}$ for $r \le \rho/2$, and thus the set A in the proof of Lemma 3 (applied with $\gamma = \varsigma$ rather than 1 as in [SS]) must contain *all* of $B(y,\rho/2)$. Therefore $T \llcorner C(y,\rho/4)$ is the current corresponding to the graph of that Lipschitz function (with Lipschitz constant ς) with multiplicity m. The "furthermore" part of the corollary is then applied exactly as in [SS] to show that in fact that Lipschitz function is $C^{1,1/2}$ on $B(y,\rho/4)$. None of this uses ellipticity except through its use in Lemma 1.

5. Discussion of "ellipticity."

We examine here the various possible characterizations of local ellipticity. We need only consider parametric integrands which satisfy (2) rather than (2*) in this discussion, since F_Υ is convex if and only if F is convex, and F_Υ satisfies (5*) if and only if F does (with the appropriate changes of $(X_0, \boldsymbol{\xi}_0)$).

We consider the following possible conditions:

(C1) F is convex and conditions (5*), (8*), and (9*) hold.

(C1A) Conditions (5*), (8*), and (9*) hold.

(C2) There exists a parametric integrand G satisfying (2) for which (5*), (8*) and (9*) (with G replacing F) hold everywhere, and $F_X(\boldsymbol{\xi}) \geq G_X(\boldsymbol{\xi})$ for all $\boldsymbol{\xi}$ whereas $F = G$ on ζ neighborhoods of both X_0 and $\boldsymbol{\xi}_0$.

(C3) Conditions (4*), (8*), (9*) hold.

(C4) For each $\boldsymbol{\xi}$ within ζ of $\boldsymbol{\xi}_0$, $\mathbf{F}(S) - \mathbf{F}(\mathbf{p}_{\boldsymbol{\xi}\#}S) \geq \mathbf{M}(S) - \mathbf{M}(\mathbf{p}_{\boldsymbol{\xi}\#}S)$ for every recifiable current S such that $\mathbf{p}_{\boldsymbol{\xi}\#}(\partial S) = \partial S$; here $\mathbf{p}_{\boldsymbol{\xi}}$ is the projection of $\mathbf{R}^n$ onto the plane whose direction is $\boldsymbol{\xi}$.

Note: (C1A) and (C3) are precisely the conditions used in the proof of regularity.

We show that it is sufficient to assume any of (C1), (C2), or (C3) to get regularity, and in codimension 1 (i.e. if $n = k+1$) then (C4) is sufficient.

(C1) implies (C1A) trivially.

(C1) implies (C2), since convexity of F_X is equivalent to the convexity of the unit ball of F_X (the set in $\wedge_k \mathbf{R}^n \cong \mathbf{R}^N$ bounded by $\{\boldsymbol{\eta} : F_X(\boldsymbol{\eta}) = 1\}$) and (5*) says that all radii of curvature of the unit ball are positive in a neighborhood of $\mathbf{e}^k$; one need only find a convex set with positive curvatures everywhere containing this unit ball and having the same boundary in a slightly smaller neighborhood of $\boldsymbol{\xi}_0$ and then let G_X be the parametric integrand with this unit ball.

(C2) implies (C1A), trivially.

(C2) implies that (C3) holds everywhere with G replacing F, as stated in [SS] (the proof, incidentally, does require that (5) hold everywhere), and therefore, since $DG_X(\boldsymbol{\xi}) = DF_X(\boldsymbol{\xi})$ for $\boldsymbol{\xi}$ near $\boldsymbol{\xi}_0$ and $G_X(\boldsymbol{\eta}) \leq F_X(\boldsymbol{\eta})$ for all $\boldsymbol{\eta}$, (C3) holds for F.

(C3) implies (C1A), since one can do a Taylor expansion for $F_X(\boldsymbol{\xi} + t\boldsymbol{\eta})$ at $\boldsymbol{\xi}$ (by homogeneity, $\langle DF_X(\boldsymbol{\xi}), \boldsymbol{\xi}\rangle = F_X(\boldsymbol{\xi})$).

Thus any of the conditions (C1), (C2), or (C3) alone is sufficient to imply that the Regularity Theorem conclusions hold. But what is the role of convexity of F? If $\boldsymbol{\xi}_0$ is a direction such that there exist $\boldsymbol{\xi}_1$ and $\boldsymbol{\xi}_2$ with $\boldsymbol{\xi}_0 = \boldsymbol{\xi}_1 + \boldsymbol{\xi}_2$ and $F(\boldsymbol{\xi}_0) > F(\boldsymbol{\xi}_1) + F(\boldsymbol{\xi}_2)$, then $\boldsymbol{\xi}_0$ cannot occur as a tangent plane direction to an $\mathbf{F}$-minimizing current, since a corrugation using many long thin plane segments with directions

$\xi_1/|\xi_1|$ and $\xi_2/|\xi_2|$ would decrease the integral of F, ; therefore the hypotheses of the theorem can never be satisfied and the theorem is true trivially. But if ξ_0 is a direction such that there exist ξ_1 and ξ_2 with $\xi_0 = \xi_1 + \xi_2$ and $F(\xi_0) = F(\xi_1) + F(\xi_2)$, then the situation is unclear: (5*) no longer implies (4*), and yet a minimizing current might well have such a tangent direction.

In [F 5.1.2] it is shown that (C1) holding everywhere implies that (C4) holds everywhere regardless of codimension; its proof works equally well in the localized versions. Similarly, if the codimension is 1 (so that all k-vectors are simple) and if (C4) holds everywhere then (C1) and (C3) hold everywhere, and the proof works equally well in the localized version. (The proof involves building a thin tent-like surface S with directions $(\xi \pm t\eta)/|\xi \pm t\eta|$ and widths $\epsilon|\xi \pm t\eta|$ respectively, plus patches at the ends whose areas are of order ϵ^2, which has its boundary the same as (and which projects onto) a rectangle of width 2ϵ in the plane with direction ξ.) Condition (C4) holding everywhere is the original definition of a parametric integrand being elliptic, [SS] notwithstanding, and is sufficient for regularity ([A] and [F]). The localization of this condition as in (C4) may also be sufficient for the conclusions of the Regularity Theorem to hold here, but it seems that the condition cannot be used only locally in the proof in [F]. In any case, applications tend to be in the case where the codimension is 1.

6. Examples.

Of significance in materials science are surface energies whose values depend on the tangent plane direction to an interface. Such surface energies are naturally parametric integrands F on $\wedge_2 \mathbf{R}^3$. The equilibrium crystal shape W_F for a parametric integrand (the shape of the region which has the least possible value of $\mathbf{F}(\partial W)$ for the volume it contains) is the Wulff shape (see [T2] for a proof)

$$W_F = \{X \in \mathbf{R}^n : X \bullet \nu \leq F(*\nu) \text{ for each unit vector } \nu \in \mathbf{R}^n\};$$

here $*\nu$ is the direction of the plane whose oriented unit normal is ν. Statisical mechanical computations [RW] lead one to expect that for certain materials in certain temperature regimes, F is such that W_F would be as shown in Figure 1.

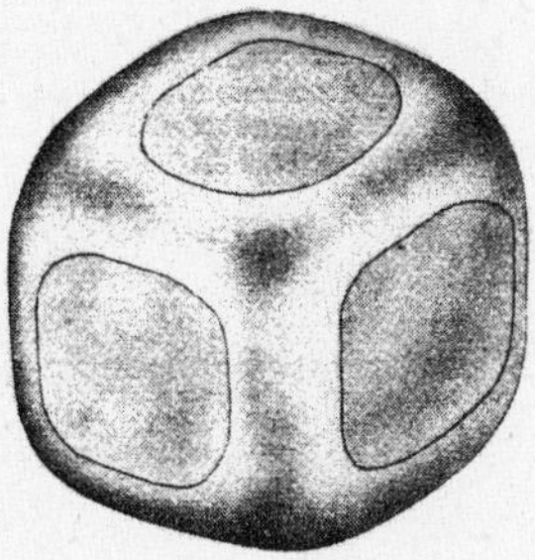

Figure 1

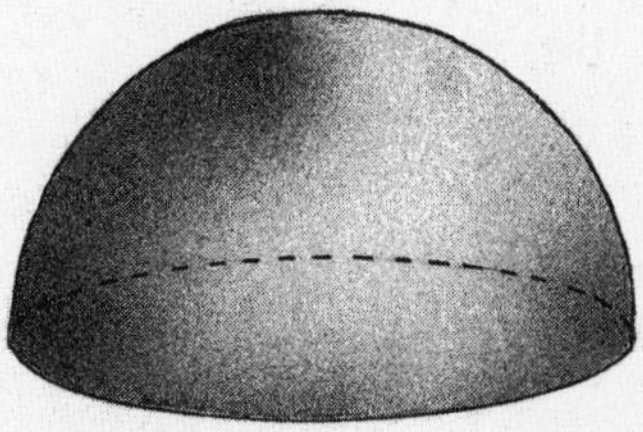

Figure 2

If W_F has a facet, then F cannot be continuously differentiable at the direction of the facet, as can be seen by the following 2-dimensional computation (the proof extends to any k, not just $k = 1$): Suppose the boundary of W_F contains the interval $\{1\} \times [x_2 - \epsilon, x_2 + \epsilon]$. Write $f(\theta)$ for F on the direction with oriented normal $(\cos\theta, \sin\theta)$. Then

$$\lim_{\theta\to 0+} (f(\theta) - f(0))/\theta \geq \lim_{\theta\to 0+} (\cos\theta + (x_2 + \epsilon)\sin\theta - 1)/\theta = x_2 + \epsilon$$

$$\lim_{\theta\to 0^-} (f(\theta) - f(0))/\theta \leq \lim_{\theta\to 0^-} (\cos\theta + (x_2 - \epsilon)\sin\theta - 1)/\theta. = x_2 - \epsilon$$

Therefore local ellipticity is a very relevant concept.

The author puts forward the following conjecture as to the nature of surfaces minimizing the integrals of integrands such as the one producing the Wulff shape of figure 1. It is proposed that (at least away from their boundaries) they consist of plane segments whose directions are those of the facets of the corresponding W_F and which have boundaries that are at least $C^{1,\alpha}$ except for isolated cusp-type singularities as in [TC], together with smoothly curving pieces of surface which meet the planar pieces at the angles that the corresponding directions meet in W_F; the curvature of the intersection C of the smooth surface and a planar piece is related to that of the surface in the direction perpendicular to that boundary via the elliptic PDE which arises from the smooth part of ∂W_F at the corresponding intersection in ∂W_F.

In particular, let F be the unique convex integrand whose Wulff shape is the upper half of the unit ball, as shown in Figure 2 above. Then $F \equiv 1$ for all directions whose normals are in the upper half sphere, and $F = 0$ on $-e_1 \wedge e_2$. If we define Υ to be the constant differential form which sends $\boldsymbol{\xi}$ to $-(1/2)(e_1 \wedge e_2) \bullet \boldsymbol{\xi}$ for each $\boldsymbol{\xi} \in \wedge_2 \mathbf{R}^3$, then $2F_\Upsilon$ satisfies condition (2) as well as conditions (4*), (5*), (7*), (8*), and (9*) with any X_0 and with $\boldsymbol{\xi}_0$ equal to any direction whose unit normal is in the open upper half sphere; the corresponding ζ is the distance from that normal to the equator of the unit sphere. Thus by the codimension 1 regularity proved in this paper, an **F**-minimizing surface S is an ordinary minimal surface in a neighborhood of any point where the surface has its oriented normal in the upper hemisphere. Away from its boundary, S is conjectured to consist of horizontal plane segments, with normals pointing down, together with pieces of minimal surfaces whose normals are in the upper hemisphere, these pieces meeting at right angles along smooth curves (except at isolated points) with the curvature of the intersection curve at each point being of the same magnitude but opposite in direction to the curvature of the minimal surface in the perpendicular direction at that point. The basis for this conjecture is the main theorem of [T1] concerning local smoothness of the singular set of "$(\nu, P) - (\mathrm{M}, \epsilon, \delta)$ minimal sets", together with the proof that cusps can exist in surfaces minimizing a different surface energy in [TC].

An example of a possible such **F**-minimizing surface is shown, viewed from above, in Figure 3a below. One can think of this as being like a free boundary soap film problem – think of the fixed outside boundary as being a wire, and of there being a thin circular glass plate suspended horizontally in the middle (Figure 3b). If the radius of the glass plate were correct, the soap film should have the shape of the smoothly curving pieces of surface, and the contact curves of film with glass should be the four horizontal curves (alternating above and below the plate, with the cusps on the edge of the plate as in [HN]). The region of the horizontal plane bounded by the four curves contributes nothing to the total energy in the soap film problem, and similarly the value of F on its direction $(-e_1 \wedge e_2)$ is zero.

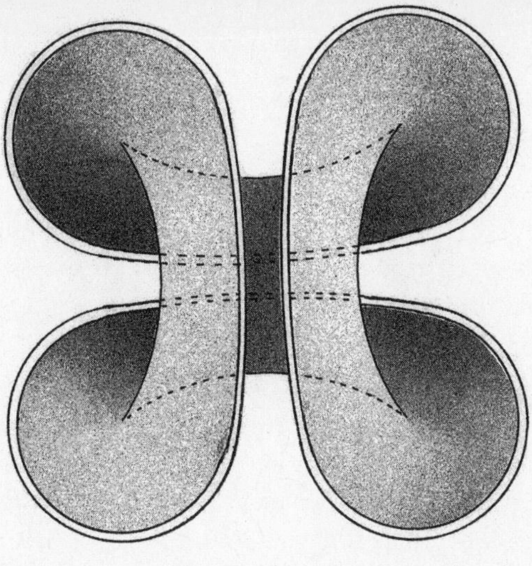

Figure 3a

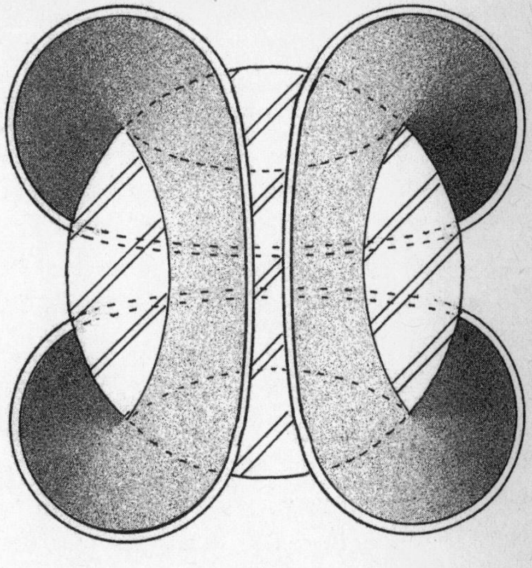

Figure 3b

REFERENCES

[A1] F. J. Almgren, *Existence and regularity almost everywhere of solutions to elliptic variational problems among surfaces of varying topological type and singularity structure,* Ann. of Math. **87** (1968), 321-391.

[A2] F. J. Almgren, *Existence and regularity almost everywhere of solutions to elliptic variational problems with constraints,* Mem. Amer. Math. Soc. **4** No. 165 (1976).

[A3] F. J. Almgren, *Q-valued functions minimizing Dirichlet's integral and the regularity of area minimizing rectifiable currents up to codimension two,* preprint.

[ABL] F. J. Almgren, W. Browder, and E. Lieb, *Co-area, liquid crystals, and minimal surfaces,* this volume.

[B] J. Brothers, Ed., *Some open problems in geometric measure theory and its applications suggested by participants of the 1984 AMS Summer Institute,* in *Geometric Measure Theory and the Calculus of Variations,* Proc. Symp. Pure Math **44** (1986), 441-464.

[F] H. Federer, Geometric Measure Theory, Springer-Verlag, Berlin/Heidelburg/New York, 1969.

[HN] S. Hildebrandt and J. C. C. Nitsche, *A uniqueness theorem for surfaces of least area with partially free boundaries on obstacles,* Arch. Rational Mech. Anal. **79** (1982), 189-218.

[RW] C. Rottman and M. Wortis, *Equilibrium crystal shapes for lattice models with nearest- and next-nearest-neighbor interactions,* Phys. Rev. B **29** (1984), 328-333.

[SS] R. Schoen and L. Simon, *A new proof of the regularity theorem for rectifiable currents which minimize parametric elliptic functionals,* Indiana Univ. Math. J. **31** (1982), 415-434.

[S] L. Simon, Lectures on Geometric Measure Theory, Centre for Mathematical Analysis, Australian National University, Canberra, 1984.

[TC] J. E. Taylor and J. W. Cahn, *A cusp singularity in surfaces that minimize an anisotropic surface energy,* Science **233** (1986), 548-551.

[T1] J. E. Taylor, *Boundary regularity for solutions to various capillarity and free boundary problems,* Comm. in Partial Diff. Eq. **2** (1977), 323-357.

[T2] J. E. Taylor, *Existence and structure of solutions to a class of nonelliptic variational problems,* Symp. Math. XIV (1974), 499-508.

A GEOMETRIC PROOF OF THE MUMFORD COMPACTNESS THEOREM

Friedrich Tomi and A. J. Tromba

Mathematisches Institut
der Universität
Im Neuenheimer Feld 288
D-6900 Heidelberg

Department of Mathematics
University of California
Santa Cruz, California 95064
USA

In [3] Mumford presented a criterion for the compactness of the moduli space $R(M)$ of a closed Riemann surface M. We take $R(M) = \mathcal{M}/\mathcal{D}$ where $\mathcal{M}$ is the space of C^∞ Riemannian metrics on M and $\mathcal{D}$ is the group of orientation preserving diffeomorphisms. This theorem has become quite important in the theory of minimal surfaces [4], [5]. We would like to present our own version and proof which differs substantially from Mumford's. He uses the uniformization theorem which allows one to represent Riemann surfaces as quotients of the upper half plane whereas ours employs very basic geometric notions and works for unorientable surfaces as well. We find our proof more natural inasmuch both the theorem and its proof employ the same language.

Theorem. Let M be a closed connected smooth surface and g^n, $n \in \mathbb{N}$, a sequence of smooth metrics of curvature -1 or 0 on M such that all their closed geodesics are bounded below in length by a fixed positive bound. In the flat case we assume furthermore that the area of (M,g^n) is independent of n. Then there exist smooth diffeomorphisms f^n of M which are orientation preserving if M is oriented, such that a subsequence of $f^n{*}g^n$ converges in C^∞ towards a smooth metric. If M admits a symmetry S which is an isometry for all g^n then the maps f^n can also be chosen to be S-symmetric and to map each half of M to itself.

Since on a negatively curved surface there are no conjugate points along any geodesic it follows that every geodesic arc is locally minimizing (with fixed end points). Therefore, any two geodesic arcs with common end points can't be homotopic with fixed endpoints; otherwise, by a common Morse-theoretic argument [2], there would exist a nonminimizing geodesic arc joining these endpoints. Hence we may conclude that a lower bound ℓ on the lengths ℓ_n of the closed geodesics of g^n implies a bound on the injectivity radius ρ_n of $M^n = (M,g^n)$, $\rho_n \geq \ell/2$.

It follows that on each open disc $B_R(z)$ where $z \in M^n$ and $R \leq \rho$

one can introduce a geodesic polar coordinate system. By a classical result in differential geometry [1] the metric tensor associated with g^n in these coordinates assumes the form

$$(g^n_{ij}) = \begin{pmatrix} 1 & 0 \\ 0 & f(r) \end{pmatrix}, \quad f(r) = \begin{cases} (\sinh r)^2 & \text{if } R(g^n) = -1 \\ r^2 & \text{if } R(g^n) = 0 \end{cases} \tag{1}$$

where r denotes the polar distance.

For the area of $B_R(z)$ we obtain from (1) the simple estimate

$|B_R(z)| \geqq \pi R^2$.

The genus of the manifolds M^n being fixed, the total area of M^n is determined by the Gauss-Bonnet formula if $R(g^n) = -1$. It follows that there is an upper bound, only depending on R , for the number of disjoint open discs $B_R(z)$ in M^n . Let us now take $R = \frac{1}{4}\rho$ and let $N(n)$ be the maximal number of open disjoint disks of radius $\frac{1}{4}R$ in M^n . By passing to a subsequence we can assume that $N(n) = N$ independent of n. If follows that for each $n \in \mathbb{N}$ we can find points $z_i^n \in M^n$, $i=1,\dots,N$, with the property that the discs $B_{\frac{1}{4}R}(z_i^n)$ are disjoint whilst the balls $B_{\frac{1}{2}R}(z_i^n)$ cover M^n . Let us now denote by H the Poincaré upper half plane in the hyperbolic case[1)] and the Euclidean plane in the flat case. We pick an arbitrary point $\zeta_0 \in H$, e.g. $\zeta_0 = i$, the imaginary unit, and introduce geodesic polar coordinates on $B_{4R}(z_i^n) \subset M^n$ and on $B_{4R}(\zeta_0)$ H, respectively. The corresponding metric tensors assume of course the same form (1) in each of both cases and we may therefore conclude that there exist isometries

$$\varphi_i^n : B_{4R}(z_i^n) \longrightarrow B_{4R}(\zeta_0) \ , \ \varphi_i^n(z_i^n) = \zeta_0 \ .$$

Let then I^n denote the set of all pairs (i,j), $1 \leqq i, j \leqq N$, such that

$$B_{2R}(z_i^n) \cap B_{2R}(z_j^n) \neq \emptyset \ .$$

By passing to a subsequence we can assume that $I^n = I$ independent of n. For $(i,j) \in I$ the transition mappings

$$\tau_{ij}^n := \varphi_i^n \circ (\varphi_j^n)^{-1} \ : \ \varphi_j^n(B_{4R}(z_i^n) \cap B_{4R}(z_j^n)) \longrightarrow \varphi_i^n(B_{4R}(z_i^n) \cap B_{4R}(z_j^n))$$

are well defined local isometries of H . Before proceeding further with the proof we first want to show that any such local isometry in fact extends to a global one. We only consider the hyperbolic case, the flat one being trivial.

---

1) This is the half plane $(x,y) \in \mathbb{R}^2$ with $y \geqq 0$ endowed with the metric $\{(dx)^2 + (dy)^2\}/4y^2$. The curvature of this metric is $\equiv -1$.

Lemma 1. Let $f: U \longrightarrow H$ be a C^1 isometry on an open connected subset U of the hyperbolic plane. Then

$$f(z) = \frac{Az + B}{Cz + D}, \quad A, B, C, D \text{ are real },$$

and $AD - BC = 1$.

Proof. The class of maps $z \longrightarrow \frac{Az + B}{Cz + D}$, $AD - BC = 1$ and all real are the group of isometries of the Poincaré metric. Thus we must show that a local isometry is also a global isometry.

It is clear that we can take f to be orientation preserving. Then an easy calculation shows that f must be holomorphic and satisfy the non-linear condition

$$|f'| = \frac{\operatorname{Im} f}{\operatorname{Im} z} . \tag{2}$$

One can check that every map of the form $z \longrightarrow \frac{Az + B}{Cz + D}$ as above satisfies the condition and that the set of maps satisfying (2) from a fixed domain to itself forms a group. Therefore, by composition with an appropriate element of the three dimensional conformal group of H we may assume that f satisfies the additional conditions: f is defined in a neighborhood of $i \in H$, and

$$f'(i) = \operatorname{Im} f(i) . \tag{3}$$

Now writing $z = x + iy$ we have

$$(\log f')' = \{\operatorname{Re}(\log f')\}_z = \{\log|f'|\}_z = \frac{(\operatorname{Im} f)_z}{\operatorname{Im} f} - \frac{y_z}{y} = \frac{-if'}{\operatorname{Im} f} + \frac{i}{y} .$$

By (3), $(\log f')'(i) = 0$. Similarly

$$(\log f')'' = -\left(\frac{if'}{\operatorname{Im} f}\right)_z + \left(\frac{i}{y}\right)_z = \frac{-if''}{\operatorname{Im} f} - \frac{if'}{(\operatorname{Im} f)^2}(if') + \frac{i^2}{y^2} .$$

And again we see that $(\log f')''(i) = 0$. Proceeding inductively we see that

$$(\log f')^n(i) = 0$$

for all n . Thus since $\log f'$ is holomorphic in a neighborhood of i, it follows that $\log f'$ is constant, and so $z \longrightarrow f'(z)$ is constant. Therefore $f(z) = z$. Since we normalized f by the isometry group of H this proves that our initial f must be in this isometry group, and this completes the proof of the lemma. For $(i,j) \in I$ we have $z_j^n \in B_{4R}(z_i^n)$, and hence $y_{ij}^n := \varphi_i(z_j^n) \in B_{4R}(\zeta_0)$, since φ_i^n is an isometry. It is obvious from the definition that

$$y_{ij}^n = \tau_{ij}^n(y_{ji}^n) . \tag{4}$$

We are now going to construct a limit manifold of the sequence

$M^n = (M, g^n)$. For this purpose we prove

Lemma 2. The family of transition mappings $(\tau_{ij}^n)_{n \in \mathbb{N}}$ is compact for each $(i,j) \in I$.

Proof. By Lemma 1 each τ_{ij}^n is a global isometry of H and there are fixed compact subset K of H and points $y_{ij}^n \in K$ such that (4) holds. From this the assertion follows at once in the flat case, since then τ_{ij}^n decomposes into a rotation and a bounded translation. In the hyperbolic case, by composition with a conformal map of H onto the unit disc $\mathcal{B} \subset \mathbb{R}^2$ we may assume that each τ_{ij}^n is a conformal map of $\mathcal{B}$ onto itself and (suppressing the indices i,j) that there are points p^n strictly staying away from $\partial\mathcal{B}$ such that also $\tau^n(p^n)$ stays away from $\partial\mathcal{B}$. Each τ^n is of the form

$$\tau^n(z) = d_n \frac{z-a_n}{1-\bar{a}_n z} ,$$

where $|a_n| < 1$, $|d_n| = 1$. It suffices to show that $|a_n|$ stays strictly below 1. If not we can assume $a_n \longrightarrow a$, $|a| = 1$ and that $d_n \longrightarrow d$, $|d| = 1$. The "limit map"

$$\tau(z) = d \frac{z-a}{1-\bar{a}z} = ad \left\{ \frac{az-1}{1-\bar{a}z} \right\} = -ad$$

collapses the disc onto a point on $\partial\mathcal{B}$, a contradiction.

We can now continue with the proof of Mumford's theorem. Passing to a subsequence we can by Lemma 3.2 assume that

$$\tau_{ij}^n \longrightarrow \tau_{ij} \, (n \to \infty) . \tag{5}$$

We now define a limiting manifold $\hat{M}$ as the disjoint union of N discs $B_R(\zeta_0) \subset H$, labelled as $B_1, \dots, B_N$ with the identifications

$$x \in B_i \text{ equals } y \in B_j \iff (i,j) \in I \quad \text{and} \quad x = \tau_{ij}(y) .$$

It is clear that $\hat{M}$ is a differentiable manifold carrying a natural Riemannian metric which on each B_i coincides with the Poincaré or Euclidean metric, respectively. We claim that $\hat{M}$ is compact. Assume to the contrary that there is a point $y \in \partial B_R(\zeta_0)$ such that $y \notin \tau_{ij}(B_R(\zeta_0))$ for some i and all j with $(i,j) \in I$. Then it follows that, for sufficiently large n, $y \notin \tau_{ij}^n(B_{\frac{3}{4}R}(\zeta_0))$ which means that $(\varphi_i^n)^{-1}(y) \notin B_{\frac{3}{4}R}(z_j^n)$. This, however, implies that

$$B_{\frac{1}{4}R}((\varphi_i^n)^{-1}(y)) \cap B_{\frac{1}{4}R}(z_k^n) = \emptyset \quad \text{for} \quad k = 1,\dots,N$$

contradicting the choice of N as the maximal number of disjoint discs in M^n of radius 1/4 R. The remainder of the proof rests upon the following

Lemma 3. There are diffeomorphisms $f^n : \hat{M} \longrightarrow M^n$, $f^n(B_i) \subset B_{2R}(z_i^n)$ such that

$$\varphi_i^n \circ f^n \longrightarrow \mathrm{id} \quad (n \to \infty) \quad \text{in } C^\infty \text{ on each } B_i . \tag{6}$$

Let us quickly finish the proof of Mumford's theorem assuming the lemma. Denoting by g the Poincaré metric or Euclidean metric, respectively, we have from (6) that

$$f^{n*}\varphi_i^{n*}g \longrightarrow g \qquad (n \to \infty)$$

on each B_i. Since, however, φ_i^n was an isometry between g and g^n on M^n this means that

$$f^{n*}g^n \longrightarrow g \qquad (n \to \infty)$$

on $\hat{M}$. Choosing now any diffeomorphism $f : M \longrightarrow \hat{M}$, we obtain

$$(f^n \circ f)^* g^n \longrightarrow f^* g \quad (n \to \infty),$$

which proves Mumford's theorem.

We now come to the proof of Lemma 3.

For the proof let us consider the manifold M^n as the disjoint union of N balls $B_{2R}(\zeta_0) \subset H$ labelled as $B_1',\dots,B_N'$ with the identifications

$$x \in B_i' \quad \text{equals} \quad y \in B_j' \quad \text{iff}$$
$$(i,j) \in I \quad \text{and} \quad x = \tau_{ij}^n(y) .$$

We denote this model of M^n by $\hat{M}^n$. It then suffices to show that there are diffeomorphisms $f^n : \hat{M} \longrightarrow \hat{M}^n$, $f^n(B_i) \subset B_i'$, such that $f_n \longrightarrow \mathrm{id}$ (as $n \to \infty$) on each B_i. We shall do this by a Morse theoretic argument.

Since $\hat{M}$ is an oriented closed surface there is a C^∞ Morse function $\Psi : \hat{M} \longrightarrow \mathbb{R}$ with distinct critical values $c_0 > c_1 > \dots > c_m$ and such that the level sets $\Psi^{-1}(c_j)$ contain only one non-degenerate critical point w_j.

We may use a partition of unity to construct a sequence of functions $\Psi^n : M^n \to \mathbb{R}$ such that on each B_i, $\Psi^n \to \Psi$ in C^∞. To see this let φ_i be the natural coordinate charts on $\hat{M}$ induced by the inclusions of B_i into H, so that $\varphi_i \circ \varphi_j^{-1} = \tau_{ij}$. Furthermore, let $\{\eta_i\}$ be a partition of unity on $\hat{M}$ with respect to the coordinate cover $\{B_i\}$. Define $\Psi^n : M^n \to \mathbb{R}$ by

$$\Psi^n(p) = \sum_j \eta_j(\varphi_j^{-1}\varphi_j^n(p))\Psi(\varphi_j^{-1}\varphi_j^n(p)) .$$

If $p = (\varphi_k^n)^{-1}(u)$ then

$$\Psi^n(p) = \sum_j \eta_j(\varphi_j^{-1}\tau_{ij}^n(u))\Psi(\varphi_j^{-1}\tau_{jk}^n(u)) .$$

As $n \to \infty$ this converges C^∞ to

$$\sum_j \eta_j(\varphi_j^{-1}\tau_{jk}(u))\Psi(\varphi_j^{-1}\tau_{jk}(u)) = \sum_j \eta_j(\varphi_k^{-1}(u)\Psi(\varphi_k^{-1}(u)) = \Psi(\varphi_k^{-1}(u)) .$$

which (after viewing Ψ^n as defined on $\hat{M}^n$) proves the result.

Consequently, for large n, Ψ^n has non-degenerate critical points $\{w_j^n\}$ "near" the $\{w_j\}$ on $\bigcup_{\ell=1}^N B_\ell$. By trivial modifications of Ψ^n we may further assume that Ψ^n has the same critical values $c_0,\ldots,c_m$ as does Ψ and that $w_j^n = w_j$ for all j.

Furthermore we can assume that in a small disc about each w_j {in the B_i's} Ψ_n and Ψ actually agree.

Let $\{\hat{M}^n\}^a = \{x \mid \Psi^n(x) \leq a\}$ and $\{\hat{M}^n\}_a = \{x \mid \Psi^n(x) \geq a\}$ with $\{\hat{M}\}^a, \{\hat{M}\}_a$ defined similarly in terms of Ψ. Let $\varepsilon > 0$ be small enough so that $\{\hat{M}^n\}_{c_0-2\varepsilon}$ and $\{\hat{M}\}_{c_0-2\varepsilon}$ contain only w_0 as its only critical point and $\{\hat{M}^n\}_{c_1-2\varepsilon}$ and $\{\hat{M}\}_{c_1-2\varepsilon}$ contain only w_0 and w_1.

Let G be a fixed metric on $\hat{M}$ which agrees with the Euclidean metric on a neighborhood of the $\{w_j\}$. As in contructing the Ψ^n we can easily find a sequence of metrics G_n on $\hat{M}^n$ so that G_n agrees with G on a neighborhood of the $\{w_j\}$ (in $\cup B_i$) and $G_n \to G$ as $n \to \infty$. Let $\nabla\Psi^n$ and $\nabla\Psi$ denote the gradients of Ψ^n and Ψ with respect to these metrics, and X^n and X the normalized fields $\frac{\nabla\Psi^n}{\|\nabla\Psi^n\|}$ and $\frac{\nabla\Psi}{\|\nabla\Psi\|}$, $\| \ \|$ denoting the norms w.r.t. G_n and G. Of course X^n and X are defined only on $\hat{M}^n - \cup w_j$ and $M - \cup w_j$ respectively. We shall define a mapping $f_n : \{\hat{M}_n\}_{c_1-2\varepsilon} \longrightarrow \hat{M}$ which is a diffeomorphism of a neighborhood of $\{\hat{M}_n\}_{c_1-\varepsilon}$ to a neighborhood of $\{\hat{M}\}_{c_1-\varepsilon}$.

Let D_0 be a "small disc" about w_0 for which the Morse lemma holds for Ψ^n and Ψ about w_0. Thus there exists a map Q from a neighborhood of 0 in $\mathbb{R}^2$ to a neighborhood of w_0 so that $\Psi_0^n Q(z) = c_0 - z_1^2 - z_2^2 = \Psi \circ Q(z)$. Thus we may take $D_0 = \{x \mid \Psi_n(x) \geq c_0 - \varepsilon\}$. Let $p \in \partial D_0$, and let $\sigma_p^n(t)$ and $\sigma_p(t)$ be the flows of the vector fields X^n and X respectively with $\sigma_p^n(0) = p = \sigma_p(0)$. It follows immediately that $\Psi^n(\sigma_p^n(t)) = \varepsilon + t = \Psi(\sigma_p(t))$. From Morse theory it follows that as t decreases the flow $\sigma_p^n(t)$ [resp. $\sigma_p(t)$] either converges to w_1 or drops into $\{\hat{M}^n\}^{c_1-2\varepsilon}$ [resp. $\{\hat{M}\}^{c_1-2\varepsilon}$]. Let U be the unstable manifold of w_1 for the flow of X^n. Then it follows that every $q \in \{\hat{M}^n\}_{c_1-2\varepsilon} \setminus \{U \cup w_0\}$ can be written as $\sigma_p(t)$ for some $t \in \mathbb{R}$ and $p \in \partial D_0$. Define the map

$$\tilde{f}^n : \{\hat{M}^n\}_{c_1-2\varepsilon} \setminus U \longrightarrow \hat{M}$$

by $\tilde{f}^n(\sigma_p^n(t)) = \sigma_p(t)$, $p \in \partial D_0$ and $\tilde{f}^n(w_0) = w_0$. Since X^n and X agree on D_0 (in some coordinate system) it follows that f_n is the identity in a neighborhood of w_0, with respect to the above coordinate system, and is thus smooth everywhere it is defined. It also follows from our construction that

$$\Psi^n(w) = \Psi(\tilde{f}^n(w))$$

and so $\tilde{f}^n$ takes level sets to level sets and also that $\tilde{f}^n \longrightarrow \mathrm{id}$ as $n \to \infty$ (on $\cup B_j$).

Now let us assume that we are in a coordinate neighborhood W_1 of w_1 where Morse's lemma holds for Ψ^n and Ψ and where $\Psi^n \equiv \Psi$. The situation is as depicted in figure 1.

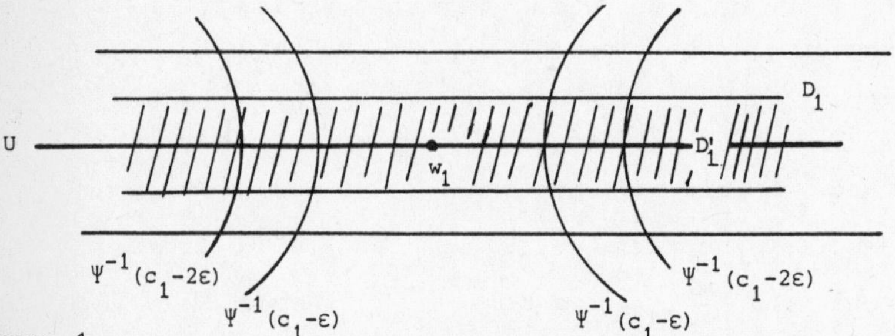

Figure 1.

Let D_1' and D_1, $\overline{D}_1' \subset D_1$ be two strips as in figure 1. Let η be a C^∞ function 1 on $W_1 - D_1$ and 0 on D_1'. Define a new map $f^n : \{\hat{M}^n\}_{c_1-2\varepsilon} \to \hat{M}$ by

$$f^n = \eta\tilde{f}_n + (1-\eta)\,\mathrm{id}$$

It is clear that for sufficiently large n, f_n is a diffeomorphism. Taking now the initial values of our trajectories to lie on $(\Psi^n)^{-1}(c_1-2\varepsilon)$ and $\Psi^{-1}(c_1-2\varepsilon)$ we can proceed inductively to extend f_n to a diffeomorphism of $\hat{M}^n$ onto $\hat{M}$. This completes the proof of Theorem 1.

REFERENCES

[1] W. BLASCHKE, Vorlesungen über Differentialgeometrie I.

[2] J. MILNOR, "Morse Theory", Annals of Math. Studies 51 (1963).

[3] D. MUMFORD, A remark on Mahler's compactness theorem, Proc. AMS 28 (1971), 289-294.

[4] F. TOMI and A. J. TROMBA, On Plateau's problem for minimal surfaces of higher genus in $\mathbb{R}^3$. Bull. AMS 13 (1985), 169-171.

[5] F. TOMI and A. J. TROMBA, Existence theorems for minimal surfaces of non-zero genus spanning a given contour in $\mathbb{R}^3$ (to appear).

HARNACK INEQUALITIES FOR FUNCTIONS IN DE GIORGI PARABOLIC CLASS

Wang, Guanglie

Department of Mathematics,
Jilin University, China.

§1. INTRODUCTION

Following the fundamental work of De Giorgi [1], Ladyzhenskaya and Ural'tseva derived that the solutions of the (elliptic, parabolic) equations in divergence form belong to certain classes of functions (which will be called "De Giorgi classes" as in [2][3]) and proved that the functions in these classes are Hölder continuous (cf., e.g., [4][5]).

Through a different approach J. Moser [6][7][8] established Harnack inequalities for these solutions and hence they are automatically Hölder continuous.

Then it is natural to ask: Do the functions in the De Giorgi classes satisfy Harnack inequalities? (cf. [3]).

Di Benedetto and Trudinger [2] gave a positive answer to the above question for the elliptic case. In the present paper we discuss the parabolic case and prove that the functions in the De Giorgi parabolic class satisfy Harnack inequalities, and thus improve the results of Ladyzhenskaya, Ural'tseva and Moser mentioned above.

For simplicity, in the present paper we only discuss the De Giorgi parabolic class in a sort of "homogeneous" case which includes the solutions of the equations studied in [7] [8] (cf. [4]), but the methods used here apply to the "general" case which may be defined as the intersection of the classes $\mathfrak{A}$ and $\mathfrak{B}$ in [4] and hence can deal with those solutions treated in [9].

Let Ω be an open set in R^n, $T>0$, $Q_T=\Omega\times(0,T)$. The space $V_2^{1,0}(Q_T)=C([0,T],L_2(\Omega))\cap L_2((0,T),W_2^1(\Omega))$, the norm of a function $u(x,t)$ in $V_2^{1,0}(Q_T)$ is defined by

$$|u|_{Q_T} = \max_{0\leq t\leq T}\|u(x,t)\|_{2,\Omega}+\|D_x u\|_{2,Q_T},$$

where $\|\cdot\|_{2,\Omega}$ and $\|\cdot\|_{2,Q_T}$ denote the L_2 norms over Ω and Q_T respectively, $D_x u = (u_{x_1}, \ldots, u_{x_n})$.

<u>Definition 1.1</u> A function $u(x,t)$ is said to belong to the De Giorgi parabolic class $DG(Q_T,N)$, $N>0$, if $u(x,t)\in V_2^{1,0}(Q_T)$ and for any $k\in R$, the function $w(x,t)=\pm u(x,t)$ satisfies the following inequalities

$$|w^{(k)}|^2_{\hat{Q}(\rho-\sigma_1\rho,\tau-\sigma_2\tau)}$$
$$\leq N[(\sigma_1\rho)^{-2} + (\sigma_2\tau)^{-1}] \, \| w^{(k)} \|^2_{2,\hat{Q}(\rho,\tau)}, \tag{1.1}$$

$$\max_{t_0 \leq t \leq t_0+\tau} \| w^{(k)}(x,t) \|^2_{2,B_{\rho-\sigma_1\rho}}$$
$$\leq \| w^{(k)}(x,t_0) \|^2_{2,B_\rho} + N(\sigma_1\rho)^{-2} \| w^{(k)} \|^2_{2,Q(\rho,\tau)}, \tag{1.2}$$

$$| w^{(k)}|^2_{Q(\rho-\sigma_1\rho,\tau-\sigma_2\tau)}$$
$$\leq N[(\sigma_1\rho)^{-2} + (\sigma_2\tau)^{-1}] \, \| w^{(k)} \|^2_{2,Q(\rho,\tau)}. \tag{1.3}$$

where $w^{(k)}(x,t) = [w(x,t)-k]^+ = \max [w(x,t)-k,0]$, ρ and τ are arbitrary positive numbers, $\hat{Q}(\rho,\tau)$ and $Q(\rho,\tau)$ are arbitrary cylinders belonging to Q_T, σ_1 and σ_2 are arbitrary numbers from $(0,1)$, and $\hat{Q}(\rho-\sigma_1\rho,\tau-\sigma_2\tau)(Q(\rho-\sigma_1\rho,\tau-\sigma_2\tau)$ resp.,) are coaxial with and have a common vertex with $\hat{Q}(\rho,\tau)(Q(\rho,\tau)$ resp.,) of the form

$$\hat{Q}(r,s) \triangleq \hat{Q}(x_0,t_0,r,s) = B_r(x_0) \times (t_0-s,t_0),$$
$$(Q(r,s) \triangleq Q(x_0,t_0,r,s) = B_r(x_0) \times (t_0,t_0+s) \text{ resp.,})$$

and

$$B_r \triangleq B_r(x_0) = \{x=(x^1, \ldots, x^n); \, [\sum_{i=1}^n (x^i-x^i_o)^2]^{\frac{1}{2}}<r\}.$$

The Harnack type inequalities that we will extablish in this paper are the following

<u>Theorem 1.2</u> Let $u(x,t) \in DG(Q_T,N)$, θ, $R>0$, $B_R(\bar{x}) \times (\bar{t},\bar{t}+\theta R^2) \subset Q_T$, $\sigma \in (0,])$. Then for any $p>0$ there exists a constant $C>0$ such that

$$\sup_{B_{\sigma R}(\bar{x})\times(\bar{t}+(1-\sigma^2)\theta R^2,\bar{t}+\theta R^2)} u(x,t) \geq C\{ \int_{B_R(\bar{x})\times(\bar{t},\bar{t}+\theta R^2)} [(u)^+]^p \, dxdt\}^{1/p}$$

where C depends only on p,n,N, σ and θ.

We always set

$$\fint_S v \, dxdt = \frac{1}{|S|} \int_S v \, dxdt$$

and $|S|$ denotes the $(n+1)$ dimensional Lebesgue measure of S.

<u>Theorem 1.3</u> Let $u(x,t) \in DG(Q_T,N)$, $u(x,t) \geq 0, \theta, R>0, B_R(\bar{x}) \times (\bar{t},\bar{t}+\theta R^2) \subset Q_T$, then for any σ_1, $\sigma_2 \in (0,1)$, $0<\theta_1<\theta_2<\theta$, there exist positive constans p and C such that

$$\inf_{B_{\sigma_1 R}(\bar{x})\times(\bar{t}+\theta_2 R^2,\bar{t}+\theta R^2)} u(x,t) \geq C \, [\fint_{B_{\sigma_2 R}(\bar{x})\times(\bar{t},\bar{t}+\theta_1 R^2)} (u)^p \, dxdt \,]^{1/p},$$

where p and C depend only on n, N, θ_1, θ_2, θ, σ_1 and σ_2.

Combining Theorem 1.2 and Theorem 1.3 we have the following full Harnack inequality

Theorem 1.4 Let $u(x,t) \in DG(Q_T,N)$, $u(x,t) \geq 0$, θ, $R>0$, $B_R(\bar{x})x(\bar{t},\bar{t}+\theta R^2) \subset Q_T$, then for any σ_1, $\sigma_2 \in (0,1)$, $0<\theta_1<\theta_2<\theta$, there exists a constant $C>0$ such that

$$\sup_{B_{\sigma_1 R}(\bar{x})x(\bar{t},\bar{t}+\theta_1 R^2)} u(x,t) \leq C \inf_{B_{\sigma_2 R}(\bar{x})x(\bar{t}+\theta_2 R^2,\bar{t}+\theta R^2)} u(x,t)$$

where C depends only on n, N, θ_1, θ_2, θ, σ_1 and σ_2 .

As pointed out by Di Benedetto and Trudinger the main tools used in [2] are the De Giorgi iteration technique and the measure argument in obtaining the Harnack inequalities for the solutions of elliptic equations in non-divergence form in [10]. It is well known that the De Giorgi iteration technique is applied to the parabolic case and the results by employing the measure argument in obtaining the Harnack inequalities for the solutions of parabolic equations in non-divergence form are also established (cf. [11], [12]). But there are still some essential difficulties for extending the Harnack inequalities from elliptic case to parabolic one as can be seen from [6], [7] and [8]. In fact, in order to do the measure argument, the method of comparison function was used to get the "diffusion lemma" (i.e. Lemma 1.3 in [11], Lemma 3.1 in [12]) in the case of equations in non-divergence form, but in the case of equations in divergence form the method of comparison function does not apply, moreover there do not appear any equations at all in the De Giorgi class.

In this paper we combine the De Giorgi iteration technique with a method of "double enlargement" to obtain the "diffusion lemma" in §3, §2 is mainly devoted to the proof of Theorem 1.2, Theorem 1.3 is derived in §4, the measure lemma (Lemma4.3) used there appeared originally in [13], for the sake of completeness we repeat the proof of the lemma with a slight modification in §5.

§2. THE MAXIMUM PRINCIPLE FOR FUNCTIONS IN $DG(Q_T,N)$

Lemma2.1 Assume $u(x,t) \in V_2^{1,0}(Q_T)$. $\theta, R_o>0$, $\hat{Q}(R_o,\theta R_o^2) \triangleq \hat{Q}(x_o,t_o,R_o,\theta R_o^2) \subset Q_T$, If the inequality (1.1) is valid for $w(x,t)=u(x,t)$, then for any $k_o \in R$, $\mu \in (0,1)$, there exists a positive constant C such that

$$\sup_{\hat{Q}(\mu R_o,\theta\mu^2 R_o^2)} u(x,t) \leq k_o$$

$$+ C\{\frac{1}{(1-\mu)^{n+2}} \fint_{\hat{Q}(R_o,\theta R_o^2)} [(u-k_o)^+]^2 dxdt\}^{1/2}\{\frac{\hat{Q}^+(k_o,R_o,\theta R_o^2)}{(1-\mu)^{n+2}|\hat{Q}(R_o,\theta R_o^2)|}\}^{\alpha/2} \tag{2.1}$$

where α is the positive solution of the equation $\alpha(1+\alpha)=2/(n+2)$, $\hat{Q}^+(k_o,R_o,\theta R_o^2)=$

$=\{(x,t)\in\hat{Q}(R_o,\theta R_o^2), u(x,t)>k_o\}$, the costant C depends only on n, N, and θ.

Lemma 2.2 Assume $u(x,t)\in V_2^{1,0}(Q_T)$, θ, $R_o>0$, $\hat{Q}(R_o,\theta R_o^2)\subset Q_T$. If the inequality (1.2) is valid for $w(x,t)=-u(x,t)$, then for any $k_o\in\mathbb{R}$, $\mu\in(0,1)$, there exists a positive constant C such that

$$\inf_{\hat{Q}(\mu R_o,\theta\mu^2R_o^2)} u(x,t) \geq k_o - C\left\{\frac{1}{(1-\mu)^{n+2}} ⨍_{\hat{Q}(R_o,\theta R_o^2)}[(k_o-u)^+]^2dxdt\right\}^{\frac{1}{2}}\left\{\frac{|\hat{Q}^-(k_o,R_o,\theta R_o^2)|}{(1-\mu)^{n+2}|\hat{Q}(R_o,\theta R_o^2)|}\right\}^{\alpha/2} \qquad (2.2)$$

where $\hat{Q}^-(k_o,R_o,\theta R_o^2)=\{(x,t)\in\hat{Q}(R_o,R_o^2\theta), u(x,t)<k_o\}$, α and C are of the same properties as the ones in Lemma 2.1.

Proof of Lemma 2.1 Let $R>r$, $\tau_R>\tau_r$, set $\bar{r}=\frac{1}{2}(R+r)$, and

$$a(k,r) = \int_{t_o-\tau_r}^{t_o} \text{mes } B_{k,r}^+(t)dt,$$

$$b(k,r) = \int_{t_o-\tau_r}^{t_o}\int_{B_{k,r}^+(t)} (u-k)^2dxdt,$$

where and in the sequel mes denotes the n-dimensional Lebesgue measure, and

$$B_{k,r}^+(t) = \{x\in B_r^+;\ u(x,t)>k\}.$$

Obviously

$$a(k,r) \leq \int_{t_o-\tau_r}^{t_o} \text{mes } B_{k,\bar{r}}^+(t)\ dt \leq a(k,R),$$

$$b(k,r) \leq \int_{t_o-\tau_r}^{t_o}\int_{B_{k,\bar{r}}^+(t)} (u-k)^2dxdt \leq b(k,R).$$

Choose $\zeta(x)\in C_o^\infty(B_{\bar{r}})$, $0\leq\zeta(x)\leq 1$, $\zeta(x)=1$ on B_r, $|D_x\zeta|\leq\frac{8}{R-r}$, then for any $h>k$ we have

$$b(h,r) \leq b(k,r) \leq \int_{t_o-\tau_r}^{t_o}\int_{B_{k,\bar{r}}^+(t)} (u-k)^2\zeta^2dxdt, \qquad (2.3)$$

using the inequality (3.7) of Ch.II in [4] to estimate the right-hand side of (2.3), we have

$$\int_{t_o-\tau_r}^{t_o}\int_{B_{k,\bar{r}}^+(t)} (u-k)^2\zeta^2dxdt \leq \beta^2[a(k,R)]^{\frac{2}{n+2}}|u^{(k)}\zeta|^2_{\hat{Q}(\bar{r},\tau_r)}$$

$$\leq \beta^2[a(k,R)]^{\frac{2}{n+2}}\{\max_{t_o-\tau_r\leq t\leq t_o}\int_{B^+_{k,\bar{r}}(t)}(u-k)^2\zeta^2dx +$$

$$+ 2\int_{t_o-\tau_r}^{t_o}\int_{B^+_{k,\bar{r}}(t)}[|D_xu|^2\zeta^2+(u-k)^2|D_x\zeta|^2]dxdt\}$$

$$\leq 2\beta^2[a(k,R)]^{\frac{2}{n+2}}\{\left|u^{(k)}\right|^2_{\hat{Q}(\bar{r},\tau_r)} + \frac{8}{(R-r)^2}b(k,R)\}, \tag{2.4}$$

where β is a constant depending only on n. In order to use (1.1) to estimate the first term in parenthesis on the right-hand side of the above inequality, in (1.1) we take $\rho-\sigma_1\rho=\bar{r}$, $\tau-\sigma_2\tau=\tau_r$, and $\rho=R$, $\tau=\tau_R$, hence $\sigma_1\rho=R-\bar{r}=\frac{1}{2}(R-r)$, $\sigma_2\tau=\tau_R-\tau_r$. And then choosing $\tau_R=\theta R^2$, $\tau_r=\theta r^2$, we thus have

$$\sigma_2\tau = \theta(R^2-r^2) \geq \theta(R-r)^2,$$

hence

$$|u^{(k)}|^2_{\hat{Q}(\bar{r},\tau_r)} \leq N[\frac{4}{(R-r)^2} + \frac{1}{\theta(R-r)^2}] \left\|u^{(k)}\right\|^2_{2,\hat{Q}(R,\tau_R)}$$

$$= \frac{N(4+1/\theta)}{(R-r)^2} b(k,R).$$

Combining this inequality with (2,3) and (2.4) we obtain

$$b(h,r)\leq\frac{2\beta^2[N(4+1/\theta)+8]}{(R-r)^2}[a(k,R)]^{\frac{2}{n+2}}b(k,R),\ \forall\ h>k, \tag{2.5}$$

On the other hand,

$$b(k,R) \geq b(k,r) \geq (h-k)^2a(h,r). \qquad \forall\ h>k,$$

hence for $\alpha>0$ (to be determined later) we have

$$[a(h,r)]^\alpha \leq \frac{1}{(h-k)^{2\alpha}}b^\alpha(k,R), \qquad \forall\ h>k, \tag{2.6}$$

Multiplying (2.5) by (2.6) we have

$$[a(h,r)]^\alpha b(h,r) \leq \frac{C}{(R-r)^2(h-k)^{2\alpha}}[a(k,R)]^{\frac{2}{n+2}}[b(k,R)]^{1+\alpha},$$

where the constant C depends only on n,N and $1/\theta$. Therefore if we choose α to be the positive solution of the equation $(1+\alpha)\alpha=2/(n+2)$, and denote

$$Y(h,r) = [a(h,r)]^\alpha b(h,r),$$

then the previous inequality becomes

$$Y(h,r) \leq \frac{C}{(R-r)^2(h-k)^{2\alpha}}[Y(k,R)]^{1+\alpha} \tag{2.7}$$

For given R_o, k_o and μ, taking $d>0$ (to be determined later), setting $R_j=R_o\mu+$ $+(1-\mu)2^{-j}R_o$, $\tau_j=\theta R_j^2$, $k_j=k_o+d-d2^{-j}$, and denoting $Y_j=Y(k_j,R_j)$, then from (2.7) we have

$$Y_j \leq \frac{C2^{2j(1+\alpha)}}{[R_o(1-\mu)d^\alpha]^2}Y_{j-1}^{1+\alpha},$$

if we choose

$$d = \frac{1}{[R_o(1-\mu)]^{1/\alpha}}\, 2^p C^{1/2\alpha} Y_o^{\frac{1}{2}},$$

where p is a fixed constant depending only on n, then by induction it is easy to prove that $Y(k_o+d,R_o\mu)=0$, hence we obtain

$$u(x,t) \leq k_o+d \qquad \forall (x,t) \in \hat{Q}(\mu R_o,\theta\mu^2R_o^2).$$

Consequently,

$$\sup_{\hat{Q}(\mu R_o,\theta\mu^2R_o^2)} u(x,t) \leq k_o$$

$$+\frac{C}{[R_o(1-\mu)]^{1/\alpha}}\{\int_{\hat{Q}(R_o,\theta R_o^2)}[(u-k_o)^+]^2dxdt\}^{\frac{1}{2}}|\hat{Q}^+(k_o,R_o,\theta R_o^2)|^{\alpha/2}$$

Thus (2.1) follows from the above inequality by the definition of α.

The proof of Lemma 2.2 is similar.

<u>Proof of Theorem 1.2</u> Denote $t_1=\bar{t}+\theta R^2$, $r=\sigma R$,

$$Q_R = \hat{Q}(\bar{x},t_1,R,\theta R^2),$$

$$Q_r = \hat{Q}(\bar{x},t_1,r,\theta r^2).$$

Then it suffices to prove

$$\sup_{Q_r} u \leq C[\fint_{Q_R}(u^+)^p dxdt]^{1/p} \tag{2.8}$$

Since $\sigma\in(0,1)$, there exists a finite number of cylinders of the form

$$\hat{Q}_{(1-\sigma)R,\theta} = B_{(1-\sigma)R}(x_o)\times(t_o-\theta((1-\sigma)R)^2,\ t_o),$$

such that $\hat{Q}_{(1-\sigma)R,\theta}\subset Q_R$, $\{\ \hat{Q}_{\frac{1}{2}(1-\sigma)R,\theta}\}\supset Q_r$

Therefore, by (2.1) with $\mu=\frac{1}{2}$, $k_o=0$ applied over $\hat{Q}_{\frac{1}{2}(1-\sigma)R,\theta}$ and the corresponding $Q_{(1-\sigma)R,\theta}$, we then have

$$\sup_{\hat{Q}_{\frac{1}{2}(1-\sigma)R,}} u \leq C[\fint_{\hat{Q}_{(1-\sigma)R,}}(u^+)^2dxdt]^{\frac{1}{2}} \leq C[\frac{1}{\theta((1-\sigma)R)^{n+2}}\int_{Q_R}(u^+)^2dxdt]^{\frac{1}{2}},$$

hence,

$$\sup_{Q_r} u \leq C[\frac{1}{(R-r)^{2+n}}\int_{Q_R}(u^+)^2dxdt]^{\frac{1}{2}}, \quad \forall r<R, \tag{2.9}$$

where C depends only on n,N,θ.

For any $p\in(0,2]$, from (2.9) we then have

$$\sup_{Q_r} u \leqq C(\sup_{Q_R} u)^{1-p/2}[\frac{1}{(R-r)^{n+2}}\int_{Q_R}(u^+)^p dxdt]^{\frac{1}{2}}$$

$$\leqq \frac{1}{2}\sup_{Q_R} u + C[\frac{1}{(R-r)^{n+2}}\int_{Q_R}(u^+)^p dxdt]^{1/p} ,$$

where the inequality $ab \leqq \varepsilon a^{q'} + C(\varepsilon,q)b^q$ with $1/q+1/q'=1$ was used for $q=2/p$ and $\varepsilon=\frac{1}{2}$.

Form the above inequality (2.8) follows if we use Lemma 2.3 below and the Hölder inequality.

Lemma 2.3 Assume $0 \leqq f(t) \leqq L < +\infty$, for $t \in [T_o,T_1]$. If $0<\varepsilon<1$, $q>0$, $A \geqq 0$, and

$$f(s) \leqq \varepsilon f(t) + \frac{A}{(t-s)^q} \quad \forall s<t ,$$

then for any $r<R$ with $r,R \quad [T_o,T_1]$, it is valid that

$$f(r) \leqq C(\varepsilon)\frac{A}{(R-r)^q} .$$

Proof Set $t_o=r$, $t_{i+1}=t_i+(1-a)a^i(R-r)$, $i=0,1,\ldots$, $a \in (0,1)$ to be determined. By induction we can find

$$f(t_o) \leqq \varepsilon^k f(t_k) + \frac{A}{(1-a)^q(R-r)^q}\sum_{i=0}^{k}\varepsilon^i a^{-iq} ,$$

therefore by taking a satisfying $\varepsilon a^{-q}<1$ the lemma follows immediately.

Now we derive a lemma which will be used in §3.

Using the inequality (2.2) with $\mu=\frac{1}{2}$, for $u \geqq 0$ and $k_o>0$ we have

$$\inf_{\hat{Q}(\frac{1}{2}R_o,\frac{1}{4}\theta R_o^2)} u \geqq k_o - C_o k_o[\frac{|\hat{Q}(k_o,R_o,\ R_o^2)|}{|\hat{Q}(R_o,\theta R_o^2)|}]^{\frac{1+\alpha}{2}} \tag{2.10}$$

where C_o is a constant depending only on n, N and θ. If by $\gamma=\gamma(n,N,\theta)$ we denote the constant determined by the following inequality

$$\gamma \leqq (\frac{1}{2C_o})^{\frac{2}{1+\alpha}} , \tag{2.11}$$

then form (2.10) we obtain

Lemma 2.4 Assume $u \in DG(Q_T,N)$, $u \geqq 0$. If for some $k_o>0$ we have

$$|\hat{Q}^-(k_o,R_o,\theta R_o^2)| \leqq \gamma|\hat{Q}(R_o,\theta R_o^2)|$$

where the constant γ is determined by (2.11), then it is valid that

$$\inf_{\hat{Q}(\frac{1}{2}R_o,\frac{1}{4}\theta R_o^2)} u \geqq \frac{1}{2}k_o .$$

§3. THE DIFFUSION PROPERTIES FOR FUNCTIONS IN $DG(Q_T,N)$

Lemma3.1 Assume $u\in V_2^{1,0}(Q_T)$, (1.2) is valid for $w(x,t)=-u(x,t)$. If $u\geqq 0,\zeta>0$, and $\operatorname{mes} B^-_{\zeta,\rho}(t_o)\leqq\frac{1}{M}\operatorname{mes} B_\rho$, $M>1$, then for any $\xi\in(\sqrt{\frac{1}{M}},1)$, there exist positive numbers $\bar\theta=\bar\theta(\xi)<1$ and $b=b(\xi)<1$ such that

$$\operatorname{mes} B^-_{(1-\xi)\zeta,\rho}(t)\leqq b(\xi)\operatorname{mes} B_\rho,\qquad \forall\, t\in[t_o,t_o+\bar\theta\rho^2],\tag{3.1}$$

where $B^-_{k,\rho}(t)=\{x\in B_\rho;u(x,t)<k\}$.

Proof By the condition of the lemma, taking $k=-\zeta$ we have

$$\int_{B^-_{\zeta,\rho-\sigma_1\rho}(t)}[\zeta-u(x,t)]^2dx\leqq\zeta^2\left(\frac{1}{M}+\frac{N\bar\theta}{\sigma_1{}^2}\right)\operatorname{mes} B_\rho,$$

On the other hand for any $\xi\in(0,1)$

$$(\xi\zeta)^2\operatorname{mes} B^-_{(1-\xi)\zeta,\rho-\sigma_1\rho}(t)\leqq\int_{B^-_{\zeta,\rho-\sigma_1\rho}(t)}[\zeta-u(x,t)]^2dx.$$

Therefore

$$\operatorname{mes} B^-_{(1-\xi)\zeta,\rho-\sigma_1\rho}(t)\leqq\xi^{-2}\left(\frac{1}{M}+\frac{N\bar\theta}{\sigma_1{}^2}\right)\operatorname{mes} B_\rho.$$

For any $\xi>\sqrt{1/M}$, it is obviously possible to select positive numbers $\sigma_1,\bar\theta(\xi)$ and $b(\xi)$ so that

$$n\sigma_1+\xi^{-2}\left(\frac{1}{M}+\frac{N\bar\theta}{\sigma_1{}^2}\right)=b(\xi)\in(0,1).$$

The lemma is proved.

To derive Lemma 3.3 below we need the following lemma which can be proved in the same way as in the proof of Inequality (5.5) of Ch. II in [4].

Lemma 3.2 Assume $u\in W_1^1(B_\rho)$. If $\operatorname{mes} B^-_{\eta,\rho}<b\operatorname{mes} B_\rho$, $b\in(0,1)$. Then for any $h<k<\eta$ we have

$$(k-h)\operatorname{mes} B^-_{h,\rho}\leqq\frac{\beta\rho}{1-b}\int_{B^-_{k,\rho}\setminus B^-_{h,\rho}}|D_xu|\,dx\tag{3.2}$$

where β is a constant depending only on n.

Lemma 3.3 Let $u(x,t)\in V_2^{1,0}(Q_T)$, inequalities (1.2) and (1.3) be valid for $w(x,t)=-u(x,t)$. Assume $u(x,t)\geqq 0,\eta>0$, $\operatorname{mes} B^-_{\eta,\rho}(t_o)\leqq\frac{1}{M}\operatorname{mes} B_\rho$ and $M>1$, hence for any $\xi\in(\sqrt{1/M},1)$, $\bar\theta=\bar\theta(\xi)$ can be determined by Lemma 3.1. Then for any $\gamma>0$ and any $\theta\in(0,\bar\theta]$. there exists $s=s(\xi,\gamma,\theta)>0$ such that

$$|Q^-(\eta(1-\xi)^s,\rho,\theta\rho^2)|\leqq\gamma|Q(\rho,\theta\rho^2)|,\tag{3.3}$$

where $Q^-(k,\rho,\theta\rho^2)=\{(x,t)\in Q(\rho,\theta\rho^2);\ u(x,t)<k\}$.

Proof By the assumption mes $B^-_{\eta,\rho}(t_o)\leq\frac{1}{M}$ mes B_ρ, we have

$$\text{mes } B^-_{(1-\xi)^i\eta,\rho}(t_o)\leq\frac{1}{M}\text{ mes } B_\rho,\quad i=0,1,\dots,s-1,$$

(s to be determines). Therefore Lemma 3.1 is applicable to $u(x,t)$ in the $Q(\rho,\theta\rho^2)$ for the level $\xi=(1-\xi)^i\eta$, it guarentees that

$$\text{mes } B^-_{(1-\xi)^{i+1}\eta,\rho}(t)\leq b(\xi)\text{ mes } B_\rho\ ,$$

for

$$t\in[t_o,t_o+\theta\rho^2].$$

Let us apply inequality (3.2) to the function $u(x,t)$ and the levels $h=(1-\xi)^{i+1}\eta$, $k=(1-\xi)^i\eta$ for $t\in[t_o,t_o+\theta\rho^2]$. This gives

$$\xi(1-\xi)^i\eta\text{ mes } B^-_{(1-\xi)^{i+1}\eta,\rho}(t)\leq\frac{\beta\rho}{1-b(\xi)}\int_{\mathcal{D}_i(t)}|D_xu(x,t)|dx,$$

where

$$\mathcal{D}_i(t)=B^-_{(1-\xi)^i\eta,\rho}(t)-B^-_{(1-\xi)^{i+1}\eta,\rho}(t).$$

We integrate both sides of this inequality with respect to t over $[t_o,t_o+\theta\rho^2]$; then we square both sides, after which we estimate the right-hand side by the Cauchy inequality

$$[\xi(1-\xi)^i\eta]^2\,|Q^-((1-\xi)^{i+1}\eta,\rho,\theta\rho^2)|^2$$

$$\leq\frac{\beta^2\rho^2}{(1-b(\xi))^2}\Big[\int_{t_o}^{t_o+\theta\rho^2}\int_{\mathcal{D}_i(t)}|D_xu|dxdt\ \Big]^2$$

$$\leq\frac{\beta^2\rho^2}{[1-b(\xi)]^2}\int_{t_o}^{t_o+\theta\rho^2}\int_{\mathcal{D}_i(t)}|D_xu|^2dxdt\int_{t_o}^{t_o+\theta\rho^2}\text{mes }\mathcal{D}_i(t)\ dt,\tag{3.4}$$

For an estimate of the first integral on the right-hand side we use inequality (1.3), choosing for $Q(\rho,\tau)$ the cylinder $Q(2\rho,4\theta\rho^2)$, and for $Q(\rho-\sigma_1\rho,\tau-\sigma_2\tau)$ the cylinder $Q(\rho,\theta\rho^2)$. After obvious simplification it gives

$$\int_{t_o}^{t_o+\theta\rho^2}\int_{\mathcal{D}_i(t)}|D_xu|^2dxdt\leq N(1+\tfrac{1}{3\theta})\rho^{-2}[(1-\xi)^i\eta]^2\ |Q(2\rho,4\theta\rho^2)|,\tag{3.5}$$

For $i=0,1,\dots,s-1$ from (3.4) and (3.5) it follows that

$$|Q^-[(1-\xi)^{i+1}\eta,\rho,\theta\rho^2]|^2 \leq \frac{N\beta^2(1+\frac{1}{3\theta})}{\xi^2[1-b(\xi)]^2}\, 2^{n+2}|Q(\rho,\theta\rho^2)| \int_{t_0}^{t_0+\theta\rho^2} \operatorname{mes}\mathcal{D}_i(t)dt,$$

and all the more so that

$$|Q^-[(1-\xi)^s\eta,\rho,\theta\rho^2]|^2 \leq \frac{2^{n+2}N\beta^2(1+\frac{1}{3\theta})}{\xi^2[1-b(\xi)]^2} \;|Q(\rho,\theta\rho^2)| \int_{t_0}^{t_0+\theta\rho^2} \operatorname{mes}\mathcal{D}_i(t)\,dt.$$

Let us sum these inequalities with respect to i from 0 to s-1. This gives

$$s|Q^-[(1-\xi)^s\eta,\rho,\theta\rho^2]|^2 \leq \frac{2^{n+2}\beta^2N(1+\frac{1}{3\theta})}{\xi^2[1-b(\xi)]^2}\,|Q(\rho,\theta\rho^2)|^2,$$

from which it is seen that for

$$s = s(\xi,\gamma,\theta) \;\geq \frac{1}{\gamma^2}\,\frac{2^{n+2}\beta^2N(1+\frac{1}{3\theta})}{\xi^2[1-b(\xi)]^2}\,, \tag{3.6}$$

we will have (3.3).

Remark 3.4 In the remaining part of this section we need to use Lemma 3.1, Lemma 3.3 and Lemma 2.4 in the following way. Assume $u(x,t)\in DG(Q_T,N)$ and $u(x,t)\geq 0$, $\eta_\circ > 0$. If

$$u(x,t) \geq \eta_\circ, \qquad\qquad \text{for } (x,t) \in B_r \times \{t=\tau\}, \tag{3.7}$$

then for given $m\geq 1$ there obviously exists $M(m)>1$ such that

$$\operatorname{mes} B_{\eta_o,2mr} \leq \operatorname{mes} B_{2mr} - \operatorname{mes} B_r \leq \frac{1}{M(m)} \operatorname{mes} B_{2mr}.$$

Set $\xi(m)= \frac{1}{2}\,[1+\sqrt{\frac{1}{M(m)}}\,]$. Then by Lemma 3.1 there exist $\bar\theta(m)=\bar\theta(\xi(m))$ and $b(m)= b(\xi(m)) \in (0,1)$ such that (3.1) holds, i.e.,

$$\operatorname{mes} B^-_{(1-\xi(m))\eta_\circ,2mr}(t) \leq b(m) \operatorname{mes} B_{2mr}\,,$$

for

$$t \in [\tau,\tau+\bar\theta(m)(2mr)^2].$$

Thus for any $\theta_m\in(0,\bar\theta(m)]$, if we take $\theta=\theta_m$ in Lemma 2.4, then by (2.11) $\gamma(m)=\gamma(n,N,\theta_m)$ is determined. Now from Lemma 3.3 there exists $s(m,\theta_m)=s(\xi(m),\gamma(m),\theta_m)>0$ such that (3.3) holds, i.e.,

$$|Q^-[\eta_\circ(1-\xi(m))^{s(m,\theta_m)},\, 2mr,\, \theta_m(2mr)^2]|$$
$$\leq \gamma(m)\,|Q(2mr,\theta_m(2mr)^2|$$

It hence follows from Lemma 2.4 that

$$u(x,t) \geq \frac{1}{2}\eta_\circ(1-\xi(m))^{s(m,\theta_m)} \tag{3.8}$$

for

$$(x,t) \in B_{mr} \times (\tau + \frac{3}{4}\theta_m(2mr)^2,\ \tau + \theta_m(2mr)^2).$$

where $\xi(m)\in(0,1)$ and $s(m,\theta_m)>0$ depend only on m and (m,θ_m) respectively (besides n,N), provied that all cylinders that appeared in the above discussion belong to Q_T .

In order to derive the main result of this section-Proposition 3.7, we need two kinds of diffusion lemmas: "given multiple" diffusion lemma-Proposition 3.5 and "given ratio" diffusion lemma-Lemma 3.6. Roughly speaking, the former says that if the density of the points at which $u(x,t)\geq\eta>0$ is sufficiently large in a fixed cube Q, and Q' is the cube adjacent to Q at the direction $t>0$ and enlarged to the "given multiple" of Q then u has a positive lower bound in Q' which can be estimated by the "given multiple"; the latter says that if $u(x,t_\circ)\geq\eta>0$ in a fixed ball $B_r(x_\circ)$ and $t_1>t_0$, then $u(x,t_1)$ has a positive lower bound in the ball $B_r(x_\circ)$ which can be estimated by the "given ratio" r/R.

<u>Proposition 3.5</u> Assume $u(x,t)\in DG(Q_T,N)$, $u(x,t)\geq 0$, $\eta>0$, $R>0$, $\theta>0$. If $\gamma=\gamma(n,N,\theta)$ is defined by (2.11) and

$$|\hat{Q}^-(\eta,x_0,t_0,R,\theta R^2)| \leq \gamma|\hat{Q}(x_0,t_0,R,\theta R^2)| \tag{3.9}$$

Then for any $m_1\geq 1$, $m_2\geq 0$ there exists a constant $C(n,N,m_1,m_2,\theta)>0$ such that

$$\left.\begin{aligned} &u(x,t) \geq C(n,N,m_1,m_2,\theta)\eta, \\ &\text{for } (x,t)\in B_{m_1R}(x_0)\times(t_0+\theta R^2,\ t_0+\theta R^2+m_2\theta(m_1R)^2, \end{aligned}\right\} \tag{3.10}$$

proviede $B_{2m_1R}(x_0)\times(t_0,t_0+\theta R^2+m_2\theta(m_1R)^2) \subset Q_T$.

<u>Proof</u> Noticing (3.9), by Lemma 2.4 we have

$$u(x,t_1)\geq\frac{\eta}{2}\ ,\quad \text{for } (x,t_1)\in B_{R/2}(x_0)\times(t_0+\frac{3}{4}\theta R^2,\ t_0+\theta R^2), \tag{3.11}$$

hence (3.7) is valid for $\eta_0=\eta/2$, $\tau=t_1$, $r=\frac{1}{2}R$. First applying Remark 3.4 for $m=2m_1$, then according to $\bar{\theta}(2m_1)$ there taking

$$\theta_1 = \theta_{m_1} = \min(\theta/12m_1^2,\ \bar{\theta}(2m_1)), \tag{3.12}$$

by (3.8) we have, for any t_1 from (3.11)

$$u(x,t)\geq\frac{1}{2}[\frac{1}{2}\eta(1-\xi(m_1))^{s(m_1)}]\ ,$$

for $(x.t)\in B_{m_1R}(x_o) \times (t_1+\frac{3}{4}\theta_1(2m_1R)^2,\ t_1+\theta_1(2m_1R)^2)$.

Therefore, repeating the above procedure at most for m_o times, where

$$m_o = \text{the integeral part of } [\frac{\frac{1}{4}\theta R^2+\theta m_1^2 m_2 R^2}{\frac{1}{4}\theta_1(2m_1R)^2} + 2] \text{ , and noticing (3.12),}$$

we have

$$u(x,t) \geqq \frac{1}{2}\eta[\frac{1}{2}(1-\xi(m_1))^{s(m_1)}]^{m_o} ,$$

$$\text{for } (x,t) \in B_{m_1R}(x_o) \times (t_o+\theta R^2,\ t_o+\theta R^2+m_2\theta(m_1R)^2),$$

which proves (3.10).

<u>Lemma 3.6</u> Let $u(x,t)\in DG(Q_T,N)$, $u(x,t)\geqq 0,\eta>0$, $R>r_o>0$, $T_1>T_o\geqq 0$, and

$$u(x,T_o)\geqq\eta, \quad \text{for } x\in B_{r_o}(x_o).$$

In Remark 3.4 with m=2, denote $\xi(m)=\xi_2$, $\bar{\theta}(\xi(m))=\bar{\theta}_2$, and $s(m,\theta_2)=s_2$ for given $\theta_2\in(0,\bar{\theta}_2]$. If for the above θ_2 and some $k\in N$ we have

$$R = 2^k r_o, \qquad T_1 = T_o + \frac{16}{3}(4^k-1)\theta_2 r_o^2 , \tag{3.13}$$

then

$$u(x,t) \geqq \eta(\frac{r_o}{R})^{\delta}, \qquad \text{for } (x,t)\in B_r(x_o) \times (T_1-\theta_2R^2,\ T_1)$$

where the constant

$$\delta = 1- \frac{s_2}{\log 2}\log(1-\xi_2) > 0$$

depends only on θ_2, n and N, provided $B_{2R}(x_o)\times(T_o,T_1)\subset Q_T$.

<u>Proof</u> Set $t_o=T_o$,

$$r_i=2r_{i-1}, \quad t_i=t_o+\sum_{j=1}^{i}\theta_2(2r_j)^2, \qquad i=1,2,\ldots,k.$$

An application of Remark 3.4 for $\tau=t_{i-1}$, $r=r_{i-1}$, $i=1,2,\ldots,k$, leads to

$$u(x,t) \geqq \eta[\frac{1}{2}(1-\xi_2)^{s_2}]^i, \quad \text{for } (x,t)\in B_{r_i}(x_o)\times(t_{i-1}+\frac{3}{4}\theta_2(2r_i)^2,t_i).$$

Hence

$$u(x,t) \geqq [\frac{1}{2}(1-\xi_2)^{s_2}]^k\eta, \quad \text{for } (x,t)\in B_R(x_o)\times(T_1-\theta_2R^2,T_1).$$

But from (3.13) we have

$$k = -\log\frac{r_o}{R}/\log 2.$$

Substituting this into the previous inequality we condlude the proof of the lemma.

Proposition 3.7 Assume $u(x,t)\in DG(Q_T,N)$, $u(x,t)\geqq 0$. If $R\geqq r>0$, $\alpha_2>\alpha_1>\alpha_0\geqq 0$ and

$$u(x,t)\geqq\eta>0,\quad \text{for } (x,t)\in B_r(x_o)\times\{t=\alpha_o R^2\}.$$

Then there exist constants C, $\delta>0$ such that

$$u(x,t)\geqq C\left(\frac{r}{R}\right)^{\delta}\eta,\quad \text{for } (x,t)\in B_R(x_o)\times(\alpha_1 R^2,\alpha_2 R^2), \tag{3.14}$$

where the constants C, $\delta>0$ depend only on n, N, $\alpha_1-\alpha_0$ and α_2, provided $B_{2R}(x_o)\times$ $\times(\alpha_0 R^2,\alpha_2 R^2)$ Q_T.

Proof For $\bar{\theta}_2$ defined in Lemma 3.6 we take

$$\theta_2 = \min\left(\bar{\theta}_2,\frac{3}{64}(\alpha_1-\alpha_0)\right). \tag{3.15}$$

For given R and r there exists a fixed $k\in\mathbb{N}$ such that

$$\frac{1}{2^k}\leqq\frac{r}{R}<\frac{1}{2^{k-1}}.$$

Set $r_0=r$, $r_i=2r_{i-1}$, for $i=1,2,\ldots,k$, $T_o=\alpha_o R^2$ and $T_1=\sum_{i=1}^{k}\theta_2(2r_i)^2+T_o$. Then a simple calculation shows that

$$T_1=\frac{16}{3}\theta_2 r^2(4^k-1)+T_o<\alpha_1 R^2.$$

Now by Lemma 3.6 we have

$$u(x,t)\geqq\eta\left(\frac{r}{R}\right)^{\delta},\quad \text{for } (x,t)\in B_R(x_o)\times(T_1-\theta_2 R^2,\ T_1),$$

where δ and θ_2 depend only on n, N, and $\alpha_1-\alpha_0$. An application of Proposition 3.5 then leads to (3.14).

§4. THE PROOF OF THEOREM 1.3

By means of the results in §3, Theorem 1.3 can be proved in the same way as in [12] using the measure lemma in [11]. But for completeness, an alternative proof is presented here in which we use Lemma 4.3 below which apeared first in [13].

In order to do the measure argument, it is needed to consider cubes instead of balls, we denote

$$K_R(x_o)=\{\ x=(x^1,\ \ldots,\ x^n);\quad \max_{1\leqq i\leqq n}|x^i-x_o^i|<R\}.$$

$$Q_{R,\theta}=Q_{R,\theta}(x_o,t_o)=K_R(x_o)\times(t_o,t_o+\theta R^2).$$

Noticing

$$B_R(x_o)\subset K_R(x_o)\subset B_{\sqrt{n}R}(x_o),$$

$$2^n \text{mes } B_R=\omega_n \text{mes } K_R, \quad (\omega_n=\text{volume of unit ball in } \mathbb{R}^n),$$

we may reformulate Proposition 3.5 and Proposition 3.7 as the following two lemmas respectively.

Lemma 4.1 Assume $u(x,t)\in DG(Q_T,N)$, $u(x,t)\geq 0, \eta>0$, $r>0$, $\theta>0$, $B_{6\sqrt{n}r}(\bar{x})\subset\Omega$.

Then there exist positive constants μ and λ depending only on n, N and θ, such that

$$|\{(x,t)\in Q_{r,\theta}(\bar{x},\bar{t}),\ u(x,t)\geq\eta\}|\geq\mu|Q_{r,\theta}(\bar{x},\bar{t})|$$

implies

$$u(x,t)\geq\lambda\eta, \quad \text{for } (x,t)\in\{K_{3r}(\bar{x})\times(\bar{t}+\theta r^2,\ \bar{t}+8\theta r^2)\}\cap Q_T .$$

Lemma4.2 Assume $u(x,t)\in DG(Q_T,N)$, $u(x,t)\geq 0$, $\eta>0,\rho\geq\varepsilon>0$, $\alpha_2>\alpha_1>\alpha_o\geq 0$, $B_{2\sqrt{n}\rho}(x_o)\times$ $\times(\alpha_o\rho^2,\alpha_2\rho^2)\subset Q_T$. If

$$u(x,\alpha_o R^2)\geq\eta \quad , \qquad \text{for } x\in K_\varepsilon(x_o),$$

then there exist positive constants C and δ depending only on n, N, $\alpha_1-\alpha_o$ and α_2 such that

$$u(x,t)\geq C\left(\frac{\varepsilon}{\rho}\right)^\delta\eta, \quad \text{for } (x,t)\in K_\rho(x_o)\times(\alpha_1\rho^2,\alpha_2\rho^2).$$

Lemma 4.3 Let $\Gamma\subset Q_{\rho,\theta}(\bar{x},\bar{t})\triangleq Q_\rho$ be a given measurable set, $|\Gamma|\neq 0$, for fixed $\mu\in(0,1)$, set

$$\tilde{\Gamma}=\Gamma\bigcup_{Q_{r,\theta}(x,t)\subset Q_\rho}\{K_{3r}(x)\times(t+\theta r^2,t+8\theta r^2)\cap Q;\ |\Gamma\ Q_{r,\theta}(x,t)|>\mu|Q_{r,\theta}(x,t)|\}.$$

Then either for $\delta=(1+\frac{1-\mu}{12})^{-1}$ it holds that

$$|\tilde{\Gamma}|\geq\delta^{-1}|\Gamma| \tag{4.1}$$

or there exists some $Q_{r_o,\theta}(x_o,t_o)\subset Q_\rho$ such that

$$\left.\begin{aligned} &r_o\geq\frac{1}{4}\Big[\frac{|\Gamma|}{(2\rho)^n\theta}\Big]^{\frac{1}{2}}, \\ &|\Gamma\cap Q_{r_o,\theta}(x_o,t_o)|\geq\mu|Q_{r_o,\theta}(x_o,t_o)|. \end{aligned}\right\} \tag{4.2}$$

By covering and changing arguments Theorem 1.3 can be derived from the following

Theorem 4.4 Assume $u(x,t)\in DG(Q_T,N)$, $u(x,t)\geq 0$, $R,\theta_3>0$, $B_{2\sqrt{n}R}(\bar{x})\times(\bar{t},\bar{t}+\theta_3 R^2)\subset Q_T$.

Then for any $\sigma_1,\sigma_2\in(0,1)$, $0<\theta_1<\theta_2<\theta_3$, there exist constants p, C>0 such that

$$\inf_{K_{\sigma_1 R}(\bar{x})\times(\bar{t}+\theta_2 R^2,\bar{t}+\theta_3 R^2)} u(x,t) \geq C\Big(\fint_{K_{\sigma_2 R}(\bar{x})\times(\bar{t},\bar{t}+\theta_1 R^2)} u^p\,dxdt\Big)^{1/p}$$

where p and C depend only on n, N, $\sigma_1,\sigma_2,\theta_1,\theta_2$ and θ_3 .

<u>Proof</u> Set $\sigma_2 R=\rho$, $\theta_1/\sigma_2^2=\theta$. Denote $Q_\rho \underline{\Delta} Q_{\rho,\theta}(x,t)=K_{\sigma_2\rho}(\bar{x})\times(\bar{t},\bar{t}+\theta_1 R^2)$,

$\Gamma_\eta=\{(x,t)\in Q_\rho,\ u(x,t)\geq\eta\}$ for any $\eta>0$.

If $|\Gamma_\eta|\neq 0$, then there exists $s\geq 0$ such that

$$\delta^s|Q_\eta| \geq |\Gamma_\eta| > \delta^{s+1}|Q_\rho|,$$

where δ is the same δ as that indicated in Lemma 4.3. Denote

$$\Gamma_\eta^j = \{(x,t)\in Q_\rho\ ;\ u(x,t) \geq \eta\lambda^j\}, \qquad j=0,1,\dots,s.$$

Then by Lemma 4.3 with fixed μ as indicated in Lemma 4.1, we know that there are only two possibilities:

Case 1. For every $j=0,1,\dots,s$, if we denote $\Gamma=\Gamma_\eta^j$, then

$$|\tilde{\Gamma}| \geq \delta^{-1}|\Gamma| = \delta^{-1}|\Gamma_\eta^j|,$$

Case 2. For some $0\leq j_0\leq s$ and $\Gamma=\Gamma_\eta^{j_0}$, there exists $Q_{r_0,\theta}(x_0,t_0)\subset Q$ such that

$$r_0 \geq \frac{1}{4}\,[|\Gamma|/\theta(2\rho)^n\,]^{\frac{1}{2}} = \frac{1}{4}[|\Gamma_\eta^{j_0}|/\theta(2\rho)^n]^{\frac{1}{2}}$$

and

$$|Q_{r_0,\theta}(x_0,t_0)\cap\{u\geq\eta\lambda^{j_0}\}| \geq \mu|Q_{r_0,\theta}(x_0,t_0)|\ .$$

In Case 1, for every fixed j, from Lemma 4.1 we know that if $Q_{r,\theta}(\hat{x},\hat{t})\subset Q_\rho$ and

$$|Q_{r,\theta}(\hat{x},\hat{t})\cap\Gamma_\eta^j| \geq \mu|Q_{r,\theta}(\hat{x},\hat{t})|,$$

then $\quad u(x,t) \geq \lambda^{j+1}\eta,\ \text{ for } (x,t)\in K_{3r}(\hat{x})\times(\hat{t}+\theta r^2,\hat{t}+8\theta r^2).$

Then for every $\Gamma=\Gamma_\eta^j$ it holds that $\Gamma_\eta^{j+1}\supset\tilde{\Gamma}$, hence $|\Gamma_\eta^{j+1}| \geq |\tilde{\Gamma}| \geq \delta^{-1}|\Gamma_\eta^j|$, therefore $|\Gamma_\eta^s|\geq\delta|Q_\rho|$. By Lemma 4.1 we then have

$$u(x,t) \geq \eta\lambda^{s+1},\ \text{ for } (x,t)\in K_{3\rho}(\bar{x})\times(\bar{t}+\theta\rho^2,\bar{t}+8\theta\rho^2)\cap Q_T$$

Consequently, by Lemma 4.2,

$$u(x,t) \geq C_1\lambda^{s+1}\eta,\ \text{ for } (x,t)\in K_{\sigma_1 R}(\bar{x})\times(\bar{t}+\theta_2 R^2,\bar{t}+\theta_3 R^2). \tag{4.3}$$

where and in the sequel capital C with different subscripts denotes the constant with the same property as formulated in the theorem.

In Case 2, an application of Lemma 4.1 gives

$$u(x,t)\geqq\lambda^{j_\circ+1}\eta,\quad \text{for } (x,t)\in K_{3r_o}(x_o)\times(t_o+\theta r_o^2,t_o+8\theta r_o^2)\cap Q_T,$$

Then by Lemma 4.2, taking $\varepsilon=r_o$, $\alpha_1=\theta_2/\sigma_2^2$, $\alpha_2=\theta_3/\sigma_2^2$, we obtain

$$\left.\begin{aligned} &u(x,t) \geqq C_2(\frac{r_o}{\rho})^{\delta}\lambda^{j_\circ+1}\eta\geqq C_3(|\Gamma_\eta^{j_\circ}|/\theta 2^n\rho^{n+2})^{\delta/2}\lambda^{s+1}\eta,\\ &\text{for } (x,t)\in K_{\sigma_1 R}(\bar{x})\times(\bar{t}+\theta_2R^2,\ \bar{t}+\theta_3R^2).\end{aligned}\right\} \tag{4.4}$$

From $|\Gamma_\eta| \leqq \delta^s|Q_\rho|$ we have

$$s \leqq \log\frac{|\Gamma_\eta|}{|Q_\rho|} \,/\, \log\delta,$$

hence

$$\lambda^{s+1} \geqq \lambda \exp\left[\frac{\log\lambda}{\log\delta}\log\frac{|\Gamma_\eta|}{|Q_\rho|}\right] = \lambda\left[\frac{|\Gamma_\eta|}{|Q_\rho|}\right]^{1/\alpha},\quad \frac{1}{\alpha} = \frac{\log\lambda}{\log\delta} > 0.$$

Consequently, combining (4.3) and (4.4) we have

$$u(x,t) \geqq C_4\eta\left[\frac{|\Gamma_\eta|}{|Q_\rho|}\right]^{1/\beta},\quad \text{for } (x,t)\in K_{\sigma_1 R}(\bar{x})\times(\bar{t}+\theta_2R^2,\bar{t}+\theta_3R^2), \tag{4.5}$$

where $\frac{1}{\beta} = \frac{1}{\alpha} + \frac{\delta}{2} > 0$.

Obviously, (4.5) is also true as $|\Gamma_\eta|=0$.

Denote

$$q = \inf_{K_{\sigma_1 R}(\bar{x})\times(\bar{t}+\theta_2R^2,\bar{t}+\theta_3R^2)} u(x,t), \tag{4.6}$$

Then from (4.5), taking $p\in(0,\beta)$, we have

$$\begin{aligned}\int_{Q_\rho} u^p\,dxdt &= p\int_o^q \eta^{p-1}|\Gamma_\eta|\,d\eta + p\int_q^\infty \eta^{p-1}|\Gamma_\eta|d\eta\\ &\leqq C_5(\frac{1}{\beta-p}+1)\,q^p|Q_\rho|.\end{aligned}$$

Noticing (4.6) we thus conclude the proof.

§5. APPENDIX

<u>Proof of Lemma 4.3</u> Without loss of generality we way assume $(\bar{x},\bar{t})=(o,o)$. Denote $K_1=K_{\rho,\theta}(o,o)$.

If $|\Gamma\cap K_1| \geqq \mu|K_1|$, then (4.2) holds, otherwise we proceed as follows

1. First Subdivision. Now $|\Gamma\cap K_1|<\mu|K_1$. We subdivide K_1 into 2^{n+2} congruent subcubes $K_{\frac{1}{2}\rho,\theta}(y_i,t_i)$, where $y_i=(y_i^1, \ldots, y_i^n)$ with $y_i^k=\frac{1}{2}\rho$, for $k=1,2,\ldots,n$; $t_i=0, \frac{1}{4}\rho^2, \frac{1}{2}\rho^2, \frac{3}{4}\rho^2$. If $|\Gamma\cap K_{2^{-1}}|<\mu|K_{2^{-1}}|$ holds for every $K_{\frac{1}{2}\rho,\theta}(y_i,t_i)=K_{2^{-1}}$, then we subdivide $K_{2^{-1}}$ into 2^{n+2} congruent subcubes $K_{2^{-2}},\ldots$

2. First Estimation. If the contrary inequality apears, that is, for some $K^o_{2^{-k}}$ we have

$$|\Gamma\cap K^o_{2^{-k}}| \geq \mu|K^o_{2^{-k}}|. \tag{5.1}$$

Denoting the ancestor cube of $K^o_{2^{-k}}$ by $\tilde{K}^o_{2^{-k+1}}$ we then have

$$|\Gamma\cap \tilde{K}^o_{2^{-k+1}}|<\mu|\tilde{K}^o_{2^{-k+1}}|.$$

Assume the cubes with the above properties adjacent with each other in the t direction are denoted by $\tilde{K}^o_{2^{-k+1}}, \tilde{K}^1_{2^{-k+1}}, \ldots, \tilde{K}^L_{2^{-k+1}}$, in the order of increasing t, i.e.,

$$|\Gamma\cap \tilde{K}^i_{2^{-k+1}}| < \mu|\tilde{K}^i_{2^{-k+1}}|, \qquad i=0,1,\ldots,L,$$

and there exists at least one $K^{o,i}_{2^{-k}}\subset\tilde{K}^i_{2^{-k+1}}$ such that

$$|\Gamma\cap K^{o,i}_{2^{-k}}| \geq \mu|K^{o,i}_{2^{-k}}| , \qquad i=0,1,\ldots, L. \tag{5.2}$$

By $\tilde{K}^{-1}_{2^{-k+1}}$(resp., $\tilde{K}^{L+1}_{2^{-k+1}}$) we denote the cube possessing the same size as and adjacent with $\tilde{K}^o_{2^{-k+1}}$ (resp., $\tilde{K}^L_{2^{-k+1}}$) from below (resp., above). Then for $K^j_{2^{-k}}\subset\tilde{K}^{-1}_{2^{-k+1}}$ or for $K^j_{2^{-k}}\subset\tilde{K}^{L+1}_{2^{-k+1}}$ $(j=1,2,\ldots,2^{n+2})$ we have

$$|\Gamma\cap K^j_{2^{-k}}| < \mu|K^j_{2^{-k}}|.$$

Denote those $K_{2^{-k}}\subset\tilde{K}^{-1}_{2^{-k+1}}$ which has common surface with $\tilde{K}^o_{2^{-k+1}}$ by $K^{-1,j}_{2^{-k}}(j=1,2,\ldots,2^n)$, similarly $K^{L+1,j}_{2^{-k}}(j=1,2,\ldots,2^n)$ has common surface with $\tilde{K}^L_{2^{-k+1}}$.

By definition,

$$\tilde{\Gamma}\cap\tilde{K}^i_{2^{-k+1}} = \tilde{K}^i_{2^{-k+1}}, \quad (i=1,2,\ldots,L),$$

$$\tilde{\Gamma}\cap K^{L+1,j}_{2^{-k}} = K^{L+1,j}_{2^{-k}} \qquad (j=1,2,\dots,2^n).$$

Noticing (5.2) we then have

$$|\tilde{\Gamma}\cap\{(\bigcup_{j=1}^{2^n} K^{L+1,j}_{2^{-k}})(\bigcup_{i=o}^{L} \tilde{K}^i_{2^{-k+1}})\}| - \frac{1-\mu}{4}|\bigcup_{j=1}^{2^n} K^{-1,j}_{2^{-k}}|$$

$$\geq |\Gamma\cap\{(\bigcup_{j=1}^{2^n} K^{L+1,j}_{2^{-k}})(\bigcup_{i=o}^{L} \tilde{K}^i_{2^{-k+1}})\}| + (1-\mu)|(\bigcup_{j=1}^{2^n} K^{L+1,j}_{2^{-k}})(\bigcup_{i=1}^{L} \tilde{K}^i_{2^{-k+1}}) -$$

$$- \frac{1-\mu}{4}|\bigcup_{j=1}^{2^n} K^{-1,j}_{2^{-k}}|$$

$$\geq \{1+\frac{\frac{3}{16}+L}{\frac{5}{4}+L}(1-\mu)\}|\Gamma\cap\{(\bigcup_{j=1}^{2^n} K^{L+1,j}_{2^{-k}})(\bigcup_{i=o}^{L} \tilde{K}^i_{2^{-k+1}})\}| . \qquad (5.3)$$

When the lower surface of $\tilde{K}^o_{2^{-k+1}}$ lies on t=0, we drop the term $\bigcup_{j=1}^{2^n} K^{-1,j}_{2^{-k}}$.

When the upper surface of $\tilde{K}^L_{2^{-k+1}}$ lies on $t=\theta\rho^2$, $K^{L+1,j}_{2^{-k}}$ lies outside of K_1, in this case instead of (5.3) we have

$$(1-\mu)|\bigcup_{j=1}^{2^n} K^{L+1,j}_{2^{-k}}| + |\tilde{\Gamma}\cap(\bigcup_{i=0}^{L} \tilde{K}^i_{2^{-k+1}})| - \frac{1-\mu}{4}|\bigcup_{j=1}^{2^n} K^{-1,j}_{2^{-k}}|$$

$$\geq [1+\frac{\frac{3}{16}+L}{1+L}(1-\mu)]|\Gamma\cap(\bigcup_{i=o}^{L} K^i_{2^{-K+1}})|. \qquad (5.4)$$

For the convenience of the following discussion those cubes $\tilde{K}^i_{2^{-k+1}}$ (i=o,1,..,L) and $K^{L+1,j}_{2^{-k}}$ $(j=1,2,\dots,2^n)$ which apear at the left-hand side of (5.4) will be called "having been added", and $\tilde{K}^{-1,j}_{2^{-k}}$ $(j=1,2,\dots,2^n)$ — "having been subtracted".

3. <u>Following Subdivision</u>. All the subcubes obtained from above steps, except the "having been added" ones, are continuously subdivided in the same way as in 1.

4. <u>Following Estimation</u>. When appears again an inequality of (5.1) type (which may also be denoted by $|\Gamma\cap K_{2^{-k}}|\geq\mu|K_{2^{-k}}|$, WLOG), we do the similar estimation as in 2. with the difference: If $\bigcup_{j=1}^{2^n} K^{-1,j}_{2^{-k}}$ is contained in the "having been added"

cubes, or the lower surface of $\tilde{K}^{o}_{2^{-k+1}}$ lies on t=0, then we drop the term $\bigcup_{i=1}^{2^n} K^{-1,j}_{2^{-k}}$ from (5.3); If $\bigcup_{j=1}^{2^n} K^{L+1,j}_{2^{-k}}$ is contained in the "having been subtracted" cubes or the upper surface of $\tilde{K}^{L}_{2^{-k+1}}$ lies on $t=\theta\rho^2$, then we have (5.4) instead of (5.3).

For the cubes appeared on the left-hand side of (5.4), we also use the same names as at the end of step 2.

5. Repeat the procedures in 3., 4. and the process is continued. Finally, summing up all the inequalities (5.3) and (5.4) obtained in 3., 4. with respect to k, noticing

$$\frac{(1-\mu)}{4}\left|\bigcup_{j=1}^{2^n} K^{-1,j}_{2^{-k}}\right| \geqq \sum_{k_1>k} (1-\mu)\left|\bigcup_{j=1}^{2^n} K^{L+1,j}_{2^{-k_1}}\right|$$

where the lower surface of $K^{L+1,j}_{2^{-k_1}}$ lies on the upper surface of those $\bigcup_{j=1}^{2^n} K^{-1,j}_{2^{-k}}$ which are "having been added" before, we then have

$$(1-\mu)|\bar{\Gamma}| + |\tilde{\Gamma}| \geqq [1+\frac{3}{20}(1-\mu)]\,|\Gamma\cap(\bigcup_k \tilde{K}_{2^{-k+1}})|. \tag{5.5}$$

where $\bar{\Gamma}$ is the union of all cubes in (5.4) with lower surface on $t=\theta\rho^2$.

Since Γ is measurable, by Lebesgue lemma we have

$$\begin{aligned} &|\Gamma - (\bigcup_k \tilde{K}_{2^{-k+1}})| = 0, \\ &|\Gamma \cap (\bigcup_k \tilde{K}_{2^{-k+1}})| = |\Gamma|. \end{aligned} \tag{5.6}$$

If the maximum height of $\bar{\Gamma}$ is not less than $\dfrac{|\Gamma|}{16(2\rho)^n}$, then (4.2) is true; otherwise by (5.5) and (5.6) we have (4.1).

REFERENCES

[1] De Giorgi, E. : Sulla differenziabilita e l'analiticita degli integrali multipli regolari, Mem. Accad. Sci. Torino Cl. Sci. Fis. Mat. Natur. (3), t.3, 1957, p. 25-43.

[2] Di Benedetto, E. and Trudinger, N.S. : Harnack inequalities for quasi-minina of variational integrals, Ann. d'Inst. Henri Poincare, Analyse Non Lineaire, Vol. 1, no.4, 1984, p.295-308.

[3] Giaquinta, M. and Giusti,E. : Quasi-Minima, Ann. d'Inst. Henri Poincare, Analyse Non Lineaire, Vol.1, no.2, 1984, p.79-107.

[4] Ladyzenskaja, O.A., Solonnikov, V.A. and Ural'ceva, N.N. : Linear and quasilinear equations of parabolic type, "Nauka", Moscow, 1967; English Transl., Amer. Math. Soc., Providence, R.I., 1968.

[5] Ladyzenskaja, O.A. and Ural'ceva, N.N.a, Linear and quasilinear elliptic equations, Academic press, New York, 1968.

[6] Moser,J., : On Harnack's theorem for elliptic differential equations, Comm. Pure Appl. Math., t. 14, 1961, p.577-591.

[7] Moser, J., : A Harnack inequality for parabolic differential equations, Comm. Pure Appl. Math., t.17, 1964, p.101-134.

[8] Moser, J., : Correction to "A Harnack inequality for parabolic differential equations" , Comm. Pure Appl. Math., t.20, 1967, p.231-236.

[9] Trudinger, N.S., : Pointwise estimates and quasilinear parabolic equations, Comm. Pure Appl. Math., t.21, 1968, p.205-226.

[10] Trudinger, N.S., : Local estimates for subsolutions and supersolutions of general second order elliptic quasilinear equations, Inventions Math. t.61, 1980, p.67-69.

[11] Krylov, N.V. and Safonov, M.V., : Certain properties of solutions of parabolic equations with measurable coefficients. Izvestia Akad. Nauk SSSR, t.40, 1980, p.161-175, English Transl., Math. USSR Izv., t.16, 1981.

[12] Reye, S.J., : Harnack inequalities for parabolic equations in general form with bounded measurable coefficients. Australian National Univ., Centre for Mathematical Analysis, Reseach report R44, 1984.

[13] Dong, Guangchang, : Local estimates for subsolutions and supersolutions of general second order parabolic quasilinear equations, to appear.

Equivariant Morse theory for isolated critical orbits and its applications to nonlinear problems

Wang Zhi-qiang

Institute of Mathematics
Academia Sinica, Beijing
People's Republic of China

Introduction

In recent years, the Morse theory has become more and more important in the studies of nonlinear problems. The classical Morse theory was only founded for nondegenerate functions. In this situation, both local and global theory are accurate and beautiful. However, applications were limited (cf. [Mil], [Pal1] etc) because it is difficult to check the nondegeneracy. In [GM1], Gromoll and Meyer studied the properties of differential functions with isolated critical points. Then, K.C.Chang studied systematically the Morse theory for this class of functions (cf. [Ch 1]), and he defined the critical gorups which were used to describe the local topological properties of isolated critical points. In [Ch 1], the author also applied his theory directly to the studies of nonlinear partial differential equations.

When space is a G-space, critical points of a G-invariant function, which appear as orbits, are generally not isolated. Therefore, the above theory does not work well for the problems with symmetries. Bott in [Bo1] introduced the concept of nondegenerate critical manifold. And by virtue of this concept, Bott, Wasserman (cf.[Bo2], [Was 1]) discussed the Morse theory for nondegenerate G-invariant functions, the so-called equivariant Morse theory. Nevertheless, the verification of nondegeneracy in this case becomes all the more complicated in applications (cf.[Pac 1] also for references).

Based on the above consideration, it seems necessary to develop a kind of Morse theory for G-invariant functions with isolated critical orbits. Our present paper is precisely intended to serve such a purpose. We also consider a series of applications to nonlinear problems. Our work, on the one hand, generalizes both the theory in [Ch 1] to the equivariant case and the theory in [Was 1] to the degenerate case. On the other hand, our work unifies the previous results in [GM 1], [Bo 2], [Was 1] and [Ch 1] to some extent. In addition, our frame of the work is suitable for the nonlinear problems with symmetries.

There are four sections in this paper. In §1, for our requirements in later sections we study the deformation properties of the equivariant gradient flow. §2 is devoted to our main abstract theory. We define the critical groups and the normal critical groups for an isolated critical orbit, by means of which we obtain the relation between the critical groups of an isolated critical orbit and the topology of the orbit itself. Some useful formulas are given, which make the calculation of the critical groups simple. Then, we consider several examples. In §3, we discuss some multiple critical orbit theorems. An easy example is given which shows that the previous three solution theorem (cf.[Ch 1]) can not be well applied to the G-invariant functionals. And we prove two theorems which guarantee the existence of at least three critical orbits. At the end of this section, a five solution theorem is given by using the critical groups of a Mountain-Pass critical orbit. In the last section, through the application of our abstract theorems we study some concrete nonlinear problems. We think the methods in this paper can be applied to some more interesting problems.

Throughout this paper, we use the following notations. G denotes a compact Lie group. A manifold M (or space) is called a G-manifold (G-space), if there is a smooth isometric action of G on M. For fixed x, $G(x) = \{gx | g \in G\}$ is called a G-orbit, which is a compact submanifold of M. The normal bundle of $G(x)$ is denoted by $\nu G(x)$. The closed subgroup of G defined by $G_x = \{g \in G | gx = x\}$ is called the isotropy group of x. $\mathrm{Fix}_G = \{x \in M | gx = x, \forall g \in G\}$ is called fixed point space. Given a function $f : M \to R$, we say that f is G- invariant if

$f(gx) = f(x)$ for each $x \in M$ and $g \in G$. If X, Y are two G-space, we say that a mapping $F : x \to Y$ is G-equivariant if $F(gx) = gF(x)$ for each $x \in X$ and $g \in G$. The concepts of tubular and slice are often used, we refer these concepts and other terminology on compact Lie transformation group to [Br 1]. If f is a smooth function, we write $K(f) = \{x \in M | df(x) = 0\}$, $f_a = \{x \in M | f(x) \leq a\}$, $K_b = f_b \bigcap K(f)$, for any real numbers a, b.

The author wishes to express his sincere thanks to Prof. K.C.Chang for his foresighted suggestion to consider this problem and for his guidance and encouragement in preparation of this work.

§1 On the deformation properties of equivariant gradient flow

For our requirements below, we discuss the deformation properties of equivariant gradient flow in this section. The main result is an equivariant generalization of the so-called second deformation theorem in [Ch 1].

Theorem 1.1: Let M be a C^2 G-Finsler manifold. Suppose that $f \in C^{2-0}(M, R)$ is G-invariant, and that f satisfies the (P.S.) condition. Assume that c is the only critical value of f in $[c, b) \subset R$ and any connected component of K_c is always a part of a certain critical orbit. Then f_c is a strong deformation retract of $f_b \setminus K_b$, i.e., there is a continuous mapping τ: $[0,1] \times f_b \setminus K_b \to f_b \setminus K_b$, such that

(1) $\tau(0, \cdot) = id$;

(2) $\tau(t, \cdot)|_{f_c} = id|_{f_c}$;

(3) $\tau(1, x) \in f_c$, $\forall x \in f_b \setminus K_b$;

(4) $\tau(t, \cdot)$ is a G-equivariant mapping for any fixed $t \in [0, 1]$.

In order to prove the above theorem, we first give following lemmas.

Lemma 1.1: Suppose that $N = G(x)$ is a G-orbit and $B = G(S_x)$ is a G-invariant tubular neighbourhood of N, where S_x is a slice at x. Let $\eta(t, u)$ be a flow defined by the equation

$$\begin{cases} \frac{d\eta(t,u)}{dt} = -df(\eta(t, u)) \\ \eta(0, u) = u \in B \end{cases}$$

Then the flow $\eta(t, u)$ preserves the slices, i.e., if there exists $t_0 > 0$ such that $\eta(u, t) \in B$ for any $t \in [0, t_0]$ and $\eta(0, u) \in S_y$ for a certain $y \in N$, then $\eta(t, u) \in S_y$ for any $t \in [0, t_0]$.

Proof: Since df is a G-equivariant operator, $\eta(t, u)$ is G-equivariant for u. Moreover, the action of G is transitive on the orbit N, so is it between slices. Then suffice it to consider the case of $u \in S_x$.

From the structure of G-tubular neighbourhood (cf.[Br 1]), there is a diffeomorphism $B \cong G \times_H S_x$, where $H = G_x$ the isotropy group at x, S_x is a H-space. Since $G_u = H_u \leq H$, for $u \in S_x$, it follows that $\dim G(u) \geq \dim N$. By the G- invariance of f, the component of $df(u)$ on the tangent space of $G(u)$ is zero. Therefore, $df(u) \in \nu_u G(u) \subset T_u S_x$, $\eta(t, u)$ has to preserve the slice S_x.

Lemma 1.2: Suppose that K is a compact metric space, F_1, F_2 are compact subsets of K. Then either there is a connected component in K which connects F_1 with F_2, or there are compact subsets M_1, M_2 in K such that $M_1 \bigcap M_2 = \phi$, $M_1 \bigcup M_2 = K$, $F_i \subset M_i$, $i = 1, 2$.

This is a result of general topology (ce.[Ke 1]).

Proof: 1^0. For each $x \in f_b \setminus (f_c \bigcup K_b)$, we define a flow as follows:

$$\begin{cases} \dot{\eta}(t, x) = -\frac{df(\eta(t,x))}{\|df(\eta(t,x))\|^2} \\ \eta(0, x) = x \end{cases}$$

It is easy to see that

$$f(\eta(t, x)) = f(x) - t.$$

By the (P.S) condition, $\eta(t,x)$ is well defined in $[0,\bar{t}(x))$, where $\bar{t}(x) = f(x) - c$, and

$$\lim_{t\to\bar{t}(x)} f(\eta(t,x)) = c. \tag{1.1}$$

2^0. We are going to prove that the limit $\lim_{t\to\bar{t}(x)-0}\eta(t,x)$ exists, and then $\eta(t,x)$ can be extended to $[0,\bar{t}(x)]$ such that $f(\eta(\bar{t}(x),x)) = c$.

Let

$$\alpha = \inf_{t\in[0,\bar{t}(x))} \operatorname{dist}(\eta(t,x), K_c) \tag{1.2}$$

We consider two cases: (i) $\alpha > 0$, (ii) $\alpha = 0$.

In case (i), one can easily prove that there is a $z \in K_c$ such that $\lim_{t\to\bar{t}(x)-0}\eta(t,x) = z$ (cf.[Ch 1] or [Wan 1]). So we need only to consider the case (ii) $\alpha = 0$. At first, we declare that

$$\lim_{t\to\bar{t}(x)-0} \operatorname{dist}(\eta(t,x), K_c) = 0. \tag{1.3}$$

The proof of this formula is referred to [Wan 1]. Hence, from the compactness of K_c and $\mathrm{df}|_{K_c} = 0$ it follows that

$$\lim_{t\to\bar{t}(x)-0} \mathrm{df}(\eta(t,x)) = 0 \tag{1.4}$$

By the (P.S) condition, for any sequence $t_i \to \bar{t}(x) - 0$, there is a convergent subsequence of $\eta(t_i,x)$. We declare that the set Λ of limit points of $\eta(t,x)$ for $t \to \bar{t}(x) - 0$ is a connected closed subset of K_c. This conclusion is also referred to [Wan 1].

Now, by the assumption of Theorem 1.1, Λ must be a part of a certain critical orbit, say N. We want to prove that Λ is a point. Take a G-tubular neighbourhood B of N such that $B \cong \nu N(\varepsilon)$, some $\varepsilon > 0$ and Lemma 1.1 holds on B. Obviously, we have

$$\lim_{t\to\bar{t}(x)-0} \operatorname{dist}(\eta(t,x), N) = 0 \tag{1.5}$$

So, there is a $\delta > 0$ such that $\eta(t,x) \in B$, if $t \in [\bar{t}(x) - \delta, \bar{t}(x))$. Using Lemma 1.1, we see that for $t \geq \bar{t}(x) - \delta$, $\eta(t,x)$ preserves the slice. Assume that $z \in \Lambda$, we must have $\eta(t,x) \in \nu_z N(\varepsilon)$ for $t \geq \bar{t}(x) - \delta$. Since $N \bigcap \nu_z N(\varepsilon) = \{z\}$, it follows from (1.5) that $\lim_{t\to\bar{t}(x)-0}\eta(t,x) = z$.

3^0. Now, we define the deformation mapping as follows:

$$\tau(t,x) = \begin{cases} \eta(t\bar{t}(x),x), & (t,x) \in [0,1) \times (f_b \setminus (f_c \bigcup K_b)) \\ \lim_{s\to 1-0}\eta(s\bar{t}(x),x), & (t,x) \in \{1\} \times (f_b \setminus (f_c \bigcup K_b)) \\ x, & (t,x) \in [0,1] \times f_c. \end{cases}$$

The continuity of τ can be verified for following cases:

(a) $(t,x) \in [0,1] \times \mathring{f}_c$;
(b) $(t,x) \in [0,1) \times (f_b \setminus (f_c \bigcup K_b))$;
(c) $(t,x) \in \{1\} \times (f_b \setminus (f_c \bigcup K_b))$;
(d) $(t,x) \in [0,1] \times f^{-1}(c)$.

Case (a) is trivial. Case (b) is verified by using the fundamental theorem of O.D.E. Case (c) and (d) are similar for verifications, we only consider case (c).

Take $x_0 \in f_b \setminus (f_c \bigcup K_b)$. Without loss of generality, assume that $z = \eta(\bar{t}(x_0), x_0) \in K_c$. Denoting the orbit of z by N, we take a G-tubular neighbourhood of N, $B \cong \nu N(\varepsilon_0)$, some $\varepsilon_0 > 0$.

At first, we declare that for any given $\varepsilon, \varepsilon_0 \geq \varepsilon > 0$, there is a $\delta > 0$ such that

$$\eta(t,x) \in B_\varepsilon, \text{if } t \in [\bar{t}(x) - \delta, \bar{t}(x)) \text{ and } \operatorname{dist}(x, x_0) < \delta, \tag{1.6}$$

where $B_\varepsilon = \nu N(\varepsilon)$.

Otherwise, there is a $\varepsilon_1 > 0$ and sequences $t_n \to \bar{t}(x_0) - 0$, $x_n \to x_0$, $n \to \infty$ such that

$$\eta(t_n, x_n) \not\in B_{\varepsilon_1}. \tag{1.7}$$

Now, let $F_1 = \{N\}$ and $F_2 = (M \setminus B_{\varepsilon_0/2}) \bigcap K_c$, then both F_1 and F_2 are compact subsets of K_c. If F_1 and F_2 are both nonempty, by the assumption of Theorem 1.1 and Lemma 1.2, there are two compact subsets of K_c, M_1 and M_2, such that

$$M_1 \bigcup M_2 = K_c, M_1 \bigcap M_2 = \phi, F_i \subset M_i, i = 1,2. \text{ Hence dist } (M_1, M_2) > 0.$$

If F_2 is empty, we take $M_1 = K_c$ and $M_2 = \phi$. Let $E = M_2 \bigcup (M \setminus B_{\varepsilon_0})$, then $\alpha = \operatorname{dist}(E, M_1) > 0$. Without loss of generality, assume that

$$\varepsilon_1 \leq \min\{\alpha/4, \varepsilon_0/4\}.$$

Choosing $\delta_1 > 0$, such that

dist $(\eta(t, x_0), z) < \varepsilon_1/8$, for $t \in [\bar{t}(x_0) - \delta_1, \bar{t}(x_0))$ and then we have $\delta_2 > 0$ such that

dist $(\eta(t, x_0), \eta(t, x)) < \varepsilon_1/8$, for $t \in [0, \bar{t}(x_0) - \delta_1]$ and dist $(x, x_0) < \delta_2$. These imply that

dist $(\eta(\bar{t}(x_0) - \delta_1, x), z) < \varepsilon_1/4$ for dist $(x, x_0) < \delta_2$. From this, we can find a subsequence of x_n (using the same notation) and another sequence t'_n such that

$$\eta(t'_n, x_n) \in B_{\varepsilon_1/4} \tag{1.8}$$

By (1.7), (1.8), one can find two sequences s'_n, s''_n and $s'_n \to \bar{t}(x_0) - 0$, $s''_n \to \bar{t}(x_0) - 0$, for $n \to \infty$ such that

$$\operatorname{dist} \ (\eta(s'_n, x_n), F_1) = \varepsilon_1$$
$$\operatorname{dist} \ (\eta(s''_n, x_n), E) = \varepsilon_1$$

$$\eta(t, x_n) \notin (F_1)_{\varepsilon_1} \bigcup (E)_{\varepsilon_1}, \text{for } t \in [s'_n, s''_n] \quad n = 1, 2, \cdots$$

where $(F_1)_{\varepsilon_1} = \{x \in M | \operatorname{dist}(x, F_1) < \varepsilon_1\}$.

By the (P.S) condition,

$$\beta = \inf_{x \in A} \|df(x)\| > 0, \quad A = f^{-1}([c,b)) \setminus (F_1)_{\varepsilon_1} \bigcup (E)_{\varepsilon_1}$$

Hence,

$$\begin{aligned} \varepsilon_1 &\leq \operatorname{dist}(\eta(s'_n, x_n), \eta(s''_n, x_n)) \\ &\leq \frac{1}{\beta} |s'_n - s''_n| \to 0 \end{aligned}$$

This is a contradiction.

Now, we shall prove that τ is continuous at $(1, x_0)$. Because, if not, there is a $\varepsilon_2 > 0$ and t_n, x_n satisfying $t_n \to \bar{t}(x_0) - 0$, $x_n \to x_0$ for $n \to \infty$ such that

$$\operatorname{dist}(\eta(t_n, x_n), z) \geq 2\varepsilon_2. \tag{1.9}$$

Without loss of generality, assume that $\varepsilon_2 \leq \frac{\varepsilon_0}{2}$. Then it follows from (1.6) that there is a $\delta > 0$ such that

$$\eta(t,x) \in B_{\varepsilon_2}, \text{for } t \in [\bar{t}(x_0) - \delta, \bar{t}(x)), \text{dist}(x, x_0) < \delta. \tag{1.10}$$

So, for n large enough (t_n, x_n) satisfies the above estimation, i.e., $\eta(t_n, x_n) \in B_{\varepsilon_2}$.

From this and (1.9), we must have

$$\text{dist}(\pi(\eta(t_n, x_n)), z) \geq \varepsilon_2 \tag{1.11}$$

where π: $\nu N \to N$ is the bundle projection.

Otherwise,

$$\begin{aligned}\text{dist}(\eta(t_n, x_n), z) \leq &\text{dist}(\eta(t_n, x_n), \pi(\eta(t_n, x_n))) \\ &+ \text{dist}(\pi(\eta(t_n, x_n)), z) < 2\varepsilon_2\end{aligned}$$

contradicting (1.9).

By virtue of Lemma 1.1 and (1.10), (1.11), we can obtain that for n large enough and $t \in [t_n, \bar{t}(x_n))$

$$\text{dist}(\eta(t, x_n), z) \geq \varepsilon_2 \tag{1.12}$$

But, similar to the proof of (1.8), for fixed t' we can find $\delta' > 0$, such that

$$\text{dist}(\eta(t', x_n), z) < \varepsilon_2, \text{for dist}(x_n, x_0) < \delta'. \tag{1.13}$$

So, for each n there is a $t'_n < t_n$, $t'_n \to \bar{t}(x_0) - 0$, if $n \to \infty$, and $\eta(t'_n, x_n) \notin B_{\varepsilon_0}$. We arrive at (1.7), then the contradiction follows in a similar way as above. The continuity of τ is proved.

4^0. One can easily check that the deformation retract τ defined above possesses all properties in Theorem 1.1. So, the proof of Theorem 1.1 is completed.

Remark 1.1: In [Wan 1] this theorem was proved for C^1-smooth function with isolated critical points. It seems that the above theorem should hold for C^1 G-invariant functions. (cf.[Ch 1] [Ro 1], etc.)

§2. Critical group of isolated critical orbits and its calculations

Let M be a C^2 Hilbert G-manifold, $f \in C^{2-0}(M, R)$ be a G-invariant function. Let $p \in M$ be a critical point of f, then we call the orbit $N = G(p)$ an isolated critical orbit if there is a neighbourhood B of N such that $B \bigcap K(f) = \{N\}$.

For a given isolated critical orbit N, we shall assign to it a series of groups which describe the local properties of f on a neighbourhood of N.

Definition 2.1: Let N be an isolated critical orbit of f. The critical group F are difined as follows:

$$C_q(f, N) = H_q(f_c \bigcap U, (f_c \setminus \{N\}) \bigcap U; F) \tag{2.1}$$

where $c = f|_N$ and U is a neighbourhood of N such that $K(f) \bigcap (f_c \bigcap U) = \{N\}$, $H_*(X, Y; F)$ stands for the singular relative homology groups with Ablian coefficient group F.

Remark 2.1: By the excision property of the singular relative homology groups, the critical groups are well defined, i.e., they do not depend upon the special choise of the neighbourhood U.

Example 2.1: Suppose that N is an isolated critical orbit corresponding to a minimum of f, then

$$C_q(f, N) = H_q(N), q = 0, 1, 2, \cdots.$$

Lemma 2.1: Let $f \in C^2(M,R)$ be a G-invariant function, then df is a G-equivariant mapping and $d^2f(p)$ a bounded selfadjoint operator satisfying

$$(2.2) \qquad < d^2f(gp)gX, Y >=< gd^2f(p)X, Y >, \forall X \in T_pM, Y \in T_{gp}M$$

Proof: Differentiate the equality $f(gp) = f(p)$ directly.

Remark 2.2: Suppose that M is a G-manifold, we introduce a G-action on TM as follows:

$$gX = dg_p(X), \quad \forall X \in T_pM$$

then, TM is a G-manifold.

Now, let $f \in C^2(M,R)$ be a G-invariant function and N an isolated critical orbit of f. From G-tubular neighbourhood theorem (cf.[Br 1]), we take a G-tubular neighbourhood B_ε of N such that B_ε is diffeomorphic to $\nu N(\varepsilon)$, where $B_\varepsilon = \{x \in M | \mathrm{dist}(x,N) < \varepsilon\}, \nu N$ is the normal bundle of N and $\nu N(\varepsilon) = \{(x,v) \in \nu N | x \in N, \|v\| < \varepsilon\}$. The diffeomorphism between B_ε and $\nu N(\varepsilon)$ is G-equivariant, and the bundle projection π: $\nu N \to N$. For any $p \in N$, $\nu_p N$ is a G_p-space, where G_p is the isotropy group of p. Since we always consider the local homological property, without loss of generality, we shall identify B_ε with $\nu N(\varepsilon)$ for the simplicity of the notations. We shall often work on the $\nu N(\varepsilon)$ if there is no confusion. We assume that 0 is isolated in $\sigma(d^2f(p)), p \in N$. By virtue of (2.2) we can easily obtain an orthogonal composition of νN,

$$\nu N = \nu^0 N \oplus \nu^+ N \oplus \nu^- N$$

where $\nu_p^0 N$ corresponds to the null space of $d^2f(p)$. $\nu_p^- N$ corresponds to the negatively definite space of $d^2f(p)$ and $\nu_p^+ N$ with similar meaning. Again by (2.2), these three bundles are all G-Hilbert vector bundles, and for any $g \in G$, $g : \nu_p^* N \to \nu_{gp}^* N$ is an isomorphism, where $* = 0, +, -$. Therefore, the following definition is well defined.

Definition 2.2: Suppose that $f \in C^2(M,R)$ and N is an isolated critical orbit. The dimension of $\nu^- N$ is called the Morse index of f at N. N is called a nondegenerate critical orbit if $\dim \nu^0 N = 0, i.e., \nu^0 N = \{N\}$.

Remark 2.3: In [Bo 1], Bott firstly introduced the concept of nondegenerate critical manifold. And in [Was 1], Wasserman developed nondegenerate equivariant Morse theory by virtue of the concept of nondegenerate critical orbit similar to the above definition.

Theorem 2.1: Suppose that $f \in C^{2-0}(M,R)$ is G- invariant, satisfying the (P.S) condition. Assume that c is an isolated critical value of f and $K_c = \{N_j\}_{j=1}^m$, where N_j is an isolated critical orbit and m finite. Then for any $\varepsilon > 0$ such that $[c-\varepsilon, c+\varepsilon]$ contains only a critical value c, we have

$$(2.3) \qquad H_*(f_{c+\varepsilon}, f_{c-\varepsilon}; F) = H_*(f_c, f_c \setminus K_c; F) = \oplus_{j=1}^m C_*(f, N_j)$$

Proof: From Theorem 1.1 in the previous section, we know that f_c is a strong deformation retract of $f_{c+\varepsilon}$ and $f_{c-\varepsilon}$, of $f_c \setminus K_c$. Hence, we have

$$H_*(f_{c+\varepsilon}, f_{c-\varepsilon}; F) = H_*(f_c, f_c \setminus K_c; F)$$

The second isomorphism in (2.3) follows from the excision theorem in singular homology theory.

Corollary 2.1: In addition, we assume that f is bounded from below and c is the minimum of f on M, then

$$(2.4) \qquad H_*(f_{c+\varepsilon}, f_{c-\varepsilon}; F) = \oplus_{j=1}^m H_*(N_j).$$

In order to describe the local property of degenerate isolated critical orbits, we introduce a spacial neighbourhood pair, GM- pair.

Definition 2.3: Assume that $f \in C^{2-0}(M,R)$ is G- invariant and satisfies the (P,S) condition. $N \subset M$ is an isolated critical orbit of f and $f|_N = c$. Let η be the negative gradient flow defined by $-df$. A pair of spaces (W, W_-) is called a GM-pair of f at N, if

(1) W is a closed neighbourhood of N, possessing the mean value property, i.e., $\forall t_1 < t_2$, that $\eta(t_i) \in W, i = 1,2$, implies $\eta(t) \in W, \forall t \in [t_1, t_2]$. And, there exists $\varepsilon > 0$ such that $W \bigcap f_{c-\varepsilon} = f^{-1}[c-\varepsilon, c) \bigcap K(f) = \phi, W \bigcap K(f) = \{N\}$.

(2) $W_- = \{x \in W | \eta(t,x) \bar{\in} W, \forall t > 0\}$ is closed in W.

(3) W_- is a piecewise submanifold and the flow η is transversal to W_-.

Theorem 2.2: Suppose that $f \in C^{2-0}(M,R)$ is G- invariant and satisfies the (P.S) condition. Let (W, W_-) be a GM-pair of f at an isolated critical orbit N. Then we have

$$C_*(f,N) = H_*(W, W_-; F). \tag{2.5}$$

Proof: By means of §1 Theorem 1.1, the proof of this theorem will be similar to that of Theorem 2.2 in [Ch 1]. So, we omit it here.

Next, we shall give the existence of GM-pair of an isolated critical orbit by constructing a special one. Suppose that $f \in C^{2-0}(M,R)$ is G-invariant, satisfying the (P.S) condition. $N \subset M$ is an isolated critical orbit of f. We take a G- invariant tubular neighbourhood O of N, which is diffeomorphic to $\nu N(\delta)$, for some $\delta > 0$ so that there is no other critical orbit in O. We shall work on the $\nu N(\delta)$. Let

$$h(x) = \lambda f(x) + \|x - \pi(x)\|^2.$$

Without loss of generality, assume that $f(p) = 0$ when $p \in N$. Choosing an $\varepsilon > 0$ such that there is no other critical value in $[-\varepsilon, \varepsilon]$, we define

$$W = f^{-1}[-\gamma, \gamma] \bigcap h_\mu \tag{2.6}$$

$$W_- = f^{-1}(-\gamma) \bigcap W \tag{2.7}$$

where λ, γ, μ are positive numbers to be determined by the following conditions.

(1) $\nu N(\frac{\delta}{2}) \subset W \subset \nu N(\delta) \bigcap f^{-1}[-\varepsilon, \varepsilon]$;

(2) $f^{-1}[-\gamma,\gamma] \bigcap h^{-1}(\mu) \subset \nu N(\delta) \setminus \nu N(\frac{\delta}{2})$;

(3) $< dh(x), df(x) >> 0, x \in \nu N(\delta) \setminus \nu N(\frac{\delta}{2})$.

From the (P.S) condition, we have

$$\beta = \inf_{x \in A} \|df(x)\| > 0, A = \nu N(\delta) \setminus \nu N(\frac{\delta}{2})$$

λ, γ and μ are determined consecutively: $\lambda > \frac{2\delta}{\beta}$, $0 < \gamma < \min\{\varepsilon, \frac{3\delta^2}{8\lambda}\}$, $\frac{\delta^2}{4} + \lambda\gamma < \mu < \delta^2 - \lambda\gamma$.

Theorem 2.3: The pair (2.6) (2.7) is a GM-pair of f at N and is G-invariant. Moreover, for any $p \in N$, $f|_{\nu_p N(\delta)}$ has a unique critical point p on $\nu_p N(\delta)$, and if we write

$$(W_p, W_p-) = (W \bigcap \nu_p N(\delta), W_- \bigcap \nu_p N(\delta)). \tag{2.8}$$

then (W_p, W_p-) is a GM-pair of $f|_{\nu_p N(\delta)}$ with respect to the isolated critical point p.

Proof: The verification that (W, W_-) is a GM-pair of f at N is almost the same as in the case of isolated critical points in [Ch 1]. We omit this part of the proof.

Next, assume that $y \in \nu_p N(\delta)$ is a critical point of $f|_{\nu_p N(\delta)}$, then $d_{\nu_p N} f(y) = 0$. Since f is G- invariant, the derivative of f along the tangent space of $G(y)$ is zero, so is it along the vertical direction of $\nu_p N$. Hence, y is a critical point of f and $y \in N$. Since $\nu_p N(\delta) \bigcap N = \{p\}$, $y = p$, i.e., p is a unique critical point of $f|_{\nu_p N(\delta)}$.

From the structure of (W, W_-) defined by (2.6), (2.7), (W, W_-) is G-invariant since f and h are G-invariant. That (W_p, W_{p-}) is a GM-pair of $f|_{\nu_p N(\delta)}$ at p follows from §1 Lemma 1.1 since (W, W_-) is a GM-pair of f at N. The proof is completed.

Definition 2.4: Suppose that $f \in C^{2-0}(M, R)$ is G- invariant, and f satisfies the (P.S) condition. N is an isolated critical orbit of f and $\nu N(\delta)$ is a tubular neighbourhood of N such that $\nu N(\delta) \bigcap K(f) = \{N\}$. The normal critical groups of f at N, $C_*^n(f, N)$, are defined as follows:

$$C_*^n(f, N) = C_*(f|_{\nu_p N(\delta)}, p) \tag{2.9}$$

where $p \in N$, $C_*(f|_{\nu_p N(\delta)}, p)$ are critical groups of $f|_{\nu_p N(\delta)}$ with repsect to isolated critical point p (cf.[Ch 1]).

Lemma 2.2: In the above definition, $C_*(f|_{\nu_p N(\delta)}, p)$ is independent of $p \in N$.

Proof: From Theorem 2.3, $C_*(f|_{\nu_p N(\delta)}, p)$ is well defined for any $p \in N$. Using Theorem 2.2 in [Ch 1] we have

$$C_*(f|_{\nu_p N(\delta)}, p) = H_*(W_p, W_{p-}; F) \tag{2.10}$$

By virtue of the structure of G-tubular neighbourhood $\nu N(\delta)$ and G-invariance of (W, W_-), one can find that for any $g \in G$,

$$(W_{gp}, W_{gp-}) = (gW_p, gW_{p-}). \tag{2.11}$$

Since g is diffeomorphism, we obtain

$$H_*(W_p, W_{p-}) = H_*(W_{gp}, W_{gp-}) \tag{2.12}$$

Since the action of G is transitive on N, the conclusion follows from (2.10) and (2.12)

Remark 2.4: From the above lemma, the definition (2.4) is well defined.

In the following, we shall see that the normal critical groups simplify the calculation of critical groups, so that the critical groups of an isolated critical orbit are related to the topology of the orbit itself.

Theorem 2.4: Suppose that $f \in C^{2-0}(M, R)$ is G- invariant, and that f satisfies the (P.S) condition. Assume that N is an isolated critical orbit of f and its normal bundle is trivial. Then we have

$$C_q(f, N) = \oplus_{i=0}^{q} C_{q-i}^n(f, N) \otimes H_i(N) \qquad q = 0, 1, 2, \cdots. \tag{2.13}$$

Proof: Assume that (W, W_-) is a GM-pair of f at N, then it follows from theorem 2.2 that

$$C_q(f, N) = H_q(W, W_-; F)$$

By the assumption that νN is a product bundle and (2.11), we know that (W, W_-) also possesses a product structure. Hence, it follows from the Kunneth formula (cf.[Gr 1]) that

$$H_q(W, W_-) = \oplus_{i=0}^{q} H_{q-i}(W_p, W_{p-}) \otimes H_i(N)$$

where $p \in N$, $q = 0, 1, 2, \cdots$. Now, the conclusion holds from Definition 2.4.

Now, assume that $f \in C^{2-0}(\nu N(\delta), R)$ is G-invariant and f satisfies the (P.S) condition. Denote $\tilde{f} = f|_{\nu^0 N(\delta)}$, then N is an isolated critical orbit of $\tilde{f}$ on $\nu^0 N(\delta)$. Hence, $C_*(\tilde{f}, N)$ and $C_*^n(\tilde{f}, N)$ are well defined.

Theorem 2.5: Suppose that $f \in C^2(M,R)$ is G- invariant, and f satisfies the (P.S) condition. Assume that N is an isolated critical orbit with the Morse index λ_N and the normal bundle of N is trivial. Then we have

$$C_q(f,N) = \oplus_{i=0}^{q} C^n_{q-i-\lambda_N}(\tilde{f},N) \otimes H_i(N) \tag{2.14}$$

$q = 0,1,2,\cdots$. $\tilde{f}$ is defined as above.

Proof: Note that the Morse index of f with respect to N is exactly the same as that of $f|_{\nu_p N(\delta)}$ with respect to p, for any $p \in N$. By virtue of Theorem 2.4 in [Ch 1] (Shifting theorem), we have

$$\begin{aligned} C_q(f|_{\nu_p N(\delta)}, p) &= C_{q-\lambda_N}(f|_{\nu_p^0 N(\delta)}, p) \\ &= C_{q-\lambda_N}(\tilde{f}|_{\nu_p^0 N(\delta)}, p) \end{aligned}$$

Hence,

$$C^n_q(f,N) = C^n_{q-\lambda_N}(\tilde{f},N)$$

The conclusion follows from this and (2.13).

Corollary 2.2: The assumptions are the same as those in Theorem 2.5, then we have

$$C_*(f,N) = C_{*-\lambda_N}(\tilde{f},N) \tag{2.15}$$

Proof: Applying (2.13) to $\tilde{f}$, we have

$$C_q(\tilde{f},N) = \oplus_{i=0}^{q} C^n_{q-i}(\tilde{f},N) \otimes H_i(N), \quad q = 0,1,2,\cdots. \tag{2.16}$$

Hence the proof will be completed by comparing (2.14) and (2.16).

Corollary 2.3: The assumptions are the same as those in the above theorem, and in addition we assume that N is a nondegenerate critical orbit, then

$$C_*(f,N) = H_{*-\lambda_N}(N) \tag{2.17}$$

Proof: Since for $p \in N$, p is a nondegenerate critical point of $f|_{\nu_p N(\delta)}$ with the Morse index λ_N, by Theorem 1.1 in [Ch 1], we have $C^n_{\lambda_N}(f,N) = F$, $C^n_q(f,N) = 0$, $q \neq \lambda_N$. Combining (2.13), we obtain (2.17).

Remark 2.5: (2.17) was proved in [Bo 1] and [Was 1] without the assumption that νN is a trivial bundle. We can also prove this result directly without the assumption of trivial normal bundle by using the Thom isomorphism theorem (cf. [Hu 1]) in the proof of Theorem 2.4 instead of the Kunneth formula, since in this situation we can reduce the calculation of $H_*(W, W_-)$ to a problem of disc bundle through a simple deformation. This motivates us to drop the assumption of trivial normal bundle in Theorems 2.4 and 2.5 and Corollary 2.2. A possible approach is to use the Leray-Hirch theorem (cf.[Hu 1]) in the proof of the theorem 2.4 instead of the Kunneth formula. However, we do not know whether or not the conditions of the Leray-Hirch theorem are all satisfied in the present situation. We tend to affirm that this is true.

Remark 2.6: The above results relate the calculation of the critical groups to the topology of the orbit itself. On the other hand, they enable us to utilize known results on isolated critical points.

Below, we shall give some examples. Take the coefficient group as Z_2, and assume that critical orbits discussed below all possess the trivial normal bundle.

Example 2.2: Let $f \in C^{2-0}(R^m, R)$ be G- invariant and N be an isolated critical orbit corresponding to a maximum of f, i.e., $\exists$ a neighbourhood O of N, such that $\forall x \in O \setminus N$, $f(x) < f(p)$, $p \in N$. If $\dim N = n$, then

$$C_q(f,N) = H_{q-m-n}(N) \quad q = 0,1,2,\cdots. \tag{2.18}$$

In fact, the result follows from (2.13) and Example 1 in [Ch 1] §1.

Example 2.3: Assume that N is an isolated critical orbit of f, and does not correspond to a minimum of f, then

$$(2.19) \qquad C_0(f,N)=0$$

In fact, from (2.13) we have

$$C_0(f,N)=C_0^n(f,N)\otimes H_0(N).$$

Now, for any $p\in N$, p is not a minimal critical point. Using Example 2 in [Ch 1] §1, we obtain $C_0^n(f,N)=0$. So the result follows.

Example 2.4: Let f be a G-invariant smooth function on R^m, and N be an isolated critical orbit of f. Assume that N does not corresponding to a maximum of f then

$$(2.20) \qquad C_m(f,N)=0$$

In fact, if we write dim $N=q_0\leq m-1$, then dim $\nu N=m-q_0$. From (2.13), we have

$$C_m(f,N)=\oplus_{i=0}^m C_{m-i}^n(f,N)\otimes H_i(N)$$

Since $H_i(N)=0$ $i>q_0$ and $C_{m-i}^n(f,N)=0$, $0\leq i<q_0$, we have $C_m(f,N)=C_{m-q_0}^n(f,N)\otimes H_{q_0}(N)$. Since for $p\in N$, p is not a maximal critical point, by Example 2 in [Ch 1] §1, we see $C_{m-q_0}^n(f,N)=0$, then $C_m(f,N)=0$ follows.

In the following, we shall calculate the critical groups of an isolated critical orbit corresponding to a Mountain-Pass critical value as an application of a series of the results above.

Let X be a Hilbert G-space and $f\in C^{2-0}(X,R)$, G- invariant. Assume that f satisfies the (P.S) condition. Let $x_0,x_1\in X$, we write

$$\Gamma=\{l\in C([0,1],X)|l(0)=x_0,l(1)=x_1\}$$

and define

$$c=\inf_{l\in\Gamma}\sup_{x\in[0,1]} f(l(x)).$$

Lemma 2.3: If $c>\text{mas}\{f(x_0),f(x_1)\}$, then c is a critical value of f, and one of the following possibilities holds:

(1) c is not isolated,

(2) K_c consists of infinitely many critical orbits,

(3) $\exists$ an isolated critical orbit $N\subset K_c$ such that the rank $C_1(f,N)>0$.

Proof: From the proof of Lemma 7.1 in [Ch 1], one can similarly obain the rank $H_1(f_{c+\varepsilon},f_c)>0$, and then c is a critical value. Now, assume that c is an isolated critical value, and that K_c consists of finite critical orbits, say $N_1,\cdots,N_m$, then by (2.4)

$$H_1(f_{c+\varepsilon},f_c)=\oplus_{j=1}^m C_1(f,N_j)$$

The lemma is proved.

Theorem 2.6: Suppose that $f\in C^2(X,RR)$ satisfies the (P.S) condition. Assume that N is an isolated critical orbit of f with a trivial normal bundle satisfying

$$(2.21) \qquad \text{rank } C_1(f,N)>0$$

Assume that the Morse index of f at N is λ_N, and that f satisfies

$$(2.22) \qquad \dim\nu^0 N\leq 1, \quad \text{if } \lambda_N=0.$$

Then $\lambda_N \le 1$. Moreover,

(1) When $\lambda_N = 1$ or rank $H_1(N) = 0$, we have
$C_q(f,N) = H_{q-1}(N) \quad q = 0,1,2,\cdots$.

(2) when $\lambda_N = 0$ and rank $H_1(N) > 0$, we have two possibilities:
(a) $C_q(f,N) = H_{q-1}(N)$
(b)$C_q(f,N) = H_q(N) \quad q = 0,1,2,\cdots$.

Proof: At first, from the rank $C_1(f,N) > 0$ and (2.15), it follows that $\lambda_N \le 1$.
If $\lambda_N = 1$, by (2.15) we know that the rank $C_0(\tilde{f},N) > 0$.

From Example 2.3, we see that N must correspond to a minimum of $\tilde{f}$. Again using Example 2.1, we obtain

$$C_q(\tilde{f},N) = \oplus_{i=0}^{q} H_q(N) \quad q = 0,1,2,\cdots.$$

Then

$$C_q(f,N) = H_{q-1}(N) \quad q = 0,1,2,\cdots.$$

If $\lambda_N = 0$ and the rank $H_1(N) = 0$, by (2.14) we see

$$C_1(f,N) = C_1^n(\tilde{f},N) \otimes H_0(N)$$

It follows that the rank $C_1^n(\tilde{f},N) > 0$. We declare that

$$\dim \nu^0 N = 1,$$

Otherwise, from (2.22) we have

$$\dim \nu^0 N = 0,$$

It follows that N is a nondegenerate critical orbit. Using (2.17) and $\lambda_N = 0$, we see that

$$\text{rank } C_1(f,N) = \text{rank } H_1(N) = 0$$

contradicting (2.21).

Now, for $p \in N$, $\tilde{f}|_{\nu_p^0 N(\delta)}$ is a one dimensional function, from rank $C_1^n(\tilde{f},N) > 0$ and Example 3 in [Ch 1] §2, we have

$$C_q^n(\tilde{f},N) = \begin{cases} Z_2 & q = 1 \\ 0 & q \neq 1. \end{cases}$$

Combining this with (2.14), we obtain

$$C_q(f,N) = \oplus_{i=0}^{q} C_{q-i}^n(\tilde{f},N) \otimes H_i(N) = H_{q-1}(N)$$

If $\lambda_N = 0$ and rank $H_1(N) > 0$ and we assume that N does not correspond to a minimum of f, then similarly as above we can prove that $\dim \nu^0 N = 1$. And from Example 2.3, we have $C_0(f,N) = 0$. By (2.14) $C_0(f,N) = C_0^n(\tilde{f},N) \otimes H_0(N) = 0$, it follows that $C_0^n(\tilde{f},N) = 0$. Again by (2.14), we see that

$$C_1(f,N) = C_1^n(\tilde{f},N) \otimes H_0(N)$$

Therefore, we obtain rank $C_1^n(\tilde{f},N) > 0$. Since for $p \in N, \tilde{f}|_{\nu_p^0 N(\delta)}$ is a one dimensional function, then

$$C_q^n(\tilde{f},N) = \begin{cases} Z_2 & q = 1 \\ 0 & q \neq 1 \end{cases}$$

Hence, the result follows from this and (2.14).

§3. Multiple critical orbit theorems

At first, we note that the three solution theorem (cf.[Ch 1]), generally, does not imply the existence of three geometrically different solutions (cf. Remark 3.1 in this section). In the first part of this section we shall consider the equivariant three solution theorems. Afterwards, we shall turn to give a five solution theorem. Below, the coefficient group is always taken as Z_2.

Lemma 3.1: Let X be a Hilbert G-space and $f \in C^{2-0}(X, R)$ be a G-invariant function. Assume that f satisfies the (P.S) condition and is bounded from below. Suppose that f has only two critical orbits N_0, N_1, corresponding to critical values $c_0 \leq c_1$ respectively. Then

(1) If $c_0 = c_1$, we have

$$H_q(N_0) \oplus H_q(N_1) = \begin{cases} Z_2, & q = 0 \\ 0, & q \neq 0 \end{cases} \tag{3.1}$$

(2) If $c_0 < c_1$, we have

$$C_q(f, N) = H_{q-1}(N_0), \ q \geq 2 \tag{3.2}$$

and a short exact sequence:

$$0 \to C_q(f, N_1) \to H_0(N_0) \to Z_2 \to C_0(f, N_1) \to 0 \tag{3.3}$$

Proof: Firstly, note that f can attain its minimum on X under the assumptions of the lemma.

(1) If $c_0 = c_1$, N_0, N_1 all correspond to the minimum of f. By Theorem 2.1 and Example 2.1, we have

$$\begin{aligned} H_q(f_{c_0+\varepsilon}) &= H_q(f_{c_0+\varepsilon}, f_{c_0-\varepsilon}) \\ &= C_q(f, N_0) \oplus C_q(f, N_1) \\ &= H_q(N_0) \oplus H_q(N_1) \end{aligned}$$

Since f has no other critical values, by the deformation property, $f_{c_0+\varepsilon}$ is a deformation retract of X, and then

$$H_q(X) = H_q(f_{c_0+\varepsilon})$$

Hence (3.1) follows from the contractibility of X.

(2) If $c_0 < c_1$, taking an $\varepsilon > 0$ such that $c_0 < c_1 - \varepsilon$ and a triad of spaces $(f_{c_0-\varepsilon}, f_{c_1-\varepsilon}, f_{c_1+\varepsilon})$, we can obtain an exact sequence (cf.[Gr 1]):

$$\cdots \to H_q(f_{c_1-\varepsilon}, f_{c_0-\varepsilon}) \to H_q(f_{c_1+\varepsilon}, f_{c_0-\varepsilon}) \to H_q(f_{c_1+\varepsilon}, f_{c_1-\varepsilon}) \to \cdots$$

By Theorem 2.1 and arguments similar to the above, we see that

$$\begin{aligned} &H_q(f_{c_1-\varepsilon}, f_{c_0-\varepsilon}) = C_q(f, N_0) = H_q(N_0) \\ &H_q(f_{c_1+\varepsilon}, f_{c_0-\varepsilon}) = H_q(f_{c_1+\varepsilon}) = H_q(X) = \begin{cases} Z_2 & q = 0 \\ 0 & q \neq 0. \end{cases} \\ &H_q(f_{c_1+\varepsilon}, f_{c_1-\varepsilon}) = C_q(f, N) \end{aligned}$$

Hence,

$$\cdots \to H_q(N_0) \to H_q(X) \to C_q(f, N_1) \to \cdots$$

and (3.2), (3.3) follow.

<u>Theorem 3.1</u>: Suppose that $f \in C^2(X, R)$ is G- invariant and bounded from below. Assume that f satisfies the (P.S) condition. Assume that θ is a nondegenerate critical point of f with the Morse index λ and $N = G(p)$ is a critical orbit of f corresponding to the minimum of f. Then f has at least three critical orbits if one of the following four conditions holds:

(1) $\dim N = 0$, $p \in \mathrm{Fix}_G$ and $\lambda \geq 1$.

(2) $\dim N = 0$, $p \notin \mathrm{Fix}_G$ and $\lambda \neq 1$.

(3) $\dim N = n \geq 1$ and N is homologically different from the n- dimensional sphere S^n.

(4) $\dim N = n \geq 1$, N is homologically the same as S^n and $\lambda \neq n+1$.

<u>Proof</u>: since θ is a nondegenerate critical point with the Morse index λ, by Theorem 1.1 in [Ch 1]

$$C_q(f, \theta) = \begin{cases} Z_2 & q = \lambda \\ 0 & q \neq \lambda \end{cases} \tag{3.4}$$

From the assumptions of this theorem, one can easily see that θ and N are different orbits. If f has only these two critical orbits, we shall deduce contradictions as follows. At first, from (3.1) we know that $c_0 < c_1$, where c_0 and c_1 are critical values corresponding to N and θ respectively. Four cases are considered respectively.

(1) since $\dim N = 0$, by (3.2)

$$C_q(f, \theta) = H_{q-1}(N) = 0 \quad \text{if } q \geq 2.$$

which contradicts (3.4) if $\lambda \geq 2$. If $\lambda = 1$, by (3.3), (3.4) and $p \in \mathrm{Fix}_G$ we obtain

$$0 \to Z_2 \to Z_2 \to Z_2 \to 0$$

which contradicts the exactness.

(2) By the assumptions, N consists of m points with $m \geq 2$. By (3.2), $C_q(f, \theta) = 0$ if $q \geq 2$. This contradicts (3.4) if $\lambda \geq 2$. If $\lambda = 0$, using (3.3) we have

$$0 \to Z_2^m \to Z_2 \to Z_2 \to 0$$

which also contradicts the exactness.

(3) In this situation, N is a compact manifold without boundary, then we see (cf.[Gr 1]):

$$H_n(N) = Z_2^m.$$

where m is the number of connected components of N. Two cases are considered: (a) $m \geq 2$ and (b) $m = 1$.

(a) $m \geq 2$. By (3.2)

$$C_{n+1}(f, \theta) = H_n(N) = Z_2^m.$$

This contradicts (3.4).

(b) By the assumption that N is homologically different from S^n and $m = 1$, there is a q_0 satisfying $1 \leq q_0 < n$ such that $H_{q_0}(N) = Z_2^k$, for some $k \geq 1$. Using (3.2) we obtain

$$C_{q_0+1}(f, \theta) = Z_2^k.$$

But again using (3.2), we also have

$$C_{n+1}(f, \theta) = Z_2.$$

Since $q_0 + 1 < n + 1$, we obtain a contradiction.

(4) By the assumptions, we have

$$H_q(N) = \begin{cases} Z_2 & q = 0, n \\ 0 & q \neq 0, n \end{cases}$$

By (3.2), $C_{n+1}(f, \theta) = Z_2$. This contradicts (3.4) for $\lambda \neq n + 1$.

Corollary 3.1: The assumptions are the same as those in the above theorem. Again assume that $\lambda \geq \dim G + 2$. Then f has at least three critical orbits.

The following theorem allows that θ is a degenerate critical point. In this situation, the topology of the critical orbit corresponding to the minimum of f will play an essential role.

Theorem 3.2: Suppose that $f \in C^2(X, R)$ is G- invariant and bounded from below, and suppose that f satisfies the (P.S) condition. Assume that θ is a critical point of f with the Morse index λ, and that N is a critical orbit corresponding to the minimum of f. Then f has at least three critical orbits if one of the following two conditions holds:

(1) N is not connected and $\lambda \geq 2$.

(2) there is a $q_0 \geq 1$ such that rank $H_{q_0}(N) > 0$ and $\lambda \geq q_0 + 2$.

Proof: Suppose that the conclusion is not true, we shall deduce contradictions for two cases respectively.

(1) By the assumption (1), $H_0(N) = Z_2^m$, where $m \geq 2$. From (3.3), we have

$$0 \to C_1(f, \theta) \to Z_2^m \to Z_2 \to C_0(f, \theta) \to 0$$

Moreover, by the shifting theorem in [Ch 1] we have $C_q(f, \theta) = 0$, if $q = 0, 1$. Combining the above exact sequence we obtain a contradiction.

(2) Again by the shifting theorem in [Ch 1],

$$C_q(f, \theta) = 0, \quad q = 0, 1, \cdots, \lambda - 1.$$

Since $2 \leq q_0 + 1 \leq \lambda - 1$, by (3.2),

$$\text{rank } C_{q_0+1}(f, \theta) = \text{rank } H_{q_0}(N) > 0$$

Again there is a contradiction. The proof is completed.

Remark 3.1: The case (1) in Theorem 3.1 exactly corresponds to the three solution theorem (cf.[Ch 1]). For the case (4), we shall give a counterexample when $\lambda = \dim N + 1$. Take $f(r)$ as a one dimensional even function, which has only three critical points $\theta, 1, -1$ such that $1, -1$ are two minimal points and $f''(\theta) < 0$. Now, let $X = R^n (n \geq 2)$, then there is a natural orthogonal action of $SO(n)$ on R^n. Let $h : R^n \to R$, $h(x) = f(\|x\|^2)$, then f is a $SO(n)$-invariant function and θ, a nondegenerate critical point with the Morse index $\lambda = n$. However, f has only one critical orbit $N = \{x \in R^n | \|x\| = 1\}$ besides θ.

As an application of Theorem 2.6, we have the following multiple critical orbit theorem. For the simplicity of statement we assume that the conditions in Theorem 2.6, such as trivial normal bundle and (2.22), are satisfied.

Theorem 3.3: Suppose that $f \in C^2(X, R)$ is G- invariant and bounded below. Assume that θ is a critical point with the Morse index λ, and that $x_0, x_1 \in \text{Fix}_G$ are two minimal critical points. Suppose that f has no other critical points in Fix_G besides θ, x_0, and x_1. Then f has at least five critical orbits if $\lambda \geq \dim G + 3$.

Proof: At first, by the Mountain-Pass lemma (cf.[Ch 1]), we obtain a critical point x_2 because we may assume that x_0, x_1 are two strictly minimal points (otherwise the theorem will hold). By the assumption, $x_2 \notin \text{Fix}_G$, we denote the orbit of x_2 by N. Without loss of generality, we write $f(\theta) = 0$, $f(x_i) = c_i$, $i = 0, 1, 2$, and $c_0 \leq c_1 < c_2 < 0$. Then, if f has no other critical orbits, we may assume that rank $C_1(f,N) > 0$ (cf.lemma 2.3). And from Theorem 1.1 and 1.2 we can show that

$$C_q(f,\theta) = H_q(f_0, f_{c_2});$$
$$C_q(f,N) = H_q(f_{c_2}, f_{c_1});$$
$$C_q(f,x_0) \oplus C_q(f,x_1) = H_q(f_{c_1}).$$

By the exact sequences of the pair (f_{c_2}, f_{c_1}) and (f_0, f_{c_2}) (see [Gr 1]), combining the above formulas we have two exact sequences:

$$\cdots \to \oplus_{i=0,1} C_q(f,x_i) \to H_q(f_{c_2}) \to C_q(f,N) \to \cdots$$
$$\cdots \to H_q(f_{c_2}) \to H_q(f_0) \to C_q(f,\theta) \to \cdots$$

Since for $i = 0, 1$, $C_q(f,x_i) = 0$, if $q \geq 1$ and $C_0(f,x_i) = Z_2$. We obtain

(3.5) $$H_q(f_{c_2}) = C_q(f,N), \text{if } q \geq 2$$

and

(3.6) $$0 \to H_1(f_{c_2}) \to C_1(f,N) \to Z_2^2 \to H_0(f_{c_2}) \to C_0(f,N) \to 0$$

On the other hand, by the deformation property, f_0 is a deformation ratract of X, then $H_q(f_0) = 0$, if $q \geq 1$, $H_0(f_0) = Z_2$. Therefore we obtain

(3.7) $$C_q(f,\theta) = H_{q-1}(f_{c_2}), \text{if } q \geq 2$$

and

(3.8) $$0 \to C_1(f,\theta) \to H_0(f_{c_2}) \to Z_2 \to C_0(f,\theta) \to 0.$$

Since $\lambda \geq \dim G + 3$, by the shifting theorem in [Ch 1],

(3.9) $$C_q(f;\theta) = 0 \quad \text{if } q = 0, 1, \cdots, \dim G + 2.$$

By virtue of Theorem 2.6, there are two possibilities:

(3.10) $$C_q(f,N) = H_{q-1}(N)$$

or

(3.11) $$C_q(f,N) = H_q(N)$$

Now, three cases are considered according to the dimension of N.

(1) $\dim N = n \geq 2$. From (3.5), (3.7) we have

(3.12) $$C_q(f,\theta) = H_{q-1}(f_{c_2}) = C_{q-1}(f,N) \quad (\text{for } q \geq 3)$$

If (3.10) holds,

$$H_n(N) = C_{n+1}(f,N) = C_{n+2}(f,\theta) = 0$$

follows from $n \leq \dim G$ and (3.9), (3.12). However, $H_n(N) = Z_2^m$, where m is the number of connected components of N (cf.[Gr 1]), we obtain a contradiction.

If (3.11) holds, we also have

$$H_n(N) = C_n(f, N) = C_{n+1}(f, \theta) = 0,$$

contradicting $H_n(N) = Z_2^m$.

(2) dim $N = 1$. If (3.10) holds, in a similar way we also have the contradiction $Z_2^m = H_1(N) = C_3(f, \theta) = 0$.

Hence, assume that (3.11) holds. By (3.8) and (3.9), $0 \to H_0(f_{c_2}) \to Z_2 \to 0$, then $H_0(f_{c_2}) = Z_2$. And by (3.7) and (3.9), $H_1(f_{c_2}) = C_2(f, \theta) = 0$. Then from (3.6), we have

$$0 \to C_1(f, N) \to Z_2^2 \to Z_2 \to C_0(f, N) \to 0. \tag{3.13}$$

Since dim $N = 1$, it is easy to see that

$$H_1(N) = H_0(N) = Z_2^m$$

Hence,

$$0 \to Z_2^m \to Z_2^2 \to Z_2 \to Z_2^m \to 0.$$

which contradicts the exactness.

(3) dim $N = 0$. In this situation, we know that $m \geq 2$ for $x_2 \notin \mathrm{Fix}_G$. And (3.13) also holds. If (3.10) holds, we obtain

$$0 \to Z_2^m \to Z_2^2 \to Z_2 \to 0$$

and if (3.11) holds we obtain

$$0 \to Z_2^2 \to Z_2 \to Z_2^m \to 0.$$

They all contradict the exactness for $m \geq 2$. Now, the theorem is proved.

§4. Applications to nonlinear partial differential equations

1^0. Consider an asymptotically linear wave equation

$$\begin{cases} u_{tt} - u_{xx} + f(t, x, u) = 0, (t, x) \in Q = [0, 2\pi) \times [0, \pi]. \\ u(t, x) = u(t + 2\pi, x) \\ u(t, 0) = u(t, \pi) = 0 \end{cases} \tag{4.1}$$

In recent years, much work has been devoted to the studies of semi-linear wave equations, some of them concern with the asymptotically linear case. K.C.Chang & al studied this problem in [CWL 1]. In their paper under some reasonable conditions they give the existence of at least three solutions. However, if the function f does not depend on the variable t, equation (4.1) has a S^1-symmetry. To be precise, if u is a solution, we define a family of functions as follows, for $\theta \in [0, 2\pi)$

$$(S_\theta u)(t, x) = u(t + \theta, x)$$

which correspond to an S^1-orbit and all are solutions of (4.1). We call two solutions geometrically different if they are in different orbits. We have to distinguish geometrically different solutions.

<u>Theorem 4.1</u>: Suppose that $f(x, u) \in C^1([0, \pi] \times R, R)$ satisfies the following conditions:

(f_1) there is $\beta > 0$ such that

$$0 < \frac{\partial f}{\partial u}(x, u) < \beta, f(x, 0) = 0, f(x, \pm\infty) = \pm\infty$$

(f_2) $\frac{f(x,u)}{u} \leq r < 3$, as $|u|$ large enough uniformly for $x \in [0, \pi]$

(f_3) $\frac{f(x,u)}{u} \geq \rho > 5$, as $|u|$ small enough uniformly for $x \in [0, \pi]$. Then the equation (4.1) has at least three geometrically different solutions.

The proof of this result will be based on the frame used in [CWL 1], and we shall apply Theorem 3.2 to the completion of the proof. So, we shall go quickly through the proof and only with careful verifications for symmetry.

For simplicity, we suppose that $f = f(u)$ only. Let $h = f^{-1}$, then $h \in C^1(R, R)$ and satisfies

(h_1) $\frac{1}{\beta} < h'(t) < \infty, h(0) = 0$

(h_2) $\frac{h(t)}{t} \geq \frac{1}{r} > \frac{1}{3}$, for $|t|$ large enough

(h_3) $\frac{h(t)}{t} \leq \frac{1}{\rho} < \frac{1}{5}$, for $|t|$ small enough.

Let $\square$ be the linear differential operator: $\frac{\partial^2}{\partial t^2} - \frac{\partial^2}{\partial x^2}$ with domain: $D(\square) = \{u \in C^2(Q) | 2\pi$-periodic in t and $u(t,0) = u(t,\pi) = 0, \forall t[0, 2\pi)\}$, and A be the self- adjoint extension of $\square$ on the Hilbert space $L^2(Q)$. We denote the range of A by $R(A)$, and the null space of A by $K(A)$. Let J be the inverse of A, defined on $R(A) = K(A)^{\perp}$. Now, the problem (4.1) is reduced to finding the critical points of the functional

$$I(v) = \frac{1}{2}\int_Q Jv \cdot v + \int_Q H(v), \tag{4.2}$$

on the Hilbert subspace $R(A)$, where $H(v) = \int_0^v h(t)dt$, or equivalently, to finding the solutions of the operator equation:

$$Jv + Ph(v) = 0 \tag{4.3}$$

where P is the orthogonal projection onto the subspace $R(A)$, and $h(\cdot)$ is the Nemytski operator $v \to h(v(x))$, from $L^2(Q)$ into itself.

Let us now introduce an S^1-action on $R(A)$ as follows:

$$(S_\theta u)(t, x) = u(t + \theta, x), \text{for any } \theta \in [0, 2\pi).$$

It is easy to see that this action is orthogonal with respect to the inner product of $L^2(Q)$, and that J and Ph are both equivariant with respect to this action. So, the operator equation (4.3) is S^1-equivariant. By means of the saddle point reduction (cf.[CWL 1]), the problem would be reduced to a finite dimensional one, and we note that the later problem is also S^1-equivariant. This can be seen as follows.

Suppose that E_λ is the spectrum resolution of J, and let

$$P^+ = \int_{-\frac{1}{\beta}}^{+\infty} dE_\lambda, P^- = \int_{-\infty}^{-\frac{1}{\beta}} dE_\lambda$$

The equation (4.3) is equivalent to the system

$$P^+Jv + P^+Ph(v) = 0$$
$$P^-Jv + P^-Ph(v) = 0$$

By the saddle reduction, we have a mapping $v_+ = v_+(v_-)$: $P^-R(A) \to P^+R(A)$ which is the unique solution of following equation for fixed $v_- \in P^-R(A)$:

$$P^+J(v_+ + v_-) + P^+Ph(v_+ + v_-) = 0$$

Since $P^+R(A)$ and $P^-R(A)$ are S^1-invariant subspace, we can check that $v_+ = v_+(v_-)$ is an S^1-equivariant mapping from the equivariance of J, P, P^+, P^-, h and the uniqueness of $v_+ = v_+(v_-)$. So, we obtain a finite dimensional problem:

$$(4.4) \qquad P^- JV(w) + P^- Ph(V(w)) = 0$$

which is also S^1-equivariant, where $w \in P^-R(A)$, $V(w) = v_+(w) + w$. Equivalently, we study the corresponding functional

$$(4.5) \qquad a(w) = \frac{1}{2}\int_Q JV(w)\cdot V(w) + \int_Q H(V(w)).$$

From (h_2), one can verify $a(w) \to +\infty$ as $\|w\| \to \infty$, so $a(w)$ is bounded below. On the other hand, one can check that θ is a critical point of a and

$$d_w a^2(\theta) = J + P^- Ph'(0)$$

By (h_3), one can easily see that the Morse index of θ is equal to or larger than the sum of dimensions of the eigenspaces corresponding to eigenvalues $-\frac{1}{3}, -\frac{1}{5}$, i.e. $\lambda \geq 4$. Now, note that $P^-R(A)$ is spanned by cosktsinjx, sinktsinjx, where, $(k, j) \in Z \times N^*$, and $0 < k^2 - j^2 < \beta$. Since $k \neq 0$, one may check that the action of S^1 on $P^-R(A)$ has no fixed points except θ. It follows that the critical orbit N corresponding to the minimum of a is homeomorphic to S^1, i.e., $H_1(N) \neq 0$. Now, the proof can be completed by applying Theorem 3.2.

2^0. Consider a bifurcation problem for potential operators

Let X be Hilbert G-space and L, a bounded self-adjoint operator on X. $\Omega \subset X$ is a G-invariant neighbourhood of θ. Assume that $F \in C^1(\Omega, X)$ is a potential operator, i.e., $\exists f \in C^2(\Omega, R)$ such that $df = F$, and that F satisfies $F(u) = o(\|u\|)$ as $\|u\| \to 0$. Assume that L and F are G-equivariant. We look for solutions of the following equation with a parameter $\lambda \in R^1$

$$(4.6) \qquad Lu + F(u) = \lambda u$$

$u = \theta$ is a trivial solution for all $\lambda \in R$. We concern with the nontrivial solutions of $(4.6)_\lambda$ with small $\|u\|$.

This problem was studied by some authors (cf.[Ra 1], [Ch 1] for references) and some multiple bifurcation results were obtained. But, if symmetry occurs, we shall face the same problem as above, that is, we have to distinguish geometrically different solutions. Our result is the following theorem which is a complement of the results in [Ra 1] and [Ch 1].

<u>Theorem 4.2</u>: Let μ be an isolated eigenvalue of L with a finite multiplicity, then (μ, θ) is a bifurcation point of $(4.6)_\lambda$. Moreover, if the multiplicity of L is equal to or larger than 2 and the action of G on the unit sphere of eigenspace of μ is not transitive, then at least one of the following alternatives occurs:

(1) (μ, θ) is not an isolated solution of $(4.6)_\mu$ in $\{\mu\} \times \Omega$.

(2) There is a one side neighbourhood Λ of μ such that for all $\lambda \in \Lambda \setminus \{\mu\}$, $(4.6)_\lambda$ possesses at least two nontrivial solutions, which are geometrically different.

(3) There is a neighbourhood I of μ such that for all $\lambda \in I \setminus \{\mu\}$, $(4.6)_\lambda$ possesses at least one nontrivial solution.

The proof depends on the Lyapunov-Schmidt reduction as in [Ch 1], and we shall only pay attention to the parts concerning with the symmtry. Let $Y = \mathrm{Ker}(L - \mu I)$, with $2 \leq \dim Y = n < +\infty$, and let P and $P^\perp$ be the orthogonal projections onto Y and $Y^\perp$ repsectively. Then (4.6) is equivalent to the following system:

$$(4.7) \qquad \mu y + PF(y + y^\perp) = \lambda y$$

$$(4.8) \qquad Ly^{\perp} + P^{\perp}F(y + y^{\perp}) = \lambda y^{\perp}$$

where $y \in Y$, $y^{\perp} \in Y^{\perp}$. As [Ch 1], by Lyapunov- Schmidt reduction, the equation (4.8) is uniquely solvable in a small neighbourhood $V \times O$ of $(\mu, \theta) \in R \times Y$, say $y^{\perp} = \varphi(\lambda, y)$ for $(\lambda, y) \in V \times O$, where $\varphi \in C^2(V \times O, Y^{\perp})$. Without loss of generality, we may assume that O is G-invariant. Since L and F are G-equivariant and Y and $Y^{\perp}$ are G-invariant subspaces, we have

$$Lg\varphi(\lambda, y) + P^{\perp}gy + g\varphi(\lambda, y) = \lambda g\varphi(\lambda, y)$$

By the uniqueness of φ,

$$\varphi(\lambda, gy) = g\varphi(\lambda, y)$$

i.e., φ is G-equivariant on $y \in Y$.

Now, the problem is reduced to a finite dimensional one:

$$(4.9) \qquad \mu y + PF(y + \varphi(\lambda, y)) = \lambda y$$

which is again a variational problem and G-equivariant. Let

$$(4.10) \qquad I_{\lambda}(y) = \frac{1}{2}(\mu - \lambda)\|y\|^2 + \frac{1}{2}(L\varphi, \varphi) + f(y + \varphi) - \frac{\lambda}{2}\|\varphi\|^2$$

It is easy to verify that (4.9) is the Euler equation of I_{λ}, and $\varphi(\lambda, y) = o(\|y\|)$, at $y = \theta$.

The problem is to find critical points of I_{λ} near $y = \theta$, for fixed λ near μ, where $I_{\lambda} \in C^2(V \times O, R)$. Obviously, θ is a critical point of I_{λ}. If θ is not an isolated critical point of I_{μ}, that is, case (1) does not hold, then there are two possibilities:

(i) either θ is a local maximum or a local minimum of I_{μ}.

(ii) θ is neither a local maximum nor a local minimum of I_{μ}.

In case (i), we suppose that θ is a local minimum of I_{μ}. Take an $\varepsilon > 0$, such that $W = (I_{\mu})_{\varepsilon} = \{y \in O | I_{\mu}(y) \leq \varepsilon\}$ is a neighbourhood of θ, containing θ as the unique critical point. Obviously, W is G-invariant and the negative gradient flow of I_{μ} preserves W. Therefore the negative gradient flow of I_{λ} preserves W for $|\mu - \lambda|$ small. If $\lambda > \mu$, it is easy to see that θ is a local maximal point, then I_{λ} has a minimal critical orbit $N \neq \{\theta\}$. W being contractible, we can use similar arguments in Theorem 3.1, and we obtain

$$C_q(I_{\lambda}, \theta) = H_{q-1}(N) \quad \text{for } q \geq 2$$

If we assume that I_{λ} has only two critical orbits θ and N. Since $n \geq 2$ and the assumption that S^{n-1} in Y is not a G-orbit, then the dimension of any G-orbit is less than $n - 1$, and we have $H_{n-1}(N) = 0$. But, since θ is a maximal point of I_{λ}, $C_n(I_{\lambda}, \theta) = Z_2$. The contradiction shows that I has at least three critical orbits.

The discussion of case (ii) is the same as [Ch 1].

As an application of the previous theorem we will study the following problem:

$$(4.11) \qquad -\Delta u = \lambda f(u), \quad u \in H^1(S^2)$$

where $S^2 \subset R^3$ is the unit sphere and $H^1(S^2)$ denotes the usual Sobolev space. Since $G = SO(3)$, or $O(3)$ has a natural orthogonal action on S^2, we then introduce a G-action on $H^1(S^2)$ as follows: for any $g \in G$,

$$(4.12) \qquad (gu)(x) = u(gx), \quad \text{any } u \in H^1(S^2).$$

Hence, (4.11) can be handled by the frame (4.6). We suppose that

(f_1) $f \in C^1(R, R), f(0) = 0, f'(0) = 1$

(f_2) $f(t) = o(|t|)$ as $|t| \to 0$

By the result in [CH 1], the eigenvalues of $(-\Delta)$ on S^2 are $\lambda_n = n(n+1)$, $n = 0, 1, 2, \cdots$, and the eigenfunctions corresponding to λ_n are

$$\cos(h\varphi)P_{n,h}(\cos\theta), \sin(h\varphi)P_{n,h}(\cos\theta), h = 0, 1, 2, \cdots n. \tag{4.13}$$

where $P_{n,h}(z)$ are Legendre polynomials. It is easy to see that the dimension of the eigenspace of λ_n is $2n+1$ and the G-action on the unit sphere of the eigenspace of $\lambda_n (n \geq 1)$ is not transitive. Hence, we have the alternative bifurcation result of Theorem 4.2 for the problem (4.11) when $\lambda_n = n(n+1)$, $n \geq 1$. The precise statements will be left to the reader.

3^0. Consider a semilinear elliptic equation on S^2

$$-\Delta u = f(u) \quad \text{on } S^2 \tag{4.14}$$

Obviously, the zero points of f are solutions of (4.14), called trivial solutions. We concern with multiple nontrivial solutions to (4.14). Consider the sublinear situation. Suppose that $f \in C^1(R, R)$ satisfies

(f_1) there is a $\delta > 0$ such that $\lim_{t\to\infty} \frac{f(t)}{t} \leq -\delta$,

(f_2) f has only three zero points $0, x_1 > 0, x_2 < 0$.

Therefore, (4.14) possesses only three trivial solution $0, x_1$, and x_2, which are constant functions. Note that there is an orthogonal action of $O(3)$ on the Sobolev space $H^1(S^2)$ defined by (4.12). Hence, (4.14) has a G-symmetry, where G denotes $O(3)$. Considering the multiple solution problem of the equation (4.14) we have to distinguish solutions by G-orbits. It is evident that $\text{Fix}_G = \{u \in H^1(S^2) | u = \text{constant}\}$. Therefore, $0, x_1$ and x_2 belong to Fix_G, and every other solution to (4.14) will correspond to a nontrivial G-orbit. Now, our result is stated as follows.

<u>Theorem 4.3</u>: Suppose that f satisfies $f'(x_1) < 0$, $f'(x_2) < 0$ and $f'(0) > 6$ besides (f_1) and (f_2), then at least one of the following two conclusions holds:

(1) (4.14) has at least one nontrivial solution and the normal bundle of the orbit of this solution is not a trivial bundle.

(2) (4.14) has at least two nontrivial geometrically different solutions.

The proof of this result will be based on Theorem 3.3. Define a functional

$$I(u) = \frac{1}{2}\int_{S^2} |\nabla u|^2 - \int_{S^2} F(u) \tag{4.15}$$

on $H^1(S^2)$, where $F(u) = \int_0^u f(t)dt$.

At first, one verifies that $I \in C^2(H^1(S^2), R)$ since $f \in C^1(R, R)$. The eigenvalues of $(-\Delta)$ on S^2 and the corresponding eigenfunctions are given by (4.13) (cf.[CH 1] also). Then, the eigenspace corresponding to $\lambda_0 = 0$ is just $\text{Fix}_G = \{u \in H^1(S^2) | u = \text{constant}\}$. Denote Fix_G by Y, there is an orthogonal composition of the space $Y \oplus Y^\perp$. Next, we declare that

$$I(u) \to +\infty \quad \text{as } \|u\|_{H^1} \to \infty \tag{4.16}$$

In fact, denote $u_n = y_n + y_n^\perp$ and assume $\|u_n\|_{H^1} \to \infty$. If $\|y_n^\perp\|_{H^1} \to \infty$, then $\|\nabla y_n^\perp\|_{L^2} \to \infty$. This is because we can apply the Sobolev imbedding theorem (cf.[Au 1]) to the functions in $Y^\perp$. From (f_1), it is easy to see that there is a $C > 0$ such that

$$\int_{S^2} F(u) \leq C \quad \text{for any } u \in H^1(S^2) \tag{4.17}$$

Hence,

$$I(u_n) = \frac{1}{2}\int_{S^2} |\nabla y_n^\perp|^2 - \int_{S^2} F(u_n) \geq \frac{1}{2}\int_{S^2} |\nabla y_n^\perp|^2 - C \to +\infty$$

If $\|y_n^\perp\|_{H^1} \leq M$, for some $M > 0$, then $\|y_n\|_{H^1} \to \infty$. By the imbedding theorem again, $\|y_n^\perp\|_{C(S^2)} \leq M'$, for some $M' > 0$ constant. Without loss of generality, assume $y_n \to +\infty$, therefore,

$$F(y_n + y_n^\perp) \leq \int_0^{y_n - M'} f(t)dt \to -\infty, \text{as } n \to \infty \tag{4.18}$$

follows from (f_1). Hence,

$$I(u_n) = \frac{1}{2}\int_{S^2} |\nabla y_n^\perp|^2 - \int_{S^2} F(u_n) \geq -\int_{S^2} F(u_n) \to +\infty.$$

Thus, we have proved that $I(u)$ is bounded from below. In the following, verify that f satisfies the (P,S) condition. Assume $u_n = y_n + y_n^\perp$ such that $I(u_n)$ are bounded and $I'(u_n) \to 0$. From the above proof, $\|y_n\|_{H^1}$ and $\|y_n^\perp\|_{H^1}$ all are bounded. y_n has a convergent subsequence and $-\nabla y_n^\perp - Pf(y_n + y_n^\perp) \to 0$. Then $P - (-\Delta)^{-1}$ is defined on $Y^\perp$ and P is the projection operator onto $Y^\perp$. It follows that $y_n^\perp$ has a convergent subsequence since $(-\Delta)^{-1}$ is a compact operator.

Now, because $f'(x_1) < 0$ and $f'(x_2) < 0$, it can be easily verified that x_1, x_2 are two non-degenerate minimal critical point of $I(u)$. Applying the Mountain-Pass lemma, there is another critical point of $I(u)$, say x_3. Denote its orbit by N, then by Lemma 2.3, rank $C_1(f; N) > 0$. Since $f'(0) > 6$, where 6 is just the eigenvalue λ_2 of $(-\Delta)$ on S^2. Noting the multiplicity of λ_n is $2n+1$ (cf. (4.13)), it is easily verified that the Morse index of θ is equal to or larger than 9. By the shifting theorem in [Ch 1], $C_1(f, \theta) = 0$. Hence, N is different from θ. Now, we have obtained a nontrivial solution of (4.14). Since dim $O(3) = 3$ and the Morse index of $\theta \geq 9 >$ dim $G + 3$, if we want to apply Theorem 3.3 we only need to check (2.22) and N possesses a trivial normal bundle. If the normal bundle of N is nontrivial, case (1) holds. Therefore, assume that case (i) does not hold, case (2) will hold if (2.22) is verified. Since x_3 is a Mountain-Pass point, inf $\{(d^2I(x_3)u, u) | u \in H^1(S^2), \|u\| = 1\} \leq 0$. If $\lambda_N = 0$ it follows that $\mu = 0$ is the minimal eigenvalue of elliptic operator $-\Delta - f'(x_3)$. Obviously, the corresponding eigenfunction u_0 is not constant, then $\int_{S^2} f'(x_3)u_0^2 = \int_{S^2} |\nabla u_0|^2 > 0$. So there is a $z_0 \in S^2$, $f'(x_3(z_0)) > 0$. By virtue of a similar method in [HK 1], one can prove that $\mu = 0$ is a simple eigenvalue. Now, we can apply Theorem 3.3 to obtain case (2). The proof is completed.

<u>Remark 4.1</u>: The same problem as (4.14) on $S^n (n \geq 3)$ can be studied by the above method. The precise statements are left to the reader.

References

[Au 1] T.Aubin, Nonlinear analysis on manifold. Monge-Ampère equations. Springer-Verlag, (1982).

[Bo 1] R.Bott, Nondegenerate critical manifold, Ann. of Math. 60, (1954) 248-261.

[Bo 2] R.Bott, lectures on Morse theory, old and new, Bull. Am. Math. Soc. 7(1982) 331-358.

[Br 1] G.Bredon, Introduction to compact transformation group, Academic press, New York (1972).

[Ch 1] K.C.Chang, Infinite dimensional Morse theory and its applications, Montreal (1983).

[CH 1] R.Courant & D.Hilbert, Methods of mathematical physics, Vol.I (1953).

[CWL 1] K.C.Chang, S.P.Wu & S.J.Li, Multiple periodic solutions for an asymptotically linear wave equation, Indiana Math. J.31 (1982) 721-729.

[GM 1] D.Gromoll & W.Meyer, On differentiable functions with isolated critical points, Topology 8(1969) 361-369.

[Gr 1] M.J.Greenberg, Lectures on algebraic topology, W.A.Benjamin, INC. New York (1967).

[HK 1] P.Hess & T.Kato, On some linear and nonlinear eigenvalue problems with an indefinite weight function, Comm. in P.D.E. 5(10), 999-1030.

[Hu 1] D.Husemoller, Fibre bundles, Springer-Verlag, (1966).

[Kc 1] J.L.Kelly, General topology. Van Nostrand, (1955).

[Mi 1] J.Milnor, Morse theory, Ann. Math. Stud. No.51 (1963).

[Pac 1] P.Pacella. Morse theory for flows in presence of a symmetry group, MRC Technical Summary Report No.2530.

[Pal 1] P.S.Palais, Morse theory on Hilbert manifolds, Topology, 2(1963), 299-340.

[Ra 1] P.H.Rabinowitz, A bifurcation theorem for potential operators, J.Funct. Analy. 25(1977) 412-424.

[Ro 1] E.H.Rothe. Critical point theory in Hilbert space under general boundary conditions. J.Math. Anal. Appl. 36(1971), 377- 431.

[Wan 1] Z.Q.Wang, A note on the deformation theorem, (to appear in Acta Math. Sinica. Vol 29(1986) No.5).

[Was 1] G.Wasserman, Equivariant differential topology, Topology Vol 8, 127-150, 1969.

A CLASS DIFFRACTIVE BOUNDARY VALUE PROBLEM WITH MULTIPLE CHARACTERISTIC

Wu Fontong
Dept. of Math., Wuhan Univ.,
Wuhan, China

Propagation of the singularity of solutions to boundary value problems for second order partial differential operator, with grazing rays has been studied by R. Melrose [8],[9],[10],[11], M.Taylor [12],[13] and V.Ivrii [6],[7] and others. Agian L.Hörmander has perfectly explained this problem in his new book [5]. The result is that the singularty is invariant under the bicharacteristic flow of the second partial differential operator.

In this paper we will consider the boundary value problem for higher order partial differential operator with constant multiple characteristic near diffractive point. Under the Levi condition the equation can be reduced to a system, and we can construct two microlocal (forwards and backwards) parametrices near diffractive point. Hence we can prove that the singularity of solutions to a boundary value problem is invariant under the bicharacteristic flow near diffractive point. Section 1 contains some preliminary knowledge concerning boundary and Fourier-Airy integral etc.. In Section 2 a few equivalent forms of the Levi condition are considered. In Section 3 we first reduce the problem to a system, and then solve the eikonal equation and transport equations. The microlocal parametrices of the boundary value problem are constructed in the last section 4.

§1. Fourier-Airy integral and preliminary

Let M be an (n+1)-manifold (C^∞, orientable, paracompact) with boundary ∂M. Let $p \in \partial M$, then near P, $M=\{(x_0,x_1,\dots,x_n);x_0\geq 0\}$ under a local coordinate system. Suppose $Q(x,D)$ is a second order partial differential operator of real principal type on M with symbol $q(x,\xi)$. Further, suppose the boundary ∂M is non-characteristic, namely,

$$\text{char}\,(Q(x,D)) \cap N^*\partial M\setminus 0=\phi.$$

<u>Definition 1.1</u> The point $Z=(x_1^0,\dots,x_n^0;\xi_1^0,\dots,\xi_n^0)\in T^*\partial M\setminus 0$ is said to be diffractive for $q(x,\xi)$ and ∂M if

$$\{q(x,\xi),\ x_0\}=0,\ \{q(x,\xi),\{q(x,\xi),\ x_0\}\}>0,\quad \text{at } Z$$

Let $A_i(\zeta)$ be the usual Airy function, the Fourier transform of $e^{is^3/3}$, and $A(\zeta)=A_\pm(\zeta)=A_i(e^{\pm 2\pi i/3}\zeta)$, so it satisfies Airy equation $A''(\zeta)-\zeta A(\zeta)=0$. (ζ,θ) is a pair of phase functions, satisfying the following eikonal equations

$$\langle d\theta, d\theta\rangle - \zeta\langle d\zeta, d\zeta\rangle = 0$$
$$\langle d\theta, d\zeta\rangle = 0 \tag{1.2}$$

where $\langle\xi,\eta\rangle$ is the bilinear form polarizing $q(x,\xi)$: $\langle\xi,\xi\rangle = q(x,\xi)$.
Note that

$$\phi_\pm = \theta\pm\frac{2}{3}(-\zeta)^{\frac{3}{2}}, \quad \text{then} \quad d\phi|_z \neq 0$$

For second order boundary value problem

$Q(x,D)u = 0$ on M, $Bu = f$ on ∂M.

$B\varepsilon L^1(M)$, there exist microlocal parametrices near z of the form

$$u = \int[g(x,\xi')A(\zeta) + ih(x,\xi')A'(\zeta)]A^{-1}(\zeta_0)e^{i\theta_{\hat{F}}}\,d\xi' \tag{1.3}$$

where $\quad \zeta_0 = \xi_1^{-\frac{1}{3}}\xi_n,\ g\varepsilon s^0,\ h\varepsilon s^{-\frac{1}{3}}$.

Lemma 1.4 Suppose A is a partial differential operator with symbol $a(x,\xi)$, $\phi(x)\varepsilon C^\infty$ is any phase function with $d_x\phi\neq 0$, then there exists

$$e^{-it\phi}A(fe^{it\phi}) = \Sigma\frac{1}{\alpha!}a^{(\alpha)}(x,td\phi(\ ))D_y^\alpha(f(y)e^{it\rho_x(y)})|_{y=x}$$

for any $f\varepsilon C^\infty$, here $\rho_x(y)=\phi(y)-\phi(x)-(y-x)\phi_x'(x)$.

Proof: It is obvious that the Lemma is true when m=0. We will use induction. It is sufficient to consider only the homogeneous part of A. Let $A_m(x,D)=a_I(x)D^I$ here $a_I(x)\varepsilon C^\infty$, $I=(I_0,\cdots,I_\ell,\cdots,I_n)$, $I_\ell>0$, $\ell\varepsilon\{0,1,\cdots,n\}$, $\alpha=(\alpha_0\cdots,\alpha_\ell,\cdots,\alpha_n)$ $\triangleq(\alpha',\alpha_\ell)$. $I=(I',I_\ell)$. $A_m(x,D)=a_I(x)D^{I\setminus|\ell}D_1$ here $|\ell=(0,\cdots,0,1,0,\cdots,0)$ $||\ell|=1$.

Suppose the Lemma is true when $|\alpha|\leqq m-1$, then

$$e^{-it\phi}A_m(fe^{it\phi}) = e^{-it\phi}a_I D^{I\setminus|\ell}(D_\ell(fe^{it\phi}))$$

$$= \sum_{\alpha\leqq I\setminus|\ell}\frac{1}{\alpha!}(a_I\xi^I\xi_\ell^{-1})|_{\xi=td\phi}\cdot D_y^\alpha[t\phi_\ell'\cdot f+D_\ell f)e^{it\rho_x(y)}]|_{y=x}$$

$$= \sum_{\substack{\alpha'\leqq I'\\ \alpha_\ell=0}}\frac{1}{\alpha!}(a_I\xi^I)^{(\alpha)}|_{\xi=td\phi}D_y^\alpha[f\cdot e^{it\rho_x(y)}]|_{y=x} -$$

$$- \sum_{\substack{\alpha'\leqq I'\\ \alpha_\ell=0}}\frac{1}{\alpha!}(a_I\xi^I\xi_\ell^{-1})^{(\alpha)}|_{\xi=td\phi}\cdot t\phi_\ell' D_y^\alpha[f\cdot e^{it\rho_x(y)}]_{y=x} +$$

$$+ \sum_{\alpha_\ell=0}^{I_\ell\setminus|\ell}\sum_{\alpha\leqq I'}\frac{1}{\alpha!}(a_I\xi^I\xi_\ell^{-1})^{(\alpha)}|_{\xi=td\xi}\cdot D_y^\alpha[(t\cdot\phi_\ell'\cdot f+D_\ell f)e^{it\rho_x(y)}]|_{y=x}$$

$$= \sum_{\substack{\alpha' \leq I' \\ \alpha_\ell = 0}} \frac{1}{\alpha!}(a_I \xi^I)^{(\alpha)}\Big|_{\xi = td\phi} D_y^\alpha [f e^{it\rho_x(y)}]\Big|_{y=x}$$

$$+ \sum_{\substack{\alpha' \leq I' \\ \alpha_\ell = 1}} \frac{1}{\alpha!}(a_I \xi^I)^{(\alpha)}\Big|_{\xi = td\phi} \cdot D_y^\alpha [f \cdot e^{it\rho_x(y)}]\Big|_{y=x} + \cdots$$

$$+ \sum_{\substack{\alpha' \leq I' \\ \alpha_\ell = k}} \frac{1}{\alpha'!k!}(a_I \xi^I)^{(\alpha',k)}\Big|_{\xi = td\phi} \cdot D_y^{(\alpha',k)} (f \cdot e^{it\rho_x(y)})\Big|_{y=x} +$$

$$+ ((k-1)-(I_\ell - 1)) \frac{(a_I \xi^I \xi_\ell^{-1})^{(\alpha',k)}}{\alpha'!k!}\Big|_{\xi = td\phi} \cdot D_y^{(\alpha',k)} \cdot (f \cdot e^{it\rho_x(y)})\Big|_{y=x} +$$

$$+ \sum_{\alpha_\ell = k}^{I_\ell - 1} \sum_{\alpha' \leq I'} \frac{(a_I \xi^I \xi_\ell^{-1})^{(\alpha)}}{\alpha!}\Big|_{\xi = td\phi} \cdot D_y^\alpha [(t\phi_\ell' \cdot f + D_\ell f) e^{it\rho_x(y)}]\Big|_{y=x}$$

Again

$$((k-1)-(I_\ell - 1)) \frac{(a_I \xi^I \xi_\ell^{-1})^{(\alpha',k-1)}}{\alpha'!k!}\Big|_{\xi = td\phi} = - \frac{I_\ell - k}{I_\ell} \frac{(a_I \xi^I)^{(\alpha',k)}}{\alpha'! \cdot k!}\Big|_{\xi = td\phi}$$

$$= - \frac{(a_I \xi^I \xi_\ell^{-1})^{(\alpha',k)}}{\alpha'! \cdot k!}\Big|_{\xi = td\phi} \cdot t\phi_\ell'$$

Equalizing this term with the next one (namely the term in $\sum$ as $\alpha_\ell = k$)

$$e^{-it\phi} A_m(x,D)(f\, e^{it\phi})$$

$$= \sum_{\alpha_\ell = 0}^{I_\ell} \sum_{\alpha' \leq I'} \frac{1}{\alpha!}(a_I(x) \xi^I)^{(\alpha)}\Big|_{\xi = td\phi} \cdot D_y^\alpha [f e^{it\rho_x(y)}]\Big|_{y=x}$$

This proves the Lemma.

Suppose P is a partial differential operator of order m with symbol $p(x,\xi)$, and its principal $\sigma_m(P) = p_m(x,\xi)$ is a homogeneous polynomial of degree m of ξ, and $g,h \varepsilon C^\infty$. (θ,ζ) is a pair of phase functions, satisfying eikonal equation (1.2). Using Lemma 1.4 we obtain

$$e^{-it\theta} P([gA(t^{\frac{2}{3}}\zeta) + iht^{-\frac{1}{3}} A'(t^{\frac{2}{3}}\zeta)] e^{it\theta})$$

$$= \sum_{|\alpha| \leq m} \frac{1}{\alpha!} p^{(\alpha)}(x, td\theta) D_y^\alpha ([gA(t^{\frac{2}{3}}\zeta) + iht^{-\frac{1}{3}} A'(t^{\frac{2}{3}}\zeta)] e^{it\phi_x})\Big|_{y=x}$$

$$= t^m A(t^{\frac{2}{3}}\zeta) g(p_m(x,d\theta) + \sum_{k=1}^{[\frac{m}{2}]} \sum_{\substack{\alpha = \alpha_1 + \cdots + \alpha_{2K} \\ |\alpha_i| = 1}} \frac{1}{\alpha!} p_m^{(\alpha)}(x,d\theta) \cdot \zeta^K \prod_{i=1}^{2k} D^{\alpha_i} \zeta)$$

$$+ t^m A(t^{\frac{2}{3}}\zeta)(ih) \sum_{k=0}^{[\frac{m}{2}]} \sum_{\substack{\alpha = \alpha_1 + \ldots + \alpha_{2K+1} \\ |\alpha_i| = 1}} \frac{1}{\alpha!} p_m^{(\alpha)}(x,d\theta) \zeta^{k+1} \prod_{i=1}^{2k+1} D^{\alpha_i} \zeta$$

$$+\ t^{m-\frac{1}{3}}\, A'(t^{\frac{2}{3}}\zeta) g \sum_{k=0}^{[\frac{m}{2}]} \sum_{\alpha=\alpha_1+\dots+\alpha_{2K+1}} \frac{1}{\alpha!} P_m^{(\alpha)}(x,d\theta)\zeta^k \prod_{i=1}^{2k+1} D^{\alpha_i}\zeta$$

$$+\ t^{m-\frac{1}{3}}\, A'(t^{\frac{2}{3}}\zeta)(ih)(P_m(x,d\theta)+\sum_{K=0}^{[\frac{m}{2}]} \sum_{\substack{\alpha=\alpha_1+\dots+\alpha_{2K} \\ |\alpha_i|=1}} \frac{1}{\alpha!} P_m^{(\alpha)}(x,d\theta)\zeta^k \prod_{i=1}^{2k} D^{\alpha_i}\zeta)$$

$$+\ O(t^{m-1}) \qquad (t\to\infty)$$

$$\triangleq t^m A(t^{\frac{2}{3}}\zeta)(g\cdot\bar{P}_1(x,d\theta,\zeta)+ih\ \zeta\ \bar{P}_2(x,d\theta,\zeta))$$

$$+\ it^{m-\frac{1}{3}}\ A'(t^{\frac{2}{3}}\zeta)(h\cdot\bar{P}_1(x,d\theta,\zeta)-ig\bar{P}_2(x,d\theta,\zeta))$$

$$+\ O(t^{m-1}) \qquad (t\to+\infty) \qquad (1.5)$$

Theorem 1.6 Suppose P is a partial differential operator of order m with principal symbol $P_m(x,\xi)$, then (1.5) is true.

Corollary 1.7 For the operator $Q(x,D)$,

$$e^{-it\theta}Q(x,D)[(gA(t^{\frac{2}{3}}\zeta)+iht^{-\frac{1}{3}}A'(t^{\frac{2}{3}}\zeta))e^{it\theta}]$$

$$=-it(2\langle d\theta,dg\rangle+2\zeta\langle d\zeta,dh\rangle+\langle d\zeta,d\zeta\rangle h-(Q^b\theta)g-\zeta(Q^b\zeta)h)A(t^{\frac{2}{3}}\zeta)$$

$$+t^{\frac{2}{3}}(-2\langle d\zeta,dg\rangle+2\langle d\theta,dh\rangle+(Q^b\zeta)g-(Q^b\theta)h)A'(t^{\frac{2}{3}}\zeta)$$

$$+O(t^0)$$

where Q^b is obtained from $Q(x,D)$ by dropping the zero order term.

§ 2. On the Levi condition

Let P be a partial differential operator of order m and Z be k-order zero point of $\sigma_m(P)=p_m(x,\xi)$, on ξ. (i.e. $\left.\frac{\partial^i\sigma_m(P)}{\partial\xi^j}\right|_Z=0$, $j=0,\dots,k-1$. $\left.\frac{\partial q(x,\xi)}{\partial\xi}\right|_Z\neq 0$,). Assume that P satisfies the following Levi condition at Z for $q(x,\xi)$ (see [1])

Definition 2.1 Let $P\varepsilon L^m(M)$, $(x°,\xi°)$ be k-order zero point of $\sigma_m(P)$ with $q(x°,\xi°)=0$ $(2k\leq m)$, and $a\varepsilon c_0^\infty(M)$ be any phase function with $d\phi\neq 0$ in supp a, $q(x, d\phi(x))=0$ and $d\phi(x°)=\xi°$, If

$$e^{-it\phi}P(ae^{it\phi}) = O(t^{m-k}) \qquad (t\to+\infty) \qquad (2.2)$$

then we say that P satisfies the Levi condition $L_{(x°,\xi°)}$.

Definition 2.3 Let $P\varepsilon L^m(M)$, Z be a k-order zero point of $\sigma_m(P)$ $(2k\leq m)$ and (θ,ζ)

be any pair of phase functions, satisfying eikonal equation (1.2). If

$$e^{-it}\, P([gA(t^{\frac{2}{3}}\zeta)+iht^{-\frac{1}{3}}\, A'(t^{\frac{2}{3}}\zeta)]e^{it}\,)$$

$$=A(t^{\frac{2}{3}}\zeta)\; 0\;(t^{m-k}) + A'(t^{\frac{2}{3}}\zeta)\; 0\;(t^{m-k-\frac{1}{3}}) \qquad (t\to+\infty) \qquad (2.4)$$

for any $g,h\in C^{\infty}$, then we say that P satisfies the condition $L^{A}_{(z)}$ at Z for $q(x,\xi)$.

Let $\tilde{M}$ denote a Smooth extension of M across the boundary ∂M, $M \subset \tilde{M}$. It is clear that P can be extended to $\tilde{M}$, and one such extension will be denoted by P.

Theorem 2.5 Let P be a partial differential operator of order m. Let Z be diffractive for $q(x,\xi)$ and ∂M, also a k-order zero point of $\sigma_m(P)$. Then the following conditions are equivalent:

i) For any (x,ξ) contained in some conical neighbourhood of Z in char (P) $(\subset T^*\tilde{M}\setminus 0)$, P satisfies the Levi condition $L_{(x,\xi)}$ for $q(x,\xi)$.

ii) P satisfies the condition $L^{A}_{(Z)}$ for $q(x,\xi)$.

iii) P is decomposable in some conical neighbourhood Γ of Z, in $T^*M\setminus 0$, i.e., there exist partial differential operators $B_j(x,D)\in L^{m-k-j}$ (Γ) such that

$$P = \sum_{j=0}^{K} B_j(x,D)Q^j(x,D)$$

here $\sigma_{m-2k}(B_k)(Z)\neq 0$

Proof: The program of proof is as follows : i) ⟶ iii) ⟶ ii) ⟶ i).

i) — iii), Set $\sigma_m(P)=(x,\xi)$, from i), we have

$$p_m^{(\alpha)}(x,\, d\phi(x)) = 0 \qquad |\alpha|<k$$

for any phase function $\phi(x)$ $(d\phi(x)\neq 0)$ satisfying $q(x,d\phi(x))=0$. Since $p_m(x,\xi)$ is a homogenous polynomial of degree m in ξ, we can write

$$p_m(x,\xi)=p_{mo}(x,\xi')\xi_0^m +\ldots+p_{mm}(x,\xi')$$

here $p_{mk}(x,\xi')\in C^{\infty}$ are homogenous polynomials of degree k in ξ, and $q(x,\xi)=q_o(x)\xi_o^2 + q_1(x,\xi')\xi_o + q_2(x,\xi')$. Since

$$\{q(x,\xi),\ \{q(x,\xi),\ x_0\}\}\neq 0,\ \{x_0,\{x_0,\ q(x,\xi)\}\}\neq 0 \text{ near Z we have}$$

$$q_o(x)\neq 0,\ \ \frac{\partial}{\partial x_0}q_2(x,\xi')\neq 0 \qquad \text{near Z}$$

Then there exist a homogenous polynomial of degree m-2 in ξ $S(x,\xi)\in C^{\infty}$(near z) and homogenous polynomial of degree 2 in $\xi\gamma(x,\xi')\xi_0+\bar{\gamma}(x,\xi')\in C^{\infty}$(near z) such that

$$p_m(x,\zeta) = S(x,\xi)\cdot q(x,\xi) + (\gamma(x,\xi')\xi_0 + \bar{\gamma}(x,\xi')).$$

Denote the projection: $T^*M \to M$ by π. Set $H=\{(x,\xi);\exists\ \xi_0'\neq\xi_0 \text{ s.t } (x;\xi_0',\xi')\ \&\ (x;\xi)\in$

$\varepsilon\ q^{-1}(0)\}$, take x^* near Πz such that $x^*\varepsilon\pi\bar{H}$, so, for $\forall\,\bar{\xi}'\varepsilon$ a neighbourhood in $\mathbb{R}^n_{\xi'}\setminus 0$ of z, there exists $\bar{\xi}_0\neq 0$ such that $p_m(x^*,\bar{\xi})=q(x^*,\bar{\xi})=0$. i.e. $\gamma(x^*,\bar{\xi}')=\bar{\gamma}(x^*,\bar{\xi}')=0$. Since the choice of $\bar{\xi}'$ is arbitrary, we have $\gamma(x,\xi')=\bar{\gamma}(x,\xi')=0$. Again $S(x,d\phi(x))=0$ from $p_m^{(j)}(x,\ d\phi(x))=0$, so likewise $s(x,\xi)$ has $q(x,\xi)$ as a factor, repeating this process gives

$$p_m(x,\xi) = b(x,\xi)\ q^k(x,\xi)$$

where $b(x,\xi)\varepsilon C^\infty$ is a homogenous polynomial of degree m-2k in ξ. We set $p_m(x,D)= =B(x,D)\ Q^k(x,D)$, $P_{m-1}=P-P_m$, so P_{m-1} is a partial differential operator of order m-1, and it also satisfies i), namely,

$$e^{-it\phi}P_{m-1}(ae^{it\phi}) = 0\ (t^{(m-1)-(k-1)}) \qquad (t\to+\infty)$$

Thus we can obtain $P_{m-1}=B_{k-1}(x,D)Q^{k-1}(x,D)$ in the same way we did above. iii) is proved.

iii) $\longrightarrow$ ii). From the preceding section we know

$$e^{-it\theta}Q^\gamma(x,D)[(gA(t^{\frac{2}{3}}\zeta)+iht^{-\frac{1}{3}}A'(t^{\frac{2}{3}}\zeta))e^{it\theta}]$$

$$=A(t^{\frac{2}{3}}\zeta)o(t^\gamma)+A'(t^{\frac{2}{3}}\zeta)o(t^{\gamma-\frac{1}{3}}) \qquad (t\to+\infty)$$

Hence $P=\sum_{j=o}^{k}B_j(x,D)Q^j(x,D)$ satisfies $L^A_{(z)}$

ii) $\longrightarrow$ i) Since h and g in the condition $L^A_{(z)}$ are arbitrary, using Theorem 1.6 one can get

$$P_m(x,\ d\theta)+\sum_{k=1}^{[\frac{m}{2}]}\ \sum_{\substack{\alpha=\alpha_1+\ldots+\alpha_{2K}\\ |\alpha_i|=1}} \frac{1}{\alpha!}P_m^{(\alpha)}(x,d\theta)\zeta^k\prod_{i=1}^{2k}D^{\alpha_i}\zeta = 0 \qquad (2.7)$$

$$\sum_{k=1}^{[\frac{m}{2}]}\ \sum_{\substack{\alpha=\alpha_1+\ldots+\alpha_{2k+1}\\ |\alpha_i|=1}} \frac{1}{\alpha!}P_m^{(\alpha)}(x,d\theta)\zeta^k\prod_{i=1}^{2k+1}D^{\alpha_i}\zeta = 0 \qquad (2.8)$$

It is known from [8][13] that there exists $X(\xi)$, homogenous of degree zero in ξ, such that $\zeta(x,\xi)<0$ as $x>X(\xi)$. Set $\Phi^\pm(x,\xi)=\theta\pm\frac{2}{3}(-\zeta)^{3/2}$, add (2.7) to $\pm i\sqrt{-\zeta}\cdot(2.8)$, we have

$$P_m(x,\ d\Phi^\pm(x,\xi)) = 0$$

Further, it is clear that $q(x,\ d\Phi^\pm(x,\xi))=0$, $(x,d\Phi(x,\xi))$ defines a null bicharacteristic of $Q(x,D)$ near Z, which is the null bicharacteristic of P_m, too. So $(x,d\phi(x))$, for any ϕ with $q(x,\ d\phi)=0$, $d\phi(x^\circ)=\xi^\circ$, lies on the null bicharacteristic through the point (x°,ξ°), namely,

$$P_m(x,d\phi) = 0.$$

Let P_m be the operator with full symbol $p_m(x,\xi)$, then $P_{m-1} \triangleq P-P_m$ also satisfies the condition $L^A_{(z)}$, hence

$$\sigma_{m-1}(P_{m-1})(x, d\phi) = \sigma_{m-1}(P)(x, d\phi)=0.$$

Repeating similar procedure yields

$$\sigma_m(P)(x, d\phi) = \sigma_{m-1}(P)(x, d\phi) = \dots = \sigma_{m-k+1}(P)(x, d\phi) = 0$$

i.e., P satisfies the Levi condition $L_{(x,\xi)}$.

§ 3. Transport Equation

In what follows we always suppose that P satisfies the Levi condition. Let

$$P = \sum_{j=0}^{k} B_j Q^j$$

where B_k is elliptic near z because $\sigma_m(P)$ vanishes up to order-k at Z. Let

$$u_1=u,\ u_2=Q(x,D)u_1,\ \dots,\ u_{k+1}=Q(x,D)u_k.$$

we have

$$\mathbb{P}\mathcal{U} = \mathbb{F}$$

here

$$\mathbb{P} = \begin{pmatrix} I \\ 0 \end{pmatrix} Q(x,D) + \begin{pmatrix} 0 & -1 & 0 & \cdots & 0 \\ 0 & 0 & 0 & \cdots & -1 \\ B_0 & B_1 & B_2 & & B_K \end{pmatrix}$$

$$\mathcal{U} = \begin{pmatrix} u_1 \\ \vdots \\ u_K \\ u_{K+1} \end{pmatrix} = \begin{pmatrix} \tilde{\mathcal{U}} \\ u_{K+1} \end{pmatrix} \qquad \mathbb{F} = \begin{pmatrix} 0 \\ \vdots \\ 0 \\ f \end{pmatrix}$$

Now consider the boundary value problem

$$\mathbb{P}\mathcal{U}=0 \text{ on } M, \quad \tilde{\mathcal{U}}=\mathcal{U}_0 \text{ on } \partial M.$$

here $\mathcal{U}_0 = \begin{pmatrix} g_0 \\ \vdots \\ g_{K-1} \end{pmatrix}$ is given.

We first consider the solutions to the eikonal equation (1.1)

Proposition 3.4 There exists a conic neighbourhood Γ of Z in $[0,+\infty)\times(T^*\partial M\backslash 0)$ and a pair of the solutions to eikonal equation (1.1) — $(\theta(x,\xi), \zeta(x,\xi))\varepsilon C^\infty(\Gamma)$, which are homogenous of degrees 1 and 2/3, respectively, in ξ, when $\xi_n \leq 0$, $\theta(x,\xi')\big|_{\partial T^*M}=\theta_0(x',\xi')$ is a phase function on $T^*\partial M$, $\zeta(x,\xi')\big|_{\partial T^*M}=\zeta_0=\xi_1^{-\frac{1}{3}}\xi_n$, when $\xi_n \geq 0$, eikonal equation is solved to infinite order on ∂M.

Proof: M. Taylor [13] mentioned that this result has been shown in the

$w=d_xR(x,\xi')$. Set $R(x,\xi')=\frac{2}{3}(-\zeta(x,\xi'))^{\frac{3}{2}}$, properly choosing initial value gives

$\zeta(0,x',\xi')=\xi_1^{-\frac{1}{3}}\xi_n$ and

$$-\zeta\langle d\zeta,d\zeta\rangle = \langle d_xR,\ d_xR\rangle = \langle W,W\rangle$$

$$=q(x,(\frac{\partial s}{\partial\xi_0}\Big|_{\xi=\bar\xi^0}\xi_1^{\frac{2}{3}}-\xi_1^{-\frac{1}{3}}\xi_n)^{\frac{1}{2}}\ d_xs\Big|_{\xi=\bar\xi^0})$$

$$=(\frac{\partial s}{\partial\xi_0}\Big|_{\xi=\bar\xi^0}\xi_1^{\frac{2}{3}}-\xi_1^{-\frac{1}{3}}\xi_n)q(x,d_xs)\Big|_{\xi=\bar\xi^0}$$

$$=(X.\xi_1^{\frac{2}{3}}-\xi_1^{-\frac{1}{3}}\xi_n)q_0(\frac{\partial s}{\partial\xi},\xi)\Big|_{\xi=\bar\xi^0}$$

$$=-\zeta^0\cdot\xi_1^{\frac{4}{3}}=-\zeta^0\langle d\zeta^0,d\zeta^0\rangle.$$

Hence $(\theta(x,\xi'),\ \zeta(x,\xi'))$ satisfies the first equation of (1.1). Similarly, we can show that the second equation of (1.1) is satisfied.

Recall (3.3), in order that $\mathbb{P}=0$. Let

$$\mathcal{U}=\int[G(x,\xi')A(\zeta(x,\xi')+iH(x,\xi')A'[\zeta(x,\xi')]]A^{-1}(\zeta_0)\ell^{i\theta(x,\xi')}\hat{E}(\xi')d\xi'$$

here $G(x,\xi')$, $H(x,\xi')$ are $(k+1)\times k$ symbol matrices, the i-th row elements of which belong to $S^{i-1}(M\times\mathbb{R}^n)$, $S^{i-1-\frac{1}{3}}(M\times\mathbb{R}^n)$, respectively, $\hat{E}$ is k-vector value function, then

$$(\mathbb{P}\{[GA(\zeta)+iHA'(\zeta)]e^{i\theta})_{ij}$$

$$=Q(x,D)\{[g_{ij}A(\zeta)+ih_{ij}A'(\zeta)]e^{i\theta}\}$$

$$-[g_{i+1,j}A(\zeta)+ih_{i+1,j}A'(\zeta)]e^{i\theta}\qquad(i=1,\cdots,k;\ j=1,\cdots,k)$$

$$(\mathbb{P}\{[GA(\zeta)+iHA'(\zeta)]e^{i\theta}\})_{k+1,j}$$

$$=\sum_{\ell=0}^{k}B_\ell(x,D)\ [g_{\ell+1,j}(x,\xi')A(\zeta)+ih_{\ell+1,j}(x,\xi')A'(\zeta)]e^{i\theta(x,\xi')}$$

$$(j=1,\cdots,k)$$

In order to find $G=(g_{i,j}(x,\xi'))$ and $H=(h_{i,j}(x,\xi'))$ we proceed by asymptotic expansion. Set

$$G\sim\sum_{\nu=0}^{\infty}G^{(\nu)}\triangleq\sum_{\nu=0}^{\infty}(g_{i,j}^{(\nu)}),\qquad g_{i,j}^{(\nu)}\in S^{i-1-\nu}(M\times\mathbb{R}^n\setminus 0)$$

$$H\sim\sum_{\nu=0}^{\infty}H^{(\nu)}\triangleq\sum_{\nu=0}^{\infty}(h_{i,j}^{(\nu)}),\qquad h_{i,j}^{(\nu)}\in S^{i-1\frac{1}{3}-\nu}(M\times\mathbb{R}^n\setminus 0)\tag{3.10}$$

and proceed by recurrence if $f=K\cdot A+S\cdot A'$, we denote $K=\mathcal{A}_p(f)$, $S=\mathcal{A}_p'(f)$, then the following transport equation is obtained from Corollary 1.7

$$(T_1(G^{(N+1)},\ H^{(N+1)}))_{i,j}$$

R. Melrose's unpublished manuscript. Here we give a sketch. Since $Q(x,D)$ is of principle type, there exists a boundary canonical transformation near z

$$X:\ (x,\eta)\mapsto(X,\xi)$$

which maps a meighbourhood of Z into a neighbourhood of $(0,\ \cdots,\ 0;\ 0,1,0,\ \cdots,\ 0)$ such that $X^*\,q_0 = q$, here $q_0 = \xi_0^2 - X_0\xi_1 + \xi_1\xi_n$ and the solutions to eikonal equation (1.1) concerning q_0 are the following:

$$\theta^0 = X_1\xi_1 + \cdots + X_n\xi_n$$

$$\zeta^0 = \xi_1{}^{-\frac{1}{3}}\xi_1 - X_0\xi_1{}^{\frac{2}{3}}$$

This transformation can be extended as $X:\ T^*\tilde{M} \to T^*(\mathbb{R}^{n+1})\setminus 0$. Let its generating function be $S(x,\xi)$, so

$$X:\ (x,\frac{\partial s}{\partial x}) \mapsto (\frac{\partial s}{\partial \xi},\xi).$$

Denote polarizations of q_0, q by $\langle\ \rangle_0$, $\langle\ \rangle$, respectively.

Set $X = \left.\frac{\partial s}{\partial \xi}\right|_{\xi_0=o}$ then

$$\langle d\theta^o, d\theta^o\rangle_o = q_0(X;\ 0,\xi') = q_0(\frac{\partial s}{\partial \xi},\xi)\Big|_{\xi_0=o}$$

$$= q(x,\frac{\partial s}{\partial x})\Big|_{\xi_0=o} = q(x,\frac{\partial s(x;0,\xi')}{\partial x})$$

$$= \langle ds(x;\ 0,\xi'),\ ds(x;\ 0,\xi')\rangle$$

hence $\theta(x,\xi') = S(x;\ 0,\xi')$. Again donote $\bar{\xi}^0 = (-\xi^{\frac{2}{3}},\ 0,\ \cdots\ 0)$. Set

$$W = (\frac{\partial s}{\partial \xi_0}\Big|_{\xi=\bar{\xi}^0}\ \xi_1^{\frac{2}{3}} - \xi_1^{-\frac{1}{3}}\xi_n)^{\frac{1}{2}}[\sum_{i=0}^{n}\frac{\partial s}{\partial x_i}\Big|_{\xi=\bar{\xi}^0}\cdot dx_i] \in \Omega^1\tilde{M}$$

here ξ is regarded as a parameter, $\left.\frac{\partial s}{\partial \xi_0}\right|_{\xi=\bar{\xi}^0} > \xi_1\xi_n$ as $\xi_1 < 0$ near Z, and

$$\frac{\partial}{\partial x_i}[(\frac{\partial s}{\partial \xi_0}\Big|_{\xi=\bar{\xi}^0}\ \xi_1^{\frac{2}{3}}\ \xi_1^{-\frac{1}{3}}\xi_n)^{\frac{1}{2}}\cdot\frac{\partial s}{\partial x_j}\Big|_{\xi=\bar{\xi}^0}]$$

$$-\frac{\partial}{\partial x_j}[(\frac{\partial s}{\partial \xi_0}\Big|_{\xi=\bar{\xi}^0}\ \xi^{\frac{2}{3}} - \xi^{-\frac{1}{3}}\xi_n)^{\frac{1}{2}}\frac{\partial s}{\partial x_i}\Big|_{\xi=\bar{\xi}^0}]$$

$$= \frac{1}{2}(\frac{\partial s}{\partial \xi_0}\Big|_{\xi=\bar{\xi}^0}\cdot\xi_1^{\frac{2}{3}} - \xi_1^{-\frac{1}{3}}\xi_n)^{-\frac{1}{2}}(\frac{\partial^2 s}{\partial\xi_0\,\partial x_i}\Big|_{\xi=\bar{\xi}^0}\cdot\frac{\partial s}{\partial x_j}\Big|_{\xi=\bar{\xi}_o} - \frac{\partial^2 s}{\partial\xi_0\,\partial x_j}\Big|_{\xi=\bar{\xi}^0}\cdot\frac{\partial s}{\partial x_i}\Big|_{\xi=\bar{\xi}^0})$$

Since

$$\frac{\partial^2 s}{\partial\xi_0\,\partial x_j}\Big|_{\xi=\bar{\xi}^0}\cdot\frac{\partial s}{\partial x_i}\Big|_{\xi=\bar{\xi}^0} = \frac{\partial s(x;\xi_o,o)}{\partial\xi_o\,\partial x_j}\Big|_{\xi_0=-\xi_1^{\frac{2}{3}}}\cdot\frac{\partial s(x;-\xi_1^{\frac{2}{3}},o)}{\partial x_i}$$

$$= \frac{\partial s(x;1,0)}{\partial x_j}\cdot\frac{\partial s(x;1,0)}{\partial x_i}\ (-\xi_1^{\frac{2}{3}})$$

the above term is zero, hence dw=o, so there exists locally $R(x,\xi')$ such that

$$\triangleq 2\langle d\theta, dg_{i,j}^{(N+1)}\rangle + 2\zeta\langle d\zeta, dh_{i,j}^{(N+1)}\rangle + \langle d\zeta, d\zeta\rangle h_{i,j}^{(N+1)}$$

$$-(Q^b\theta)g_{i,j}^{(N+1)} - \zeta(Q^b\zeta)h_{i,j}^{(N+1)} - ig_{i+1,j}^{(N+1)}$$

$$= -i(\mathcal{A}_p(e^{-i\theta}\mathbb{P}\{\sum_{\nu=0}^{N} [G^{(\nu)}A(\zeta) + iH^{(\nu)}A'(\zeta)]e^{i\theta}\}))_{i,j}$$

$$= -iQ(x,D)g_{i,j}^{(N)} \tag{3.11$_1$}$$

$$(T_2(G^{(N+1)}, H^{(N+1)}))_{i,j} \triangleq -2\langle d\zeta, dg_{i,j}^{(N+1)}\rangle + 2\langle d\theta, dh_{i,j}^{(N+1)}\rangle + (Q^b\zeta)g_{i,j}^{(N+1)} -$$

$$-(Q^b\theta)h_{i,j}^{(N+1)} - ih_{i+1,j}^{(N+1)}$$

$$= -(\mathcal{A}'_p(e^{-i\theta}\mathbb{P}\{\sum_{\nu=0}^{N} [G^{(\nu)}A(\zeta)+iH^{(\nu)}A'(\zeta)]\ ^{i\theta}\}))_{i,j}$$

$$= -iQ(x,D)h_{i,j}^{(N)} \qquad (i=1, \ldots, k;\ j=1, \ldots, k+1) \tag{3.11$_2$}$$

For (3.9), using (1.5) gives

$$(T_1(G^{(N+1)}, H^{(N+1)}))_{k+1,j}$$

$$\triangleq \sum_{\ell=0}^{k} (\bar{b}_\ell)_1(x,d\theta,\zeta)\cdot g_{\ell+1,j}^{(N+1)} + i\zeta\sum_{\ell=0}^{k}(\bar{b}_\ell)_2(x,d\theta,\zeta)h_{\ell+1,j}^{(N+1)}$$

$$= \mathcal{A}_p(e^{-i\theta}\sum_{\ell=0}^{k} B_\ell\{\sum_{\nu=0}^{N}[g_{\ell+1,j}^{(\nu)}A(\zeta) + ih_{\ell+1,j}^{(\nu)}A'(\zeta)]e^{i\theta}\}) \tag{3.12$_1$}$$

$$(T_2(G^{(N+1)}, H^{(N+1)}))_{k+1,j}$$

$$\triangleq \sum_{\ell=0}^{K}(\bar{b}_\ell)_1(x,d\theta,\zeta)\cdot h_{\ell+1,j}^{(N+1)} - i\sum_{\ell=0}^{K}(\bar{b}_\ell)_2(x,d\theta,\zeta)\cdot g_{\ell+1,j}^{(N+1)}$$

$$= i\mathcal{A}'_p(e^{-i\theta}\sum_{\ell=0}^{k} B_\ell\{\sum_{\nu=0}^{N}[g_{\ell+1,j}^{(\nu)}A(\zeta) + ih_{\ell+1,j}^{(\nu)}A'(\zeta)]\ell^{i\theta}\}) \tag{3.12$_2$}$$

Summing up, we shall consider the transport equations

$$\begin{cases} T_1(G^{(\nu)}, H^{(\nu)}) = W^{(\nu-1)} \\ T_2(G^{(\nu)}, H^{(\nu)}) = V^{(\nu-1)} \end{cases} \tag{3.13}$$

here $W^{(\nu-1)} = (W_{i,j}^{(\nu-1)})$, $V^{(\nu-1)} = (v_{i,j}^{(\nu-1)})$, $W_{i,j}^{(\nu-1)}$, $v_{i,j}^{(\nu-1)} \in S^{i-\nu}$

are homogenous of degree $i-\nu$ in ξ. Moreover we give initial conditions

$$\tilde{G}^{(\nu)}\Big|_{\substack{\partial T^*M \\ (\xi_n<0)}} = G_0^{(\nu)}, \qquad \tilde{H}^{(\nu)}\Big|_{\substack{\partial T^*M \\ (\xi_n<0)}} = H_0^{(\nu)} \tag{3.14}$$

here the matrix $\tilde{A}$ is obtained from $(k+1)\times k$ matrix A by dropping (k+1)-th row $G_o^{(\nu)} = (g_{ijo}^{(\nu)})$, $H_o^{(\nu)} = (h_{ijo}^{(\nu)})$, $g_{ijo}^{(\nu)} \in S^{i-1-\nu}(\partial M\times\mathbb{R}^n)$,

$h_{ijo}^{(\nu)} \in S^{i-1\frac{1}{3}-\nu}(\partial M\times\mathbb{R}^n)$. Thus we have

Proposition 3.15 Let $W^{(\nu-1)}, V^{(\nu-1)}$ and $G_o^{(\nu)}, H_o^{(\nu)}$ be given symbol matrices, whose elements $W_{i,j}^{(\nu-1)}$, $V_{i,j}^{(\nu-1)} \in S^{i-\nu}(M\times\mathbb{R}^n)$, $g_{ijo}^{(\nu)} \in S^{i+\nu}(\partial M\times\mathbb{R}^n)$, $H_{ijo}^{(\nu)} \in S^{i-1\frac{1}{3}-\nu}(\partial M\times\mathbb{R}^n)$ then there exists a conic neighbourhood $\Gamma'\subset\Gamma$(prop 3.14) of z, such that the problem (3.13) and (3.14) has the unique solution $G^{(\nu)}$, $H^{(\nu)}$, in Γ', satisfying (3.10).

Proof: In $\xi_n<0$, add the second term of (3.13)(namely $(3.11)_2$ and $(3.12)_2$) multiplied by $\mp\sqrt{-\xi}$ to the first term of (3.13)(namely $(3.11)_1$ and $(3.12)_1$), denote $e_{i,j}^{\pm} = g_{i,j}^{(\nu)} \mp\sqrt{-\zeta}\, h_{i,j}^{(\nu)}$, $f_{i,j}^{\pm} = g_{ijo}^{(\nu)}\mp\sqrt{-\zeta}h_{ijo}^{(\nu)}$, $\Phi^{\pm} = \theta(x,\xi') \pm \frac{2}{3}(-\zeta(x,\xi'))^{\frac{3}{2}}$,

then

$$2\langle d\Phi^{\pm}(x,\xi'),de_{i,j}^{\pm}\rangle - (Q^b\theta\pm\sqrt{-\zeta}Q^b\zeta)e_{i,j}^{\pm} - ie_{i+1,j}^{\pm} = f_{i,j}^{\pm} \tag{3.15}$$

$$i=1, \dots, k;\quad j=1, \dots, k+1.$$

$$\sum_{\ell=0}^{K} b_\ell(x,d\Phi^{\pm})\, e_{\ell+1,j}^{\pm} = f_{k+1,j} \tag{3.16}$$

$$j=1, \dots, k+1$$

For (3.16), one can solve $e_{K+1,j}^{\pm}$ since $b_k(x,d\Phi^{\pm}) \neq 0$. Substituting this into (3.15), noting that $(x,d\Phi^{\pm})$ define a bicharacteristic of Q(x,D) and that its projection on the base space is the characteristic ourve of Q(x,D), we can reduce to a system of ordinary differential equations along the characteristic curve. Then there exists a neighbourhood $\Gamma'\subset\Gamma$ small enough(see [8], [12]) such that there is the unique solution $g_{i,j}^{(\nu)}$, $h_{i,j}^{(\nu)}$ on Γ', taking the initial data on $\partial M\times\mathbb{R}^n\setminus 0$, as $\xi_n\leq 0$, which can continue smoothly to $\xi_n\geq 0$ as solutions to infinite order on ∂M. The proposition is proved.

Solving the transport equations and taking almost holomorphic extensions of these functions in ξ_n, one can obtain G, H, such that $\mathbf{P}\mathfrak{U}=0$ (mod C^∞). E in (3.7) will be determined in the next section.

§ 4. Boundary value problem

Let us first recall some facts about Airy functions. We have the following asymptotic expansion, valid for $-\pi<\arg Z<\pi$

$$Ai(Z) = \Phi(Z)\, e^{-\frac{2}{3}Z^{\frac{3}{2}}} \tag{4.1}$$

$$(Ai(z))^{-1} = \Psi(z)e^{\frac{2}{3}z^{\frac{3}{2}}} \tag{4.2}$$

here $\Phi(z)$, $\Psi(z)$ have uniform asymptotic expansion

$$\Phi(z) \sim Z^{\frac{1}{4}}\sum_{\nu=0}^{\infty}\alpha_\nu Z^{-\frac{3}{2}\nu} \qquad (\text{as } Z\to\infty) \tag{4.3}$$

$$\Psi(z) \sim Z^{\frac{1}{4}}\Sigma\beta_\nu Z^{-\frac{3}{2}\nu} \qquad (\text{as } Z\to\infty) \tag{4.4}$$

hence

$$A_\pm(\zeta) = Ai(e^{\pm 2\pi i/3}\zeta) = Ai(e^{\mp i\pi/3}(-\zeta))$$
$$= \Phi(e^{\mp i\pi/3}(-\zeta))e^{\pm\frac{2}{3}i(-\zeta)^{3/2}} \qquad (\xi_n<0)$$
$$(A_\pm(\zeta_0))^{-1} = \Psi(e^{\mp i\pi/3}(-\zeta_0))\cdot e^{\mp\frac{2}{3}i(-\zeta_0)^{3/2}}$$

For (3.7), we rewrite $\mathcal{U}$

$$\mathcal{U} = \int[G(x,\xi')+iH(x,\xi')\,\frac{A'(\xi)}{A(\xi)}\,]Be^{i(\theta+\gamma)}\,\hat{E}d\xi' \tag{4.5}$$

here

$$B = \begin{cases} \Phi(-\zeta)\cdot\Psi(-\zeta_0) & \xi_n< 0 \\ 1 & \xi \geqq 0 \end{cases} \tag{4.6}$$

$B\,\varepsilon\, S^{o}_{1,o}$ on ∂M.

$$\gamma= \gamma_\pm= \begin{cases} \pm\frac{2}{3}\,[(-\zeta)^{\frac{3}{2}} - (-\zeta_0)^{\frac{3}{2}}] & \xi_n< 0 \\ 0 & \xi_n \geqq 0 \end{cases} \tag{4.7}$$

We consider the boundary value problem (3.3). $\tilde{A}$ denotes a submatrix from A by dropping the last row, so

$$\tilde{\mathcal{U}} = \int[\tilde{G}(x,\xi')A(\zeta)+i\tilde{H}(x,\xi')A'(\zeta)]A^{-1}(\zeta_0)e^{i\theta(x,\xi')}\hat{E}(\xi')d\xi'$$

According to proposition 3.15, we can take $\tilde{H}|_{\partial T^*M}=H_0=0$, $\tilde{G}|_{\partial T^*M}=G_0$ is non-degenerate, (we can also arrange $\xi|_{\partial T^*M}=\xi_0+0(|\xi_n/\xi_1|^\infty)$, $\tilde{G}|_{\partial T^*M}$ is non-degenerate as [12]) so

$$\mathcal{U}_0 = \tilde{\mathcal{U}}|_{\partial M} = J(E)$$

here J is an elliptic Fourier integral operator with phase function $\theta(o,x'\xi')$, J^{-1} denote a parametrix of J, Then

$$E = J^{-1}(\mathcal{U}_0)$$

$$= \int e^{i(x',\xi'-\theta(o,y',\xi'))}K(\xi',y')\mathcal{U}_o(y')d\!\!{}^-\xi'dy'$$

sc

$$\hat{E}(\xi') = \int e^{-i\theta(o,y',\xi')}K(\xi',y')\mathcal{U}_o(y')dy' \tag{4.8}$$

Hence

$$\mathcal{U}_\pm = \int [G(x,\xi')+iH(x,\xi')\frac{A'_\pm(\xi)}{A_\pm(\xi)}]\ B\cdot K\cdot e^{i[\phi^\pm(x,\xi')-\phi^\pm(o,y'\xi')]}\mathcal{U}_o(y')dy'd\xi'$$

$$= E_o^\pm \mathcal{U}_o \tag{4.9}$$

namely

$$\mathcal{U}_\pm = \sum_{\ell=0}^{K-1} E^\pm_{0,\ell}\, g_\ell$$

here $(E^\pm_{0,0}, \ldots, E^\pm_{0,K-1})$ is the first row of the matrix $E^\pm_0$.

Now we consider the inhomogenous boundary value prophem

$$\begin{cases} Pu = f & \text{in } M \\ Q^j u\big|_{\partial M} = g_j & (j=o,1,\ldots,k-1) \end{cases}$$

If P satisfies the Levi condition near Z, the problem above can be rewritten as

$$\tilde{\mathbb{P}}\,\tilde{u} = \begin{pmatrix} o \\ \vdots \\ f \end{pmatrix}$$

$$\tilde{u}\big|_{\partial M} = u_o = \begin{pmatrix} g_0 \\ \vdots \\ g_{K-1} \end{pmatrix} \tag{4.11}$$

here

$$\tilde{\mathbb{P}} = \begin{pmatrix} Q & & 0 \\ & \ddots & Q \\ 0 & & B_K Q \end{pmatrix} + \begin{pmatrix} 0 & -1 & & 0 \\ & \ddots & \ddots & \\ 0 & & 0 & -1 \\ B_o & \cdots & & B_{K-1} \end{pmatrix} \tag{4.12}$$

Using the notations in R. Melrose [11], let E be a vector bundle on M. The space $\mathcal{D}'(M, E)$ of extendable distributions is the dual of compactly supported C^∞ sections of $E^*\otimes\Omega$ vanishing to the infinite order on the boundary ∂M, where E^* is the dual bundle of E, Ω is density on M. The space $\dot{\mathcal{D}}'(M, E)$ of distribution of E supported by M is the dual of compactly supported C^∞ sections of $E^*\otimes\Omega$. Moreover $\mathcal{D}'(\partial M, E_{(m)})$ $=\{u\in\mathcal{D}'(M, E),\ \text{supp}\ u \subset \partial M \subset M,\ x_o^{m+1}u=0\ (m\geq 0)\quad C^\infty([0,+\infty); \mathcal{D}'(\mathbb{R}^n))=C^\infty(\mathbb{R}; \mathcal{D}'(\mathbb{R}^n))/$ $/\{\psi\in C^\infty(\mathbb{R}; \mathcal{D}'(\mathbb{R}^n));\ \text{supp}\ \psi \subset (-\infty,0)\times\mathbb{R}^n\}$. $\mathcal{D}'_\partial(M, E)$ is a subspace of $\mathcal{D}'(M, E)$, the elements of which belong to $C^\infty([0, +\infty); \mathcal{D}'(\mathbb{R}^n))$. Thus the map $C^\infty(M, E)\subset\dot{\mathcal{D}}'(M,E)$, $u\mapsto(u)_c$ can be extended to $\mathcal{D}'_\partial(m, E)\to\dot{\mathcal{D}}'(M, E)$.

Since P satisfies the Levi condition near Z, it follows from [2] and [3] that there exist microlocal parametrices near Z for P. Thus there exists a conic neighbourhood Γ of Z and an open neighbourhood U of πZ and maps

$$F^\pm:\ \varepsilon'(U)\to\mathcal{D}'(U) \tag{4.13}$$

such that if $f \in \mathcal{E}'(U)$ and $WF(f) \subset \Gamma$ then $PF^{\pm}(f) - f \in C^{\infty}(U)$ and $F^{\pm}$ satisfy the conditions

$$WF(F^{\pm}(f)) \subset \Omega^{\pm} \circ WF(f) \tag{4.14}$$

where Ω^{+}(resp. Ω^{-}) $= \{(Z,Z') \in T^*U \times T^*U;\ Z \neq 0,\ Z' = 0$ and either $Z = Z'$ or $p(z) = p(z') = 0$ and Z lies on the forward (resp. back ward) half-bicharacteristic in Γ through $Z'\}$.
Thus

$$\tilde{u} = \begin{pmatrix} 1 \\ Q \\ \vdots \\ Q^{k-1} \end{pmatrix} F^{\pm}(f)$$

satisfies $\tilde{\tilde{\mathbb{P}}}\tilde{u} = \begin{pmatrix} 0 \\ \vdots \\ f \end{pmatrix}$ (mod $C^{\infty}(U)$). We can also construct an operator $\tilde{A}$ according to [8] §11 and denote

$$^{c}\left(\begin{pmatrix} 1 \\ Q \\ \vdots \\ Q^{k-1} \end{pmatrix} F^{\pm}\tilde{A}(f) \right) = \begin{pmatrix} 1 \\ Q \\ \vdots \\ Q^{k-1} \end{pmatrix} F^{\pm}\tilde{A}(f) - \left(\begin{pmatrix} 1 \\ Q \\ \vdots \\ Q^{k-1} \end{pmatrix} F^{\pm}\tilde{A}(f) \right)_{c}$$

here $\begin{pmatrix} 1 \\ Q \\ \vdots \\ Q^{k-1} \end{pmatrix} F^{\pm}\tilde{A}(f) \in \dot{\mathcal{E}}'((-\infty,o] \times \mathbb{R}^n)$. Set

$$\tilde{E}^{\pm}(f, g_0, \ldots, g_{k-1})$$

$$= \left(\begin{pmatrix} 1 \\ Q \\ \cdot \\ \cdot \\ \cdot \\ Q^{k-1} \end{pmatrix} F^{\pm}\tilde{A}(f) \right) \left(E_0^{\pm}\left(W^{\pm}\begin{pmatrix} 0 \\ \vdots \\ f \end{pmatrix} + u_o\right) \right)_{c} \tag{4.15}$$

here

$$W^{\pm}\begin{pmatrix} 0 \\ \vdots \\ f \end{pmatrix} = \begin{pmatrix} W_0\ (f) \\ \cdot \\ \cdot \\ W_{k-1}^{\pm}(f) \end{pmatrix} \qquad u_o = \begin{pmatrix} g_0 \\ \vdots \\ g^{k-1} \end{pmatrix}$$

$$W_i^{\pm}(f) = \lim_{x_0 \uparrow 0} Q^{i-1} F^{\pm}\tilde{A}(f)(x_0, \cdot\ ), \tag{4.16}$$

$$(i = o, \ldots, k-1.).$$

Thus we have already proved the following proposition.

Proposition 4.17 Let the partial differential operator of order m P satisfy the Levi condition $L_{(x^{\circ},\xi^{\circ})}$ for $q(x,\xi)$ and the boundary ∂M is non-characteristic for $Q(x,D)$, then

$$u^{\pm} = E^{\pm}(f, g_0, \dots, g_k)$$

$$= {}^c[F^{\pm}\tilde{A}(f)] + [\sum_{\ell=0}^{K-1} E^{\pm}_{O,\ell}\,(W^{\pm}_{\ell}(f) + g_{\ell})]_c$$

satisfy

$$Pu^{\pm} - f \varepsilon C^{\infty}(M) \qquad \text{near } Z$$

$$Q^i(x,D)u^{\pm}\Big|_{\partial M} - g_i \ \varepsilon\ C^{\infty}(\partial M) \qquad (i=0, \dots, k-1)$$

Proof : Since $\tilde{\mathbb{P}}$ is non-characteristic, from [11] we have $\dot{\mathbb{P}}u = \tilde{\mathbb{P}}(u_c) - (\tilde{\mathbb{P}}u)_c\ \varepsilon$ $\mathcal{D}'(\partial M, E_{(1)})$. Suppose Q is a right inverse, namely, $\dot{\mathbb{P}}Q = Id$, $Q\dot{\mathbb{P}} - Id \subset \ker(\dot{\mathbb{P}})$.

Furthermore

$$\tilde{\mathbb{P}}E^{\pm}(f, g_O, \dots, g_{k+1}) - f$$

$$= \tilde{\mathbb{P}}[\begin{pmatrix} 1 \\ Q \\ \cdot \\ \cdot \\ \cdot \\ Q^{k+1} \end{pmatrix} F^{\pm}\tilde{A}(f)] - [\begin{pmatrix} 0 \\ \vdots \\ f \end{pmatrix}] - \tilde{\mathbb{P}}[\begin{pmatrix} 1 \\ Q \\ \cdot \\ \cdot \\ \cdot \\ Q^{k+1} \end{pmatrix} F^{\pm}\tilde{A}(f)]_c +$$

$$+ \ \tilde{\mathbb{P}}\ [E^{\pm}_O(W^{\pm}\begin{pmatrix} 0 \\ \vdots \\ f \end{pmatrix} + u_0)]_c$$

$$= \dot{\mathbb{P}}\ (E^{\pm}_O[W^{\pm}\begin{pmatrix} 0 \\ \vdots \\ f \end{pmatrix} - \begin{pmatrix} 1 \\ Q \\ \cdot \\ \cdot \\ \cdot \\ Q^{k-1} \end{pmatrix} F^{\pm}\tilde{A}(f)] + \dot{\mathbb{P}}(E^{\pm}_O\, u_O),\ \varepsilon\, \mathcal{D}'(\partial M, E_{(1)})$$

but

$$\mathbb{R}\cdot Q\cdot(\tilde{\mathbb{P}}\tilde{E}^{\pm}(f;g_O, \dots, g_{k-1}) - \begin{pmatrix} 0 \\ \vdots \\ f \end{pmatrix})$$

$$= \mathbb{R}\ (E^{\pm}_O u_O + \zeta_1 + \zeta_2)$$

here

$$\zeta_1 = E^{\pm}_O\ [W^{\pm}\begin{pmatrix} 0 \\ \vdots \\ f \end{pmatrix} - \begin{pmatrix} 1 \\ Q \\ \vdots \\ Q^{k-1} \end{pmatrix} F^{\pm}\tilde{A}(f)], \qquad \zeta_2 \varepsilon\ \ker(\dot{\mathbb{P}})$$

then $\mathbb{R}\zeta_2 = 0$, and $\mathbb{R}$ is the extension map :

$\mathcal{A}'(M, E) \to \mathcal{D}'(\partial M, E)$ of the restrict map : $C^{\infty}(M, E) \to C^{\infty}(\partial M, E)$ (see [11] (3.13))

but ζ_1 is zero on ∂M then $\mathbb{R}\zeta_1 = 0$. Hence

$$\mathbb{R}\ Q(\tilde{\mathbb{P}}\ \tilde{E}^{\pm} - f) = u_O \qquad \varepsilon \mathcal{D}'(M, E)$$

Thus,using porposition 3.28 of [11] we have

$$\tilde{u}^{\pm} = \tilde{E}^{\pm}(f, g_O, \dots, g_{k-1})\ \varepsilon\, \mathcal{A}'(M, E)$$

and

$$\tilde{u}^{\pm}\Big|_{\partial M} = u_0$$

completing the proof.

Note the similarity of these parametrices with parametrices constructed by R. Melrose [8]. By duality arguments these parametrices can be shown to be microlocally unique and using them one can obtain results of propagation of singularities like Theorem 3.1 in [8].

REFERENCES

[1] J.Chazarain, Operateurs hyperboliques a caracteristiques de multiplicite constante, Ann. Inst. Fourier, (Grenoble) 24,1 (1974) 173-202

[2] ______, Propagation des singularites pour une classe d'operateurs a caracteristiques multiples et resobulite locale, Ann. Inst. Fourier. (Grenoble) 24, (1974) 203-223

[3] J.Duistermaat and L.Hörmander, Fourier integral operators, II Acta Math. 128 (1972) 183-269

[4] F.Friedlander, The wavefront set of the solution of a simple initial-boundary value problem with glancing rays, Math. Proc. Camb. Phil. Soc., 79 (1976) 145-149

[5] L.Hörmander, The analysis of linear partial defferential operators III, Springer-Verlag, Berlin 1985.

[6] V.Ivrii, The propagation of singularities of a solution of the wave equation near a boundary (in Russian), English translation see in Soviet Math. Dokl. 19 (1978)

[7] ______, The propagation of singularities of solutions of nonclassical boundary value problems for second order hyperbolic equation (in Russion), English translation see in Moscow Math. Soc. (1983) 78-99

[8] R.Melrose, Microlocal parametrices for diffractive boundary value problems, Duke Math. J. 42 (1975) 605-635

[9] ______, Equivalence of glancing hypersurfaces, Invent. Math. 37, (1976) 165-191

[10] ______, Differential boundary value problems of principal type, In Seminar on singularities of solutions of linear partial differential equations, Ed. L.Homander. Princeton Univ. Press, Princeton (1978)

[11] ______, Transformation of boundary problems, Acta Math. Vol.147 (1981) 149-236

[12] M.Taylor, Grazing rays and reflection of singularities of solution to wave equations, Comm. Pure. Appl. Math. 29 (1976) 1-38

[13] ______, Airy operator calculus, Contemporary Mathe. Vol.27 (1984) 169-192

EXISTENCE, UNIQUENESS AND REGULARITY OF THE MINIMIZER OF A CERTAIN FUNCTIONAL

Wu Lancheng
Peking University, China.

§1. Introduction

In this paper we present a functional, which arises from an elastic-plastic theory and prove the existence, uniqueness and regularity of the minimizer of this functional in certain Sobolev space, described as follows.

Let Ω be an open domain in $\mathbb{R}^3$. For u: $\Omega \to \mathbb{R}^3$ consider the functional

$$E[u]: = \frac{1}{2}\int_\Omega (\partial u)^T D\partial u dx - \int_\Omega j(q)dx - \int_\Omega u^T f dx, \tag{1.1}$$

subject to Dirichlet boundary condition:

$$u = 0 \quad \text{on the boundary of } \Omega, \tag{1.2}$$

where ∂ is the differential operator matrix:

$$\partial = \begin{pmatrix} \frac{\partial}{\partial x_1} & 0 & 0 & 0 & \frac{\partial}{\partial x_3} & \frac{\partial}{\partial x_2} \\ 0 & \frac{\partial}{\partial x_2} & 0 & \frac{\partial}{\partial x_3} & 0 & \frac{\partial}{\partial x_1} \\ 0 & 0 & \frac{\partial}{\partial x_3} & \frac{\partial}{\partial x_2} & \frac{\partial}{\partial x_1} & 0 \end{pmatrix}^T \tag{1.3}$$

D is the elastic matrix, and

$$j(q) = \begin{cases} 0 & q<0 \\ \frac{1}{2}\alpha_\circ q^2 & q\geq 0 \end{cases}, \tag{1.4}$$

$$q = N^T D \partial u, \tag{1.5}$$

$$N = [N_1\ N_2\ N_3\ N_4\ N_5\ N_6]^T \tag{1.6}$$

$$\alpha_\circ = \frac{1}{h_\circ + N^T DN}, \tag{1.7}$$

where $h_\circ > 0$ and N_i (i=1,..., 6) are given constants.

The physical background of the functional E[u] will be stated in §2. The existence and uniqueness of the minimizer of E[u] in H_0^1 follows from the strict convexity of the integrand with respect to ∇u, which will be proved in §3. In §4, we consider the $C^{0,\delta}$ and $C^{1,\delta}$ -regularity of such a minimizer.

Throughout this paper we shall denote by $H^{m,p}(\Omega, \mathbb{R}^n)$ the Cartesian product

$$H^{m,p}(\Omega)\times H^{m,p}(\Omega)\times\ldots\times H^{m,p}(\Omega) = (H^{m,p}(\Omega))^n,$$

$H^{m,p}(\Omega)$ being the standard Sobolev spaces. A similar meaning holds for $C^{m,\delta}(\Omega, \mathbb{R}^n)$ and so on.

§2. Physical motivation.

Let us consider an elastic-plastic material and suppose that an admissible stress state $\sigma=[\sigma_x\ \sigma_y\ \sigma_z\ \tau_{yz}\ \tau_{zx}\ \tau_{xy}]^T$ satisfies

$$\Phi(\sigma):=N^T\sigma-a\leq 0 \qquad N\in\mathbb{R}^6,\quad a\in\mathbb{R}$$

and a critical state of stress is defined by the condition

$$\Phi(\sigma)=0$$

As we know, the surface $\Phi(\sigma)=0$ is called the yield-surface of this material. in this paper, we confine ourselves to the quasi-static process and assume that the material obeys certain hardening rule (see [7]). For this kind of elastic-plastic material K.S. Havner and H.P. Patel [7] established a mathematical model in which the displacement rate $u=[u_1\ u_2\ u_3]^T$ and the plastic multiplier γ satisfy

$$\int_\Omega (\partial\phi)^T D(\partial u-\gamma N)dx = \int_\Omega \phi^T f dx+\int_{\Gamma_2}\phi^T p dS,\ \forall\phi, \tag{2.1}$$

$$(h_\circ+N^T DN)\gamma-N^T D\partial u \geq 0,\quad \gamma\geq 0, \tag{2.2}$$

$$\gamma[(h_\circ+N^T DN)\gamma-N^T D\partial u]=0 \tag{2.3}$$

$$u=0 \quad \text{on } \Gamma_1 \tag{2.4}$$

here $\Gamma_1\cup\Gamma_2=\partial\Omega$, $\Gamma_1\cap\Gamma_2=\phi$, $h_\circ>0$ is a constant coming from the hardening rule, N is the outer normal vector of the yield surface, ∂u stands for the total strain rate and D is the elastic matrix:

$$D=\frac{E}{2(1+\nu)}\begin{pmatrix} \frac{2(1-\nu)}{1-2\nu} & \frac{2\nu}{1-2\nu} & \frac{2\nu}{1-2\nu} & 0 & 0 & 0\\ \frac{2\nu}{1-2\nu} & \frac{2(1-\nu)}{1-2\nu} & \frac{2\nu}{1-2\nu} & 0 & 0 & 0\\ \frac{2\nu}{1-2\nu} & \frac{2\nu}{1-2\nu} & \frac{2(1-\nu)}{1-2\nu} & 0 & 0 & 0\\ 0 & 0 & 0 & 1 & 0 & 0\\ 0 & 0 & 0 & 0 & 1 & 0\\ 0 & 0 & 0 & 0 & 0 & 1 \end{pmatrix} \tag{2.5}$$

where $E>0$ is Young's modulus and $-1<\nu<\frac{1}{2}$ is Poisson's ratio.

For the sake of simplicity we consider Dirichlet problem, i.e. $\partial\Omega=\Gamma_1$. Let

$$k=\{(u,\gamma)\in H_0^1(\Omega,\mathbb{R}^3)\times L^2(\Omega),\quad \gamma\geq 0\}.$$

In [9] and [11] we reduce (2.1)-(2.4) to the following problem:

$$\begin{cases} \text{Finding } (u,\gamma)\in k \quad \text{such that} \\ a(u,v-u)+C(\gamma,\lambda-\gamma)-[\gamma,v-u]-[\lambda-\gamma,u] \geqq (f,v-u), \\ \qquad \forall\, (v,\lambda)\in k\,, \end{cases} \tag{2.6}$$

where

$$a(v,u) = \int_\Omega (\partial v)^T D\partial u dx,$$

$$C(\lambda,\gamma) = \int_\Omega \lambda(h_o+N^T DN)\gamma dx,$$

$$[\gamma,v] = \int_\Omega \gamma N^T D\partial v dx,$$

$$(f,v) = \int_\Omega v^T f dx,$$

and then it is easy to check that this problem is equivalent to

$$\begin{cases} \text{Finding } (u,\gamma)\in k \quad \text{such that} \\ J(u,\gamma) \leqq J(v,\lambda), \\ \text{for every } (v,\lambda)\in k, \end{cases} \tag{2.7}$$

where

$$J(v,\lambda) = \frac{1}{2}a(v,v)+ \frac{1}{2}C(\lambda,\lambda)-[\lambda,v]-(f,v). \tag{2.8}$$

Form (2.2) and (2.3), we have

$$\gamma = \gamma(q) = \begin{cases} 0 & q<0 \\ \alpha_o q & q\geqq 0, \end{cases} \tag{2.9}$$

where q and α_o are defined in (1.5) and (1.7) respectively.

Substituting (2.9) into (2.8), we have

$$J(u, \gamma(q)) = \frac{1}{2}\int_\Omega (\partial u)^T D\partial u dx - \int_\Omega j(q)dx - \int_\Omega u^T f dx,$$

where $j(q)$ is defined in (1.4).

Let

$$E[u]: = J(u,\gamma(q)),$$

then we are led to the problem

$$\begin{cases} \text{Finding } \quad u\in H_o^1(\Omega, \mathbb{R}^3) \quad \text{such that} \\ E[u] \leqq E[u+\phi] \\ \text{for every } \quad \phi\in H_o^1(\Omega, \mathbb{R}^3). \end{cases} \tag{2.10}$$

The solution of Problem (2.10) is called the minimizer of functional $E[u]$.

§3. Existence and uniqueness theorem

In this section we prove the existence and uniqueness of the solution of Problem (2.10).

It is easy to check that functional $E[u]$ can be rewritten as follows:

$$E[u] = \frac{1}{2}\int_\Omega (\partial u)^T D_{ep}(q)\partial u dx - \int_\Omega u^T f dx, \tag{3.1}$$

where

$$D_{ep}(q) = D - DN\alpha(q)N^T D , \tag{3.2}$$

$$\alpha(q) = \begin{cases} 0 & q<0 \\ \alpha_o & q\geqq 0 , \end{cases} \tag{3.3}$$

q and α_o are defined in (1.5) and (1.7) respectively.

It is well known that D is a positive definite matrix, i.e. for every $\xi \in \mathbb{R}^6$ there exists a constant $\lambda_e>0$ such that

$$\xi^T D\xi \geqq \lambda_e|\xi|^2. \tag{3.4}$$

Then we have

<u>Lemma 3.1</u> $D_{ep}(q)$ is a positive definite matrix, i.e., for $\forall\xi\in\mathbb{R}^6$, there exists a constant $\lambda>0$, such that

$$\xi^T D_{ep}(q)\xi \geqq \lambda|\xi|^2 ,$$

where λ depends on λ_e, D, N, $h_\circ$.

<u>Proof</u> For $\xi\in\mathbb{R}^6$, we have

$$\begin{aligned}\xi^T D_{ep}(q)\xi &= \xi^T[D-DN\alpha(q)N^T D]\xi \\ &= \xi^T D\xi-\alpha(q)(N^T D\xi)^2 \\ &\geqq \xi^T D\xi-\alpha(q)(N^T DN)(\xi^T D\xi) \\ &\geqq [1-\alpha_o(N^T DN)]\xi^T D\xi \\ &\geqq \frac{h_\circ\lambda_e}{h_\circ+N^T DN}|\xi|^2 = :\lambda|\xi|^2,\end{aligned}$$

by (3.3)(3.4).

Q.E.D.

Now we use the following notations:

$$\hat{P}:= [p_{ij}] \quad i,j=1,2,3, \qquad p_{ij}\in\mathbb{R},$$

$$P:= [p_{11} \quad p_{22} \quad p_{33} \quad p_{32}+p_{23} \quad p_{13}+p_{31} \quad p_{21}+p_{12}]^T$$

$$=: [p^i] \quad i=1,2,\ldots,6,$$

$$F(u,p):= \frac{1}{2} P^T D_{ep}(N^T DP)P-u^T f,$$

$$\hat{F}(u,\hat{P}):= F(u,P),$$

then we have

<u>Theorem 3.1</u> $F(u,p)$ is strictly convex with respect to p.

<u>Proof</u> It is sufficient to prove that for every P and $P_\circ$ there exist constants

$a_i(P_\circ)$ $(i=1,\ldots,6)$ such that

$$F(u,P)-F(u,P_\circ) \geqq \sum_{i=1}^{6} a_i(P_\circ)(p^i-p_\circ^i)$$

and the equality holds if and only if $P=P_\circ$.

For every P and $P_\circ$ we have four posibilities:

1°. $N^TDP<0$ and $N^TDP_\circ<0$. In this case, we have

$$\begin{aligned}F(u,P)-F(u,P_\circ) &= \frac{1}{2}P^TDP-\frac{1}{2}P_\circ^TDP_\circ\\ &= \frac{1}{2}(P+P_\circ)^TD(P-P_\circ)\\ &= \frac{1}{2}(P-P_\circ)^TD(P-P_\circ)+P_\circ^TD(P-P_\circ)\\ &\geqq \sum_{i=1}^{6} a_i(P_\circ)(p^i-p_\circ^i)+\frac{1}{2}\lambda_e|P-P_\circ|^2.\end{aligned}$$

2°. $N^TDP\geqq0$ and $N^TDP_\circ\geqq0$. In this case, we have

$$\begin{aligned}F(u,P)-F(u,P_\circ) &= \frac{1}{2}P^T(D-DN\alpha_\circ N^TD)P-\frac{1}{2}P_\circ^T(D-DN\alpha_\circ N^TD)P_\circ\\ &= \frac{1}{2}(P-P_\circ)^T(D-DN\alpha_\circ N^TD)(P-P_\circ)+P_\circ^T(D-DN\alpha_\circ N^TD)(P-P_\circ)\\ &\geqq \sum_{i=1}^{6} \tilde{a}_i(P_\circ)(p^i-p_\circ^i)+\frac{1}{2}\lambda|P-P_\circ|^2 .\end{aligned}$$

3°. N^TDP 0 and $N^TDP_\circ$ 0. In this case, we have

$$\begin{aligned}F(u,P)-F(u,P_\circ) &= \frac{1}{2}P^TDP-\frac{1}{2}P_\circ^T(D-DN\alpha_\circ N^TD)P_\circ\\ &= P_\circ^T(D-DN\alpha_\circ N^TD)(P-P_\circ)+\frac{1}{2}P^TDN\alpha_\circ N^TDP\\ &\quad+\frac{1}{2}(P-P_\circ)^T(D-DN\alpha_\circ N^TD)(P-P_\circ)\\ &\geqq \sum_{i=1}^{6} \tilde{a}_i(P_\circ)(p^i-p_\circ^i)+\frac{1}{2}\lambda|P-P_\circ|^2 .\end{aligned}$$

4°. $N^TDP\geqq0$ and $N^TDP_\circ<0$. In this case, we have

$$\begin{aligned}F(u,P)-F(u,P_\circ) &= \frac{1}{2}P^T(D-DN\alpha_\circ N^TD)P-\frac{1}{2}P_\circ^TDP_\circ\\ &= \frac{1}{2}(P-P_\circ)^T(D-DN\alpha_\circ N^TD)(P-P_\circ)+P_\circ^T(D-DN\alpha_\circ N^TD)P\\ &\quad-\frac{1}{2}P_\circ^T(D-DN\alpha_\circ N^TD)P_\circ-\frac{1}{2}P_\circ^TDP_\circ\\ &\geqq \frac{1}{2}(P-P_\circ)^T(D-DN\alpha_\circ N^TD)(P-P_\circ)+P_\circ^TDP-P_\circ^TDP_\circ\\ &= \frac{1}{2}(P-P_\circ)^T(D-DN\alpha_\circ N^TD)(P-P_\circ)+P_\circ^TD(P-P_\circ)\\ &\geqq \sum_{i=1}^{6} a_i(P_\circ)(p^i-p_\circ^i)+\frac{1}{2}\lambda|P-P_\circ|^2.\end{aligned}$$

So in each case, there exist constants $a_i(P_\circ)$ $(i=1,\ldots,6)$, such that

$$F(u,P)-F(u,P\ ) \geq \sum_{i=1}^{6} a_i(P_\circ)(p^i-p_\circ^i)$$

and the equality holds if and only if $P=P_\circ$.

Q.E.D.

Corollary 3.1 $\hat{F}(u,\hat{P})$ is strictly convex with respect to $\hat{P}$.

Proof. From Theorem 3.1, we have

$$F(u,\mu P_1+(1-\mu)P_2) < \mu F(u,P_1)+(1-\mu)F(u,P_2)$$

for every $\mu\in(0,1)$, $P_1\neq P_2$. It follows that

$$\hat{F}(u,\mu\hat{P}_1+(1-\mu)\hat{P}_2) < \mu\hat{F}(u,\hat{P}_1)+(1-\mu)\hat{F}(u,\hat{P}_2)$$

for every $\mu\in(0,1)$, $\hat{P}_1\neq \hat{P}_2$.

Q.E.D.

Therefore, in conclusion [1], we have

Theorem 3.2 If $f\in L^2(\Omega, \mathbb{R}^3)$, then there exists one and only one solution of problem (2.10).

§4. Regularity of the minimizer of E[u].

Recently, the regularity of the minimizers of general functionals of the following type

$$\int_\Omega F(Du)dx$$

has been considered by L.C. Evans [2], who proved $C^{1,\delta}$- partial regularity of minimizers of such functionals under the assumptions that the integrand be uniformly strictly quasiconvex and F(P) is twice continuously differentiable, later on N.Fusco, J.Hutchinson [4], M.Giaquinta, G.Modica [6] and M.C.Hong [8] extended this result to more general case:

$$\int_\Omega F(x,u,Du)dx\to\inf$$

and soon after, L.C.Evans and R.F.Gariepy [3] presented a new proof of the result in [2]. But all these results can not cover the case in which F is not twice continuously differentiable in P while functional E[u] in (1.1) is, however, exactly this case.

In this section, we shall prove $C^{0,\delta}$-regularity and under an additional assumption prove $C^{1,\delta}$-regularity of the minimizer of functional E[u] by considering the Euler equation of E.

First of all, let us derive the Euler equation of E at u.

Lemma 4.1 If u is a minimizer of functional E[u], then $u\in H_0^1(\Omega, \mathbb{R}^3)$ satisfies

$$\int_\Omega(\partial\phi)^T D_{ep}(q)\partial u dx-\int_\Omega\phi^T f dx = 0,\ \forall\ \phi\in H_0^1(\Omega, \mathbb{R}^3), \qquad (4.1)$$

where $D_{ep}(q)$ is defined in (3.2).

Proof. Let u be a solution to probelm (2.10), then $u\in H_o^1(\Omega, \mathbb{R}^3)$ satisfies

$$\frac{d}{dt}E[u+t\phi]\Big|_{t=0} = 0 \tag{4.2}$$

for all $\phi\in H_o^1(\Omega, \mathbb{R}^3)$. From (1.1), we have

$$\begin{aligned} E[u+t\phi] &= \frac{1}{2}\int_\Omega(\partial u)^T D\partial u dx + t\int_\Omega(\partial\phi)^T D\partial u dx \\ &+ \frac{1}{2}t^2\int_\Omega(\partial\phi)^T D\partial\phi dx - \int_\Omega j(q+t\hat{q})dx \\ &- \int_\Omega u^T f dx - t\int_\Omega \phi^T f dx, \end{aligned}$$

where $q=N^T D\partial u$, $\hat{q}=N^T D\partial\phi$. so we have

$$\frac{d}{dt}E[u+t\phi]\Big|_{t=0} = \int_\Omega(\partial\phi)^T D\partial u dx - \int_\Omega\gamma(q)\hat{q}dx - \int_\Omega\phi^T f dx,$$

where $\gamma(q)$ is defined in (2.9). Hence

$$\begin{aligned} \frac{d}{dt}E[u+t\phi]\Big|_{t=0} &= \int_\Omega(\partial\phi)^T D\partial u dx - \int_\Omega(\partial\phi)^T DN\alpha(q)N^T D\partial u dx - \int_\Omega\phi^T f dx \\ &= \int_\Omega(\partial\phi)^T D_{ep}(q)\partial u dx - \int_\Omega\phi^T f dx. \end{aligned}$$

So u satisfies (4.1). Q.E.D.

Now for $\forall\eta\in\mathbb{R}^3$, let

$$M_{ep}(\eta_1,\eta_2,\eta_3) := \begin{pmatrix} \eta_1 & 0 & 0 \\ 0 & \eta_2 & 0 \\ 0 & 0 & \eta_3 \\ 0 & \eta_3 & \eta_2 \\ \eta_3 & 0 & \eta_1 \\ \eta_2 & \eta_1 & 0 \end{pmatrix}^T D_{ep}(q) \begin{pmatrix} \eta_1 & 0 & 0 \\ 0 & \eta_2 & 0 \\ 0 & 0 & \eta_3 \\ 0 & \eta_3 & \eta_2 \\ \eta_3 & 0 & \eta_1 \\ \eta_2 & \eta_1 & 0 \end{pmatrix} \tag{4.3}$$

we have

Lemma 4.2 For $\forall\eta\in\mathbb{R}^3$, $\eta\neq0$, $M_{ep}(\eta_1,\eta_2,\eta_3)$ is a positive definite matrix.

Proof. It is sufficient to show that for $\forall\,\xi\in\mathbb{R}^3$, there exists a constant $\bar{\lambda}>0$ such that for $\forall\eta\in\mathbb{R}^3$, $\eta\neq0$, we have

$$\mathcal{L}(\eta,\xi) := \xi^T M_{ep}(\eta_1,\eta_2,\eta_3)\xi \geq \bar{\lambda}|\xi|^2 .$$

In fact, from Lemma 3.1, we can see that for

$$\xi_\eta := [\xi_1\eta_1 \quad \xi_2\eta_2 \quad \xi_3\eta_3 \quad \xi_2\eta_3+\xi_3\eta_2 \quad \xi_1\eta_3+\xi_3\eta_1 \quad \xi_1\eta_2+\xi_2\eta_1]^T$$

we have

$$\mathcal{L}(\eta,\xi) = \xi_\eta^T D_{ep}(q)\xi_\eta \geq \lambda|\xi_\eta|^2$$

Hence $\mathcal{L}(\eta,\xi)\geq0$ and for $\forall\eta\in\mathbb{R}^3$, $\eta\neq0$, $\mathcal{L}(\eta,\xi)=0$ if and only if $\xi=0$.

Now for $\forall\eta\in\mathbb{R}^3$, $\eta\neq0$, let $\bar{\lambda} := \min_{|\xi|=1}\mathcal{L}(\eta,\xi)>0$, then it is easy to show that

$$\mathcal{L}(\eta,\xi) \geq \bar{\lambda}|\xi|^2.$$

Q.E.D.

Therefore, the Euler equation of E is a strong elliptic system in Visik-Nirenberg sense.

Now we shall prove the regularity of the solution of (4.1). For the sake of simplicity, we shall only deal with the regularity in the interior.

Theorem 4.1 If $f \in L^2(\Omega, \mathbb{R}^3)$, let $u \in H^1_o(\Omega, \mathbb{R}^3)$ be the solution of (4.1), then $u \in H^{2.2}_{loc}(\Omega, \mathbb{R}^3)$.

Proof. Following the difference-quotient method of L. Nirenberg [12], we define

$$\Delta^h_k u := \frac{1}{h}[u(x+he_k)-u(x)], \qquad k=1,2,3$$

with $e_1=(1,0,0)$, $e_2=(0,1,0)$, $e_3=(0,0,1)$. Let

$$B_R(x_\circ)=\{x \in \mathbb{R}^3, \ |x-x_\circ|<R\}.$$

For $B_{3R}(x_\circ) \subset\subset \Omega$, let ζ be a cut-off function: $\zeta \in C^\infty_o(B_{2R}(x_\circ))$, $0 \leqq \zeta \leqq 1$, $\zeta \equiv 1$ on $B_R(x_\circ)$, $|D\zeta|<\frac{C}{R}$ and for every $\phi \in H^1_o(\Omega, \mathbb{R}^3)$ Taking $\Delta^h_k(\zeta\phi)$ as the test function in (4.1), we have

$$\int_\Omega [\partial\Delta^h_k(\zeta\phi)]^T D_{ep}(q)\partial u dx - \int_\Omega [\Delta^h_k(\zeta\phi)]^T f dx = 0,$$

$$\forall \ \phi \in H^1_o(\Omega, \mathbb{R}^3).$$

Using the formula

$$\int_\Omega f\Delta^h_k g dx = -\int_\Omega g(x+he_k)\Delta^h_k f dx \tag{4.4}$$

with $\operatorname{dist}(\partial\Omega, \operatorname{supp}.g)>|h|>0$, we obtain

$$\int_\Omega [\partial(\zeta\phi)]^T_{x+he_k} \Delta^h_k[D_{ep}(q)\partial u]dx + \int_\Omega [\Delta^h_k(\zeta\phi)]^T f dx = 0, \tag{4.5}$$

$$\forall \ \phi \in H^1_o(\Omega, \mathbb{R}^3).$$

It is easy to verify that

1°. If $\zeta \in C^\infty_o(\Omega)$ and $W \in H^1_o(\Omega, \mathbb{R}^3)$, then

$$\partial(\zeta W) = \zeta\partial W + [W^T\partial^T(\zeta I)]^T \tag{4.6}$$

where I is a 6×6 unit matrix.

2°. $\Delta^h_k[D_{ep}(q)\partial u] = \hat{D}_{ep}\partial(\Delta^h_k u)$,

where
$$\hat{D}_{ep} = D - DN(\Delta\gamma(q))N^T D, \tag{4.7}$$

$$\Delta\gamma(q) = \begin{cases} \dfrac{\gamma(q(x+he_k))-\gamma(q(x))}{q(x+he_k)-q(x)} & \text{if } q(x+he_k)\neq q(x) \\ 0 & \text{if } q(x+he_k)=q(x). \end{cases}$$

Hence the first term on the left-hand side of (4.5) can be rewritten as follows:

$$(I):= \int_\Omega [\partial(\zeta\phi)]^T_{x+he_k} \Delta^h_k [D_{ep}(q)\partial u]dx$$

$$= \int_\Omega \zeta|_{x+he_k} (\partial\phi)^T_{x+he_k} \hat{D}_{ep}\partial(\Delta^h_k u)dx$$

$$+\int_\Omega [\phi^T\partial^T(\zeta I)]_{x+he_k} \hat{D}_{ep}\partial(\Delta^h_k u)dx.$$

Using (4.6) again, we have

$$\begin{aligned} (I) = &\int_\Omega (\partial\phi)^T_{x+he_k} \hat{D}_{ep}\partial(\zeta|_{x+he_k} \Delta^h_k u)dx \\ &-\int_\Omega (\partial\phi)^T_{x+he_k} \hat{D}_{ep}[(\Delta^h_k u)^T\partial^T(\zeta I)_{x+he_k}]^T dx \\ &+\int_\Omega [\phi^T\partial^T(\zeta I)]_{x+he_k} \hat{D}_{ep}\partial(\Delta^h_k u)dx, \end{aligned} \tag{4.8}$$

Substituting (4.8) into (4.5) and taking $\phi=\zeta\Delta^{-h}_k u$, we have

$$\begin{aligned} &\int_\Omega [\partial(\zeta\Delta^{-h}_k u)]^T_{x+he_k} \hat{D}_{ep}\partial[\zeta|_{x+he_k} \Delta^h_k u]dx \\ &-\int_\Omega [\partial(\zeta\Delta^{-h}_k u)]^T_{x+he_k} \hat{D}_{ep}[(\Delta^h_k u)^T\partial^T(\zeta I)_{x+he_k}]^T dx \\ &+\int_\Omega (\Delta^h_k u)^T\partial^T(\zeta I)_{x+he_k} \hat{D}_{ep}\partial(\zeta|_{x+he_k} \Delta^h_k u)dx \\ &-\int_\Omega (\Delta^h_k u)^T\partial^T(\zeta I)_{x+he_k} \hat{D}_{ep}[(\Delta^h_k u)^T\partial^T(\zeta I)_{x+he_k}]^T dx \\ &+\int_\Omega [\Delta^h_k(\zeta^2\Delta^{-h}_k u)]^T f dx = 0, \quad k=1,2,3. \end{aligned} \tag{4.9}$$

It is not difficult to show that $\hat{D}_{ep}$ is a positive definite matrix and elements of $\hat{D}_{ep}$ are uniformly bounded, whose bounds depend on D, N, $h_\circ$. Therefore, using Hölder inequality, we obtain

$$\lambda\int_\Omega |\partial(\zeta|_{x+he_k} \Delta^h_k u)|^2 dx \leq$$

$$\leq C(\|u\|_{H^1(\Omega,\, \mathbb{R}^3)}+\|f\|_{L^2(\Omega,\, \mathbb{R}^3)}).$$

where $\lambda>0$ is the same constant as that one in Lemma3.1. By Korn's inequality and the properties of ζ, we have

$$\|\Delta_k^h u\|_{H^1(B_{\frac{R}{2}}(x_o),\, \mathbb{R}^3)} \leq C(\|u\|_{H^1(\Omega,\, \mathbb{R}^3)} + \|f\|_{L^2(\Omega,\, \mathbb{R}^3)}).$$

Using a standard covering argument, we have

$$\|\Delta_k^h u\|_{H^1(\Omega',\, \mathbb{R}^3)} \leq C(\|u\|_{H^1(\Omega,\, \mathbb{R}^3)} + \|f\|_{L^2(\Omega,\, \mathbb{R}^3)})$$

for $\forall\,\Omega' \subset\subset \Omega$. The right-hand side of this inequality being independent of h, we obtain

$$u \in H^{2,2}_{loc}(\Omega, \mathbb{R}^3)$$

and we can see that u satisfies the following system

$$\partial^T(D_{ep}(q)\partial u) + f = 0 \qquad \text{a.e. in } \Omega. \tag{4.10}$$

Theorem 4.2 Suppose $f \in L^s(\Omega, \mathbb{R}^3)$, $s>2$. Let $u \in H^1_o(\Omega, \mathbb{R}^3)$ be the solution of (4.1), then there exists an exponent $r>2$ such that $u \in H^{2,r}_{loc}(\Omega, \mathbb{R}^3)$, where r depends only on λ, D, N, h_o, s.

Proof. Let us divide the proof into two steps.

1°. Rewrite (4.1) as follows:

$$\int_\Omega (\partial\phi)^T (D\partial u - DN\gamma(q))dx - \int_\Omega \phi^T f dx = 0, \tag{4.11}$$
$$\forall\ \phi \in C_o^\infty(\Omega, \mathbb{R}^3),$$

where $\gamma(q)$ is defined in (2.9).

Now for $\forall \psi \in C_o^\infty(\Omega, \mathbb{R}^3)$, inserting $\phi = \psi_{x_k} = \frac{\partial\psi}{\partial x_k}$ in (4.11)

we have

$$\int_\Omega (\partial\psi_{x_k})^T (D\partial u - DN\gamma(q))dx - \int_\Omega \psi_{x_k}^T f dx = 0. \tag{4.12}$$

Since we proved $u \in H^{2,2}_{loc}(\Omega, \mathbb{R}^3)$ in Theorem 4.1, so $[D\partial u - DN\gamma(q)] \in H^{1,2}_{loc}(\Omega, \mathbb{R}^6)$ and can be differentiated in the weak sense [13; Lemma 1.1], i.e.

$$\int_\Omega (\partial\psi_{x_k})^T (D\partial u - DN\gamma(q))dx$$
$$= -\int_\Omega (\partial\psi)^T (D\partial u_{x_k} - DN\alpha(q)N^T D\partial u_{x_k})dx. \tag{4.13}$$

Therefore, $u_{x_k} = \frac{\partial u}{\partial x_k} \in H^1(\Omega, \mathbb{R}^3)$ $(k=1,2,3)$ satisfies

$$\int_\Omega (\partial\phi)^T D_{ep}(q)\partial u_{x_k} dx + \int_\Omega \phi_{x_k}^T f dx = 0, \forall\ \phi \in C_o^\infty(\Omega, \mathbb{R}^3), \tag{4.14}$$
$$k=1,2,3.$$

2°. We are now to get the result of this theorem from (4.14).

Let Q be an n-cube, $Q\subset\subset\Omega$. Here and hereafter, we regard n as 3.

For each $x^\circ\in Q$ and each $R<\frac{1}{2}\mathrm{dist}(x^\circ,\partial Q)$, we construct a cut-off function $\eta(x)$:

$$\eta\in C_0^\infty(Q_{2R}(x^\circ)),\ 0\leqq\eta\leqq 1,\ \eta\equiv 1 \text{ on } Q_R(x^\circ),\ |D\eta|<\frac{C}{R} \tag{4.15}$$

and choose as test function in (4.14) $\phi=\phi_k=\eta^2(u_{x_k}-(u_{x_k})_{2R})$, $k=1,2,3$, where $(u_{x_k})_{2R}=\fint_{Q_{2R}(x^\circ)}u_{x_k}\,dx$ is the average of u_{x_k} on $Q_{2R}(x^\circ)$, then we have

$$\int_\Omega[\partial(\eta^2(u_{x_k}-(u_{x_k})_{2R}))]^T D_{ep}(q)\partial(u_{x_k}-(u_{x_k})_{2R})dx$$
$$+\int_\Omega[\eta^2(u_{x_k}-(u_{x_k})_{2R})]^T_{x_k} f dx = 0,\quad k=1,2,3.$$

Some computations yield

$$\int_\Omega[\partial(\eta(u_{x_k}-(u_{x_k})_{2R}))]^T D_{ep}(q)\partial(\eta(u_{x_k}-(u_{x_k})_{2R}))dx$$
$$=\int_\Omega[\partial(\eta(u_{x_k}-(u_{x_k})_{2R}))]^T D_{ep}(q)(\partial^T(\eta I))^T(u_{x_k}-(u_{x_k})_{2R})dx$$
$$-\int_\Omega(u_{x_k}-(u_{x_k})_{2R})^T\partial^T(\eta I)D_{ep}(q)\partial(\eta(u_{x_k}-(u_{x_k})_{2R}))dx$$
$$+\int_\Omega(u_{x_k}-(u_{x_k})_{2R})^T\partial^T(\eta I)D_{ep}(q)(\partial^T(\eta I))^T(u_{x_k}-(u_{x_k})_{2R})dx$$
$$-\int_\Omega\eta[\eta(u_{x_k}-(u_{x_k})_{2R})]^T_{x_k} f dx-\int_\Omega\eta_{x_k}\eta(u_{x_k}-(u_{x_k})_{2R})^T f dx,\quad k=1,2,3,$$

where I is a 6×6 unit matrix.

Since $D_{ep}(q)$ is a positive definite matrix, and the elements of $D_{ep}(q)$ are bounded, whose bounds depend only on D, N, h_0, so we have

$$\sum_{k=1}^3\int_\Omega|\partial(\eta(u_{x_k}-(u_{x_k})_{2R}))|^2dx$$
$$\leq\sum_{k=1}^3 C\int_\Omega|\partial^T(\eta I)|^2|u_{x_k}-(u_{x_k})_{2R}|^2dx+C\int_\Omega\eta^2|f|^2dx,$$

where C depends on λ, D, N, h_0 .

By Korn's inequality, we obtain

$$\sum_{k=1}^3\int_\Omega|D(\eta(u_{x_k}-(u_{x_k})_{2R}))|^2dx$$
$$\leq\sum_{k=1}^3 C\int_\Omega|D\eta|^2|u_{x_k}-(u_{x_k})_{2R}|^2dx+C\int_\Omega\eta^2|f|^2dx$$

and from (4.15) we get

$$\int_{Q_R(x^\circ)}|D^2u|^2dx\leqq\frac{C}{R^2}\int_{Q_{2R}(x^\circ)}|Du-(Du)_{2R}|^2dx+C\int_{Q_{2R}(x^\circ)}|f|^2dx. \tag{4.16}$$

Using Sobolev-Poincare inequality, we obtain

$$\int_{Q_{2R}(x^\circ)} |Du-(Du)_{2R}|^2 dx \leq C\left(\int_{Q_{2R}(x^\circ)} |D^2u|^{\frac{2n}{n+2}} dx\right)^{\frac{n+2}{n}} \tag{4.17}$$

Combining (4.16) and (4.17), we have

$$⨍_{Q_R(x^\circ)} |D^2u|^2 dx \leq C\left(⨍_{Q_{2R}(x^\circ)} |D^2u|^{\frac{2n}{n+2}} dx\right)^{\frac{n+2}{n}} + C ⨍_{Q_{2R}(x^\circ)} |f|^2 dx.$$

Now using reverse Hölder inequality with increasing support [5; Proposition 1.1], we get

$$|D^2u|^{\frac{2n}{n+2}} \in L^p_{loc}(Q), \quad p\in\left[\frac{n+2}{n}, \frac{n+2}{n}+\varepsilon\right)$$

for some $\varepsilon>0$. Moreover, for all $Q_{2R}\subset\subset Q\subset\subset\Omega$, we have

$$\left(⨍_{Q_R}|D^2u|^{\frac{2n}{n+2}p} dx\right)^{\frac{1}{p}} \leq C\left(⨍_{Q_{2R}}|D^2u|^2 dx\right)^{\frac{n}{n+2}} + C\left(⨍_{Q_{2R}}|f|^{\frac{2n}{n+2}p} dx\right)^{\frac{1}{p}},$$

where C and ε are positive constants depending only on n, D, N, $h_\circ$, s.

Let $r=\frac{2n}{n+2}p$, then $|D^2u|\in L^r_{loc}(Q)$, $r\in[2,2+\varepsilon_1)$ for some $\varepsilon_1>0$, i.e.

$$u\in H^{2,r}_{loc}(Q, \mathbb{R}^3), \quad r>2, \ \forall\ Q\subset\subset\Omega.$$

Therefore, by the standard covering argument we have

$$u\in H^{2,r}_{loc}(\Omega, \mathbb{R}^3) \quad \text{for some } r>2.$$

Q.E.D.

<u>Corollary 4.1</u> Under the assumptions of Theorem 4.2, the minimizer of functional $E[u]$ in $H^1_0(\Omega, \mathbb{R}^3)$ belongs to $C^{o,\delta}_{loc}(\Omega, \mathbb{R}^3)$, where $\delta=2-\frac{3}{r}>0$ for some $r>2$.

<u>Theorem 4.3</u> Suppose $f=[f_1\ f_2\ f_3]^T$, where $f_i=D_\alpha g_i^\alpha$, $g_i^\alpha\in C^{o,\delta}(\Omega)$, $i=1,2,3$, $\alpha=1,2,3$, $\delta>0$ and let $u\in H^1_0(\Omega, \mathbb{R}^3)$ be a minimizer of E. Then there exists a constant $\beta>0$ such that if $h_o>\beta$ then

$$u\in C^{1,\delta}_{loc}(\Omega, \mathbb{R}^3), \quad \delta>0,$$

where β depends only on μ, λe, D, N, δ ($\mu>0$ is a constant which appears in Korn's inequality).

<u>Proof</u>. For $\tilde{\Omega}\subset\subset\Omega$, $\forall x_\circ\in\tilde{\Omega}$, $\forall R$, $0<R<\text{dist}(\tilde{\Omega},\partial\Omega)$, the Euler equation (4.1) may be rewritten as follows:

$$\int_\Omega (\partial\phi)^T D\partial u dx - \int_\Omega (\partial\phi)^T DN(\gamma(q)-\gamma((q)_R))dx - \int_\Omega \phi^T f dx = 0,$$

$$\forall\, \phi\in H^1_o(\Omega, \mathbb{R}^3),$$

where $(q)_R = \fint_{B_R(x_o)} q\,dx.$

Let $v\in H^1(B_R(x_o), \mathbb{R}^3)$ be the solution to the Dirichlet problem

$$\begin{cases} \int_\Omega (\partial\phi)^T D\,\partial v\,dx = 0, & \forall\, \phi\in H^1_o(B_R(x_o), \mathbb{R}^3), \\ v-u\in H^1_o(B_R(x_o), \mathbb{R}^3) \end{cases}$$

Then we have, [see 5; Ch III, Theorem 2.1], for all $\rho<R$

$$\int_{B_\rho(x_o)} |Dv-(Dv)_\rho|^2 dx \leq C\left(\frac{\rho}{R}\right)^{n+2} \int_{B_R(x_o)} |Dv-(Dv)_R|^2 dx$$

and therefore

$$\int_{B_\rho(x_o)} |Du-(Du)_\rho|^2 dx \leq C\left(\frac{\rho}{R}\right)^{n+2} \int_{B_R(x_o)} |Du-(Du)_R|^2 dx + C\int_{B_R(x_o)} |D(u-v)|^2 dx. \tag{4.18}$$

Here and hereatter we regard n as 3.

If we set $W=u-v$, then $W\in H^1_o(B_R(x_o), \mathbb{R}^3)$ satisfies

$$\int_{B_R(x_o)} (\partial\phi)^T D\,\partial W dx = \int_{B_R(x_o)} (\partial\phi)^T DN(\gamma(q)-\gamma((q)_R))dx + \int_{B_R(x_o)} \phi^T f dx,$$

$$\forall\, \phi\in H^1_o(B_R(x_o), \mathbb{R}^3)$$

In particular we may take $\phi=W$, since D is a positive definite matrix, so that using Korn's inequality and Hölder inequality we get

$$\int_{B_R(x_o)} |DW|^2 dx \leq \frac{C(n,\mu)}{\lambda_e^2} |DNN^T D|^2 \alpha_o^2 \int_{B_R(x_o)} |Du-(Du)_R|^2 dx + \frac{C(n,\mu)}{\lambda_e^2} \int_{B_R(x_o)} |g-(g)_R|^2 dx, \tag{4.19}$$

where $g=(g^\alpha_i)$ $\alpha=1,2,3$, $i=1,2,3$.

Combining (4.18) and (4.19) we obtain

$$\int_{B_\rho(x_o)} |Du-(Du)_\rho|^2 dx \leq C(n,\mu,\lambda_e, D, N)\left[\left(\frac{\rho}{R}\right)^{n+2} + \alpha_o^2\right] \int_{B_R(x_o)} |Du-(Du)_R|^2 dx +$$

$$+C(n,\mu,\lambda_e) \|g\|^2_{C^{o,\delta}(\Omega,\, \mathbb{R}^{n\times n})} R^{n+2\delta}$$

By an iteration lemma [see 5, ChIII, Lemma 2.1], we can see that there exist two positive constants $\varepsilon_o=\varepsilon_o(n,\mu,\lambda_e,D,N,\delta)$ and $C_o=C_o(n,\mu,\lambda_e,D,N,\delta)$ such that if $\alpha_o^2<\varepsilon_o^2$ then for all $\rho<R$ we have

$$\int_{B_\rho(x_o)} |Du-(Du)_\rho|^2 dx \leqq$$

$$\leqq C_o\{(\frac{\rho}{R})^{n+2\delta}\int_{B_R(x_o)} |Du-(Du)_R|^2 dx+\|g\|^2_{C^{o,\delta}(\Omega,\, \mathbb{R}^{n\times n})}\rho^{n+2\delta}\}$$

which implies

$$Du \in C^{o,\delta}_{loc}(\Omega,\, \mathbb{R}^{3\times 3}),\quad \delta>0.$$

Here $\alpha_o^2<\varepsilon_o^2$ means $h_o+N^TDN>\frac{1}{\varepsilon_o}$ [see (1.7)], i.e.

$h_o>\frac{1}{\varepsilon_o} - N^TDN=: \beta(\mu,\lambda_e,D,N,\delta)$. Therefore, if $h_o>\beta$ then we have

$$u \in C^{1,\delta}_{loc}(\Omega,\, \mathbb{R}^3),\quad \delta>0.$$

Q.E.D.

REFERENCES

[1] I.Beju. Theorems on existence, uniqueness and stability of the solution of the place boundary-value problem, in statics, for hyperelastic materials, Arch. Ratonal. Mech. Anal., 42, 1971,1-23.

[2] L.C.Evans. Quasiconvexity and partial regularity in the calculus of variations, preprint, MD84-45LE.

[3] L.C.Evans, R.F.Gariepy. Blow-up, compactness and partial regularity in the calculus of variations, preprint, MD85-43-LERG.

[4] N.Fusco, J.Hutchinson. $C^{1,\alpha}$ partial regularity of functions minimizing quasiconvex integrals, manuscripta math. Vol. 54, Fasc 1-2, 1985,121-142.

[5] M.Giaquinta. Multiple Integrals in the Calculus of Variations and Nonliner Elliptic Systems, Princeton University Press, 1983.

[6] M.Giaquinta, G.Modica. Partial regularity of minimizers of quasiconvex integrals, Ann. Inst. H. Poincare, Analyse non lineaire, 3, 1986, 185-208.

[7] K.S.Havner, H.P.Patel. On convergence of the finite element method for a class of elastic-plastic solids, Quarterly of Appl. Math. Vol. 34, No.1, 1976, 59-68.

[8] M.C.Hong. Existence and partial regularity in the calculus of variations, preprint.

[9] L.S.Jiang, L.C.Wu, Y.D.Wang, Q.X.Ye. On the existence, uniqueness of a class of elastic-plastic problem and the convergence of the approximate solutions, Acta Mathematicae Applicatae Sinica, Vol. 4. No.2, 1981, 166-174.

[10] L.S.Jiang, L.C.Wu. A class of nonlinear elliptic systems with discontinuous coefficients, Acta Mathematicae Sinca, Vol. 26, No.6, 1983, 660-668.

[11] L.S.Jiang. On an elastic-plastic problem, Journal of Differential Equations, Vol. 51, No.1, 1984, 97-115.

[12] L.Nirenberg. Remarks on strongly elliptic partial differential equations, Comm. Pure Appl. Math. 8, 1955, 649-675.

[13] G.Stampacchia. Equations Elliptiques du Second Ordre a Coefficients Discontinus, Montreal University Press, 1966.

[14] L.C.Wu. A note on the regularity of solutions to a nonlinear system from elasticity-plasticity theory, preprint.

EVERYWHERE REGULARITY FOR SOLUTIONS TO QUASILINEAR ELLIPTIC SYSTEMS OF TRIANGULAR FORM

Yan Ziqian

Jilin University,
Changchun, China.

In recent years many mathematicians have paid attention to regularity of weak solutions to quasilinear elliptic systems

$$-D_\alpha[A_{ij}^{\alpha\beta}(x,u)D_\beta u^j + a_i^\alpha(x,u)] = B_i(x,u,Du),$$

$$i = 1, \ldots, N, \qquad x\in\Omega\subset R^n,$$

with coefficients having controllable or natural growth.

Here and in the sequel, we agree that repeated indices are to be summed for α and β from 1 to n, for i and j from 1 to N, but not for k, unless otherwise specified.

The main results in this respect may be found in Giaquinta [1], Hidebrandt [2] and the references quoted by them. It is known that any weak solution is Hölder continuous if N=1 (single equation) or n=2. However, for $N\geqq 2$ and $n\geqq 3$, the situation is quite different. An example [1,p.253] shows that there is a linear elliptic system without lower order terms and having bounded coefficients, which has a weak solution being singular in a dense set. Thus, the general problem of imposing reasonable conditions on the leading parts $A_{ij}^{\alpha\beta}$ to ensure solutions to have everywhere regularity should be interesting. This kind of conditions being known now is only

1) $A_{ij}^{\alpha\beta} = A_{ij}^{\alpha\beta}(x)\in C^\circ(\Omega)$ and $A_{ij}^{\alpha\beta}(x)\xi_\alpha\xi_\beta\eta^i\eta^j \geqq \lambda|\xi|^2|\eta|^2, \lambda>0$;

2) $A_{ij}^{\alpha\beta}(x,u) = 0$ when $j\neq i$, $A_{kk}^{\alpha\beta}(x,u) = a^{\alpha\beta}(x,u)$ for $k=1, \ldots, N$ and

$$\lambda|\xi|^2 \leqq a^{\alpha\beta}(x,u)\xi_\alpha\xi_\beta \leqq \Lambda|\xi|^2, \ \Lambda \geqq \lambda > 0;$$

or

3) $A_{ij}^{\alpha\beta} = A_{ij}^{\alpha\beta}(x)$ and $\lambda|\xi|^2 \leqq A_{ij}^{\alpha\beta}(x)\xi_\alpha^i\xi_\beta^j \leqq \Lambda|\xi|^2, \quad \lambda>0$

with Λ/λ close to 1,

which was presented in [1,p.183].

In this paper we show that weak solutions of elliptic systems of triangular form are Hölder continuous, provided that the growth of $B_i(x,u,p)$ with respect to $|p|$ is less than 2. The method used here is combining that used by Giaquinta to prove Theorem 1.4 of [1,p.184] and an iteration procedure which gives the decay of the Dirichlet integral for u^{k+1} in terms of that for u^j, j=1, ..., k.

Let Ω be an open set in R^n. Consider the quasilinear elliptic systems of triangular form, i.e., the systems

$$-D_\alpha[A_{ij}^{\alpha\beta}(x,u)D_\beta u^j + a_i^\alpha(x,u)] = B_i(x,u,Du), \qquad (1)$$
$$i=1, \ldots, N, \qquad x \in \Omega$$

with bounded $A_{ij}^{\alpha\beta}(x,u)$ satisfying

$$\begin{aligned} &A_{ij}^{\alpha\beta}(x,u) = 0 \quad \text{when} \quad j>i, \\ &A_{kk}^{\alpha\beta}(x,u)\xi_\alpha\xi_\beta \geq \lambda|\xi|^2, \qquad k=1, \ldots, N, \quad \lambda>0. \end{aligned} \qquad (2)$$

Let us first suppose that the controllable growth conditions hold, i.e., (assume $n\geq3$)

$$|a_i^\alpha(x,u)| \leq C(|u|^{\frac{n}{n-2}} + f_i^\alpha), \qquad f_i^\alpha \in L^\sigma(\Omega), \qquad (3)$$

$$|B_i(x,u,p)| \leq C(|p|^{\frac{n+2}{n}} + |u|^{\frac{n+2}{n-2}} + g_i), \qquad g_i \in L^\tau(\Omega). \qquad (4)$$

Theorem 1 Under conditions (2), (3) and (4) with $\sigma>n$ and $\tau>n/2$, any weak solution to the system (1) of triangular form in fact belongs to $C_{loc}^{0,\mu}(\Omega,R^N)$ for some $\mu>0$.

By a weak solution to (1) we mean a function $u\in H^1(\Omega,R^N)$ satisfying the integral identity

$$\int_\Omega[A_{ij}^{\alpha\beta}(x,u)D_\beta u^j + a_i^\alpha(x,u)]D_\alpha\phi^i dx = \int_\Omega B_i(x,u,Du)\phi^i dx \qquad (5)$$

for any $\phi\in H_0^1(\Omega,R^N)$.

In order to prove Theorem 1 we need the following

Proposition Suppose conditions (2), (3) and (4) with $\sigma>2$ and $\tau>\frac{2n}{n+2}$ hold. Then there exists an exponent $p>2$ such that, if $u\in H^1(\Omega,R^N)$ is a weak solution to (1), then $Du\in L_{\ell oc}^p(\Omega,R^{nN})$. Moreover, for $B_{R/2}\subset B_R\subset\Omega$ and R small enough we have

$$\left[⨍_{B_{R/2}}(|u|^{\frac{2n}{n-2}} + |Du|^2)^{\frac{p}{2}}dx\right]^{\frac{1}{p}} \leq C\left\{\left[⨍_{B_R}(|u|^{\frac{2n}{n-2}} + |Du|^2)dx\right]^{\frac{1}{2}} + \right.$$
$$\left. + \left[⨍_{B_R}\sum_{i,\alpha}|f_i^\alpha|^p dx\right]^{\frac{1}{p}} + \left[⨍_{B_R}\sum_i|g_i|^{\frac{pn}{n+2}}dx\right]^{\frac{n+2}{pn}}\right\}$$

where $B_R=B_R(x_\circ)=\{x:|x-x_\circ|<R\}$.

Proof For each $k\in\{1, \dots, N\}$ we set $\phi^k=(u^k-u^k_R)\eta^2$ and $\phi^i=0$ when $i\neq k$ into (5), where

$$u^k_R = ⨍_{B_R} u^k dx = \frac{1}{|B_R|}\int_{B_R} u^k dx$$

and $\eta\in C^\infty_\circ(B_R)$, $0\leq\eta\leq 1$, $|\nabla\eta|\leq C/R$. Then we have

$$\begin{aligned}\int_\Omega [A^{\alpha\beta}_{kk}D_\alpha u^k D_\beta u^k\eta^2 + (\sum_{j=1}^{k-1} A^{\alpha\beta}_{kj}D_\beta u^j + a^\alpha_k)D_\alpha u^k\eta^2 + \\ + (\sum_{j=1}^{k} A^{\alpha\beta}_{kj}D_\beta u^j + a^\alpha_k)(u^k-u^k_R)2\eta D_\alpha\eta]dx \\ = \int_\Omega B_k(u^k-u^k_R)\eta^2 dx.\end{aligned}$$

By (2)-(4),

$$\begin{aligned}\int_\Omega |Du^k|^2\eta^2 dx \leq C[\int_\Omega \sum_{j=1}^{k-1}|Du^j|^2\eta^2 dx + \int_\Omega (u^k-u^k_R)^2|D\eta|^2 dx + \\ + \int_\Omega \sum_\alpha |f^\alpha_k|^2\eta^2 dx] + \varepsilon[\int_\Omega |u^k-u^k_R|^{\frac{2n}{n-2}}\eta^2 dx]^{\frac{n-2}{n}} + \\ + C(\varepsilon,u)[\int_\Omega |u|^{\frac{2n}{n-2}}\eta^2 dx + (\int_\Omega |Du|^2\eta^2 dx)^{1+\frac{2}{n}} \\ + (\int_\Omega |g_k|^{\frac{2n}{n+2}}\eta^2 dx)^{1+\frac{2}{n}}], \quad \forall\ \varepsilon>0.\end{aligned}$$

Multiplying both sides by δ^{k-1} with $\delta\in(0,1)$ satisfying

$$C(\delta+ \dots +\delta^{N-1})<\frac{1}{2}$$

and summing for k from 1 to N we obtain

$$\begin{aligned}\delta^{N-1}\int_\Omega |Du|^2\eta^2 dx \leq \sum_{k=1}^{N}\delta^{k-1}\int_\Omega |Du^k|^2\eta^2 dx \\ \leq 2C[\int_\Omega |u-u_R|^2|D\eta|^2 dx + \int_\Omega \sum_{k,\alpha}|f^\alpha_k|^2\eta^2 dx] + 2\varepsilon\sum_{k=1}^{N}[\int_\Omega |u^k-u^k_R|^{\frac{2n}{n-2}}\eta^2 dx]^{\frac{n-2}{n}} + \\ + (2C(\varepsilon,u)+1)[\int_\Omega |u|^{\frac{2n}{n-2}}\eta^2 dx + (\int_\Omega |Du|^2\eta^2 dx)^{1+\frac{2}{n}} + \sum_{k=1}^{N}(\int_\Omega |g_k|^{\frac{n}{n+2}}\eta^2 dx)^{1+\frac{2}{n}}].\end{aligned}$$

Arguing as in [1,pp.140-142], one easily gets

$$⨍_{B_{R/2}} (|u|^{\frac{2n}{n-2}} + |Du|^2)dx \leq C\{[⨍_{B_R} (|u|^{\frac{qn}{n-2}} + |Du|^q)dx]^{\frac{2}{q}}$$

$$+ \fint_{B_R} \sum_{i,\alpha} |f_i^\alpha|^2 dx + \left(\fint_{B_R} \sum_i |g_i|^{\frac{2n}{n+2}} dx\right)^{1+\frac{2}{n}}\} + \frac{1}{2}\fint_{B_R} |Du|^2 dx$$

for all R small enough, where

$$q = \frac{2n}{n+2}.$$

Now the result follows from Gehring's Lemma (see [1,p.122]).

Remark Theorem 2.2 of [1,p.148] does not apply to the Proposition since here we only assume the weaker ellipticity condition (2) instead of

$$A_{ij}^{\alpha\beta}(x,u)\xi_\alpha^i\xi_\beta^j \geq \lambda|\xi|^2, \quad \lambda>0.$$

Let us now come back to

The proof of Theorem 1 Let $u=(u^1, \ldots, u^N) \in H^1(\Omega,R^N)$ be a weak solution to the system (1). For any $x_\circ\in\Omega$ and any $R<R_\circ=\min(\mathrm{dist}(x_\circ,\partial\Omega),1)$ we split u as v+w on $B_R=B_R(x_\circ)$. Here $v=(v^1, \ldots, v^N)$, $w=(w^1, \ldots, w^N)$, v^k is a solution to the Dirichlet problem

$$\begin{cases} -D_\alpha[A_{kk}^{\alpha\beta}(x,u(x))D_\beta v^k] = 0 \quad \text{in } B_R, \\ v^k-u^k \in H_\circ^1(B_R), \end{cases} \tag{6$_k$}$$

and w^k satisfies the integral identity

$$\begin{aligned} &\int_{B_R} [A_{kk}^{\alpha\beta}(x,u(x))D_\beta w^k + \sum_{j=1}^{k-1} A_{kj}^{\alpha\beta}(x,u(x))D_\beta u^j + a_k^\alpha(x,u(x))]D_\alpha w^k dx \\ &\qquad = \int_{B_R} B_k(x,u(x),Du(x))w^k dx \end{aligned} \tag{7$_k$}$$

for each $k\in\{1, \ldots, N\}$. By the De Giorgi-Nash Theorem, from $(6)_k$ one easily gets that for some $\gamma>0$ and for all $\rho<R$

$$\int_{B_\rho} |Dv^k|^2 dx \leq C\left(\frac{\rho}{R}\right)^{n-2+2\gamma}\int_{B_R} |Dv^k|^2 dx$$

(see [1,p.175]) and hence

$$\int_{B_\rho} |Du^k|^2 dx \leq C\left[\left(\frac{\rho}{R}\right)^{n-2+2\gamma}\int_{B_R} |Du^k|^2 dx + \int_{B_R} |Dw^k|^2 dx\right]. \tag{8$_k$}$$

Using the assumptions (2)-(4), the Young inequality, Hölder inequality and Sobolev inequality from $(7)_k$ we get

$$\int_{B_\rho} |Dw^k|^2 dx \leq C\left[\sum_{j=1}^{k-1} \int_{B_R} |Du^j|^2 dx + \left(\int_{B_R} |Du|^2 dx\right)^{1+\frac{2}{n}} + \int_{B_R} |u|^{\frac{2n}{n-2}} dx + R^{n-2+2\mu}\right], \tag{9$_k$}$$

where $\mu=\min(1-\frac{n}{\sigma},2-\frac{n}{\tau})$. For simplicity we assume $\mu<\gamma$.

Substituting $(9)_k$ into $(8)_k$ we find that

$$\int_{B_\rho}|Du^k|^2dx \leq C[(\frac{\rho}{R})^{n-2+2\gamma}\int_{B_R}|Du^k|^2dx + \sum_{j=1}^{k-1}\int_{B_R}|Du^j|^2dx + (\int_{B_R}|Du|^2dx)^{1+\frac{2}{n}} + \int_{B_R}|u|^{\frac{2n}{n-2}}dx + R^{n-2+2\mu}]. \tag{10$_k$}$$

By the Proposition, for small ρ,

$$\int_{B_\rho}|Du|^2dx \leq C\rho^\varepsilon \tag{11}$$

where $\varepsilon=n(1-\frac{2}{p})$. Note that

$$\int_{B_R}|u-u_R|^{\frac{2n}{n-2}}dx \leq C(\int_{B_R}|Du|^2dx)^{\frac{n}{n-2}} .$$

Therefore, (11) implies

$$\int_{B_\rho}|u|^{\frac{2n}{n-2}}dx \leq C[(\frac{\rho}{R})^n\int_{B_R}|u|^{\frac{2n}{n-2}}dx + R^{\varepsilon\frac{n}{n-2}}] .$$

Applying Lemma 2.1 of [1,p.86] to $\phi(\rho)=\int_{B_\rho}|u|^{\frac{2n}{n-2}}dx$, we get

$$\int_{B_\rho}|u|^{\frac{2n}{n-2}}dx \leq C\rho^{\varepsilon\frac{n}{n-2}} \leq C\rho^{\varepsilon(1+\frac{2}{n})}. \tag{12}$$

Rewriting $(10)_1$ in the form below:

$$\int_{B_\rho}|Du^1|^2dx \leq C[(\frac{\rho}{R})^{n-2+2\gamma}\int_{B_R}|Du^1|^2dx + (\int_{B_R}|Du|^2dx)^{1+\frac{2}{n}} + \int_{B_R}|u|^{\frac{2n}{n-2}}dx+R^{n-2+2\mu}],$$

applying the lemma just mentioned to $\phi(\rho)=\int_{B_\rho}|Du^1|^2dx$ and taking (11),(12) into account we see that

$$\int_{B_\rho}|Du^1|^2dx \leq C\rho^{\varepsilon(1+\frac{2}{n})} .$$

Substituting this estimate into $(10)_2$ and using the same lemma we get

$$\int_{B_\rho}|Du^2|^2dx \leq C\rho^{\varepsilon(1+\frac{2}{n})}$$

By iteration in this way we deduce

$$\int_{B_\rho} |Du^3|^2 dx, \ \ldots, \ \int_{B_\rho} |Du^N|^2 dx \le C\rho^{\varepsilon(1+\frac{2}{n})}$$

successively, and hence

$$\int_{B_\rho} |Du|^2 dx \le C\rho^{\varepsilon(1+\frac{2}{n})} . \tag{13}$$

Just as (12) follows from (11), it follows from (13) that

$$\int_{B_\rho} |u|^{\frac{2n}{n-2}} dx \le C\rho^{\varepsilon\frac{n}{n-2}(1+\frac{2}{n})} . \tag{14}$$

Thus, (11) and (12) are improved by (13) and (14).

Arguing step by step, after m steps where m is an integer satisfying

$$\varepsilon(1+\frac{2}{n})^{m-1} < n-2+2\mu \le \varepsilon(1+\frac{2}{n})^m,$$

we arrive at

$$\int_{B_\rho} |Du|^2 dx \le C\rho^{n-2+2\mu} , \text{ for small } \rho.$$

By a well-known Morrey's Lemma (see [1,p.64]), $u \in C^{o,\mu}_{\ell oc}(\Omega, R^N)$.

Next we consider the regularity for systems (1) with (2) under natural growth conditions:

$$a_i^\alpha(x,u) \equiv a_i^\alpha(x) \in L^\sigma(\Omega), \quad \sigma > n, \tag{15}$$

$$|B_i(x,u,p)| \le a|p|^r + g_i(x), \quad 1+\frac{2}{n}<r<2, \tag{16}$$

$$g_i \in L^\tau(\Omega), \quad \tau > \frac{n}{2}, \ a \text{ — const.}$$

<u>Theorem 2</u> <u>Under the conditions (2), (15) and (16), any bounded weak solution to the system (1) of triangular form in fact belongs to</u> $C^{o,\mu}_{loc}(\Omega,R^N)$ <u>for some</u> $\mu>0$.

<u>Proof</u>. Let $u \in H^1(\Omega,R^N)$ with $\sup_\Omega|u|=M$ be a solution to the system (1), i.e., (5) holds for any $\phi \in H^1_o(\Omega) \cap L^\infty(\Omega)$. We split $u=v+w$ as before. This time $(7)_k$ and $(8)_k$ are still true, while $(9)_k$ and $(10)_k$ become even simpler. Comparing conditions (15) and (16) with (3) and (4) we see that, instead of $(9)_k$ and $(10)_k$, now we have

$$\int_{B_R} |Dw^k|^2 dx \le C[\sum_{j=1}^{k-1} \int_{B_R} |Du^j|^2 dx + \int_{B_R} |Du|^r dx + R^{n-2+2\mu}] \tag*{$(17)_k$}$$

$$\le C[\sum_{j=1}^{k-1} \int_{B_R} |Du^j|^2 dx + R^\varepsilon \int_{B_R} |Du|^2 dx + R^{n-2+2\mu}]$$

for $\varepsilon>0$ small, say $\varepsilon=2(1-\mu)(2-r)/r$, and for k=1, ..., N, and

$$\int_{B_\rho} |Du^k|^2 dx \leq C[(\frac{\rho}{R})^{n-2+2\gamma}\int_{B_R} |Du^k|^2 dx + \sum_{j=1}^{k-1}\int_{B_R} |Du^j|^2 dx + R^\varepsilon \int_{B_R} |Du|^2 dx + R^{n-2+2\mu}] \qquad (18)_k$$

for k=1, ..., N, respectively.

Applying Lemma 2.1 of [1,p.86] successively to $(18)_k$ we find that

$$\int_{B_\rho} |Du^1|^2 dx, \; \ldots, \; \int_{B_\rho} |Du^N|^2 dx \leq C\rho^\varepsilon$$

and hence

$$\int_{B_\rho} |Du|^2 dx \leq C\rho^\varepsilon .$$

By iteration as in the proof of Theorem 1 we finally arrive at

$$\int_{B_\rho} |Du|^2 dx \leq C\rho^{n-2+2\mu}$$

for some $\mu>0$, which, by Morrey's Lemma, completes the proof.

REFERENCES

[1] Giaquinta, M.: Multiple integrals in the calculus of variations and nonlinear elliptic systems, Princeton, Princeton Univ. Press, 1983.

[2] Hildebrandt, S.: Nonlinear elliptic systems and harmonic mappings, Proceedings of the 1980 Beijing symposium on differential geometry and differential equations, Beijing, Science Press, 1982, 481-615.

ON THE DIRICHLET PROBLEM FOR A CLASS OF QUASILINEAR ELLIPTIC SYSTEMS OF PARTIAL DIFFERENTIAL EQUATIONS IN DIVERGENCE FORM

Zhang Ke-Wei

Peking University, Beijing, China

1. Introduction

In this paper we introduce the notions of "quasimonotone" mappings and "semiconvex" functions. With these notions we study the second order elliptic systems involving quasimonotone mappings in the form

(1.1) $$-\frac{\partial}{\partial x^\alpha} A^i_\alpha(x, u(x), Du(x)) + B^i(x, u(x), Du(x)) = 0, x \in \Omega$$

(1.2) $$u(x) = u_o(u), \qquad x \in \partial\Omega$$

where $u : \Omega \subset R^n \longmapsto R^N$ is a vector valued function, $i = 1, \cdots, N, N \geq 1$, Ω is a bounded Lipschitz domain with boundary $\partial\Omega$. (We use the summation convention throughout with i, j running from 1 to N and α, β running from 1 to n). We study the existence problem of weak solution of (1.1) - (1.2).

A feature of our paper is that we treat a class of problems which the classical monotone operator methods developed by Visik [19], Minty [12], Browder [5], Brezis [2], Lions [10] do not work. The study of quasimonotone mappings is not only of the interest for function theory but also for its applications. For example, in the mathematical theory of nonlinear elastostatics, equations governing the equilibrium state of general homogeneous elastic materials without external forces are

(1.3) $$\sum_{\alpha=1}^{3} \frac{\partial}{\partial x^\alpha} A^i_\alpha(Du(x)) = 0 \quad i = 1, 2, 3$$

where $(Du(x))^i_\alpha = \frac{\partial u^i(x)}{\partial x^\alpha} = u^i_{,\alpha}(x), i = 1, 2, 3, \alpha = 1, 2, 3$. As for the hyperelastic materials, equations to be solved can be reduced to finding the stationary points of the functional

$$J(u) = \int_\Omega F(Du(x))dx$$

i.e., to solving the related Euler equations:

$$\underset{\alpha}{\mathrm{div}} F_{p^i}(Du(x)) = 0 \quad i = 1, 2, 3$$

in some proper functional spaces.

Generally, system (1.3) does not permit the uniqueness of solutions, so we can not add simply on (A^i_α) and F monotonicity and convexity conditions respectively (see for example Ball [3], Truesdell [17], Truesdell & Noll [18]). Hence it is necessary for us to study more general type of mappings and functions.

An important task of nonlinear elasticity is to find the constitutive conditions which (A^i_α) and F satisfy. Ball considers the polyconvex functions (which are a special cases of Morrey's quasiconvex functions (see [14])) and studies in [3,4] the minimizing problem of the functional

$$J(u) = \int_\Omega F(x, u, Du)dx$$

and proves several existence theorems in nonlinear elasticity. However, for non-hyperelastic materials, in other words, when (A^i_α) is not a potential map, similar consideritions do not appear in present literatures. Motivated by these, we introduce our conception.

Suppose F: $R^{Nn} \longmapsto R$ is a C^1 function. F is called a quasiconvex function (see Ball [3,4], Morrey [13, 14]), if for every $P \in R^{Nn}$, every open subset G of R^n, and every $z \in C^1_0(G; R^N)$, we have

$$\int_G F(P + Dz(x))dx \geq F(P)\text{meas}(G)$$

If we define

$$h_{P,z}(t) = \int_\Omega F(P + tDz(x))dx \quad t \in R$$

then quasiconvexity means that for every P, z, G as above, $h_{P,z}(t)$ attains its minimum at $t = 0$. This, however, tells us nothing about the behavior of $h_{P,z}$ at other points $t \in R$. if we strengthen the proceeding condition on $h_{P,z}(t)$ to satisfy

$$t\frac{d}{dt}h_{P,z}(t) \geq 0, i.e., \int_G F_{P^i_\alpha}(P + tDz(x))tz^i_{,\alpha}(x)dx \geq 0$$

we could expect better properties of h. This leads us to introduce the following definitions.

Definition 1.1 A C^1 function F: $M^{N\times n} \to R$ is called a semiconvex function if for every P, z, G as above, we have

$$\int_G F_{P^i_\alpha}(P + Dz(x))z^i_{,\alpha}(x)dx \geq 0 \tag{1.4}$$

where $F_{P^i_\alpha}$ denotes the partial derivative of F with respect to P P^i_α, $M^{N\times n}$ is the set of all real $N \times n$ matrices.

Definition 1.2 A continuous map $A : M^{N\times n} \to M^{N\times n}$ is called a quasimonotone map, if for every P, z, G as above, we have

$$\int_G A^i_\alpha(P + Dz(x))z^i_{,\alpha}(x)dx \geq 0 \tag{1.5}$$

In section 2, notations, preliminary results and examples are given which play important roles in the proof of our existence theorem.

In section 3 we state and prove our main theorem.

Acknowledgment The auther is grateful to professors Wang Row-Huai, Chang Kung-Ching and Wu Lan-Cheng for their valuable help in preparing this paper.

2. Notations, Preliminaries and Examples

If $a \in R^k$, then $|a|$ is its euclidean norm. $M^{N\times n}$ is the set of all $N \times n$ real matrices with reduced R^{Nn} topology, that is, if $p \in M^{N\times n}$, then $|p|$ is the norm of p when regarded as a vector in R^{Nn}. The Lebesgue measure of a measurable set S in R^n will be denoted by meas(S).

Let $\Omega \subset R^n$ be an open set, $1 \leq p \leq +\infty, N \geq 1$, we define $L^p(\Omega; R^N)$ as the collection of all N-tuples $(f^1, \cdots, f^N)$ of functions in $L^p(\Omega)$. Analogously, we say that $u \in W^{1,p}(\Omega; R^N)$ if u belongs to $L^p(\Omega; R^N)$ together with its distribution derivatives $\frac{\partial u^i}{\partial x^\alpha}$, $1 \leq i \leq N$, $1 \leq \alpha \leq n$. The $N \times n$ matrix of these derivatives will be denoted by the symbol Du. $W^{1,p}(\Omega; R^N)$ becomes a Banach space if it is endowed with the norm

$$\|u\|_{W^{1,p}(\Omega;R^N)} = \||u|\|_{L^p(\Omega)}p + \||Du|\|_{L^p(\Omega)}$$

where

$$|u|(x) = |u(x)|, |Du|(x) = |Du(x)|$$

$u \in C_0^1(\Omega; R^N)$ if each u^i is a C^1 function on Ω with compact support, while $W_0^{1,p}(\Omega; R^N)$ is the closure of $C_0^1(\Omega; R^N)$ in the topology of $W^{1,p}(\Omega; R^N)$. However, we can introduce an equivalent norm on $W_0^{1,p}(\Omega; R^N)$, say

$$\|u\|_{W_0^{1,p}(\Omega;R^N)} = \||Du|\|_{L^p_{(\Omega)}}$$

Remark 2.1 it is easy to examine that for C^1 function F: $M^{N\times n} \to R$, we have the following implications: convexity $\Rightarrow$ semiconvexity $\Rightarrow$ quasiconvexity;
and for continuous mapping A: $M^{N\times n} \to M^{N\times n}$, we have
monotonicity $\Rightarrow$ quasimonotonicity $\Rightarrow$ ellipticity;
where ellipticity means

$$\frac{\partial A^i_\alpha(P)}{\partial P^j_\beta}\lambda^i\lambda^j\eta_\alpha\eta_\beta \geq 0 \tag{2.1}$$

for every $P \in M^{N\times n}$, $\lambda \in R^N$, $\eta \in R^n$.

To show that quasimonotonicity implies ellipticity, we can use the methods in Morrey [14], Ball [3].

Remark 2.2 Example 2.1 shows that semiconvexity does not imply convexity. Consequently quasimonotonicity does not imply monotonicity.

In what follows, if A is a map from $R^n \times R^N \times M^{N\times n}$ into $M^{N\times n}$, with meas $(I) = 0$, such that for every $\bar{x} \in R^n \setminus I$, and $\bar{s} \in R^N$, the map

$$P \longrightarrow (A^i_\alpha(\bar{x}, \bar{s}, P))$$

is quasimonotone.

Definition 2.3 $f : R^n \times R^N \times M^{N\times n} \to R$ is a Caratheodory function if the following conditions are satisfied:

for every $(s, P) \in R^n \times M^{N\times n}$, $x \to f(x, s, P)$ is measurable;

for almost all $x \in R^n$, $(s, P) \to f(x, s, P)$ is continuous.

The following result of Scoraza & Dragoni ([8, page 235]) characterizes the class of Caratheodory functions.

Lemma 2.4 $f : R^n \times R^N \times M^{N\times n} \to R$ is a Caratheodory function if and only if for every compact set $K \subset R^n$, and $\varepsilon > 0$, there exists a compact set $K_\varepsilon \subset K$ with mean $(K \setminus K_\varepsilon) < \varepsilon$, such that the restriction of f to $K_\varepsilon \times R^N \times M^{N\times n}$ is continuous.

The following lemma can be found in [7].

Lemma 2.5 let $G \subset R^n$ be measurable, with meas$(G) < \infty$. Assume (M_k) is a sequence of measurable subsets of G, such that, for some $\varepsilon > 0$, the following estimate holds:

$$\text{meas}(M_k) \geq \varepsilon, \text{for all } k \in N$$

Then a subsequence (M_{k_h}) can be selected such that $\bigcap_{h\in N} M_{k_h} \neq \phi$.

Lemma 2.6 (see Acerbi & Fusco [1]) Let (f_k) be a bounded sequence in $L^1(R^n)$. Then for each $\varepsilon > 0$, there exists a triple $(A_\varepsilon, \delta, S)$, where A_ε is measurable and meas$(A_\varepsilon) < \varepsilon$, $\delta > 0$ and S is an infinite subset of N, such that for all $k \in S$,

$$\int_B |f_k(x)|dx < \varepsilon$$

whenever B and A_ε are disjoint and meas$(B) < \delta$.

If $r > 0$, and $x \in R^n$, we set $B_r(x) = y \in R^n : |y - x| < r$ and for $f \in L^1(R^n)$, set

$$\fint_{B_r(x)} f(x)dx = \frac{1}{\text{meas}(B_r(x))} \int_{B_r(x)} f(x)dx$$

Definition 2.7 Let $u \in C_0^1(R^n)$, we define

$$(M^* u)(x) = (Mu)(x) + \sum_{\alpha=1}^{n} (MD_\alpha u)(x)$$

where we set

$$(Mf)(x) = \sup_{r>0} \fint_{B_r(x)} f(x)dx$$

for every locally summable f, and $D_\alpha u = \frac{\partial u}{\partial x^\alpha}$.

The following two lemmas are contained in [11].

Lemma 2.8 If $u \in C_0^\infty(R^n)$, then $M^* u \in C^0(R^n)$ and

$$|u(x)| + \sum_{\alpha=1}^{n} |D_\alpha u(x)| \le (M^* u)(x)$$

for all $x \in R^n$. Moreover (see [16]) if $p > 1$, then

$$\|M^* u\|_{L^p(R^n)} \le C(n,p) \|u\|_{W_0^{1,p}(R^n)}$$

and if $p = 1$, then

$$\text{meas}(\{x \in R^n : (m^* u)(x) \le \lambda\}) \le \frac{C(n)}{\lambda} \|u\|_{W^{1,1}(R^n)}$$

for all $\lambda > 0$.

Lemma 2.9 Let $u \in C_0^\infty(R^n)$ and put

$$U(x,y) = \frac{|u(y) - u(x) - \sum_{\alpha=1}^{n} D_\alpha u(x)(y^\alpha - x^\alpha)|}{|y - x|}$$

Then for every $x \in R^n$ and $r > 0$

$$\int_{B_r(x)} U(x,y)dy \le 2\text{meas}(B_r(x))(M^* u)(x)$$

Lemma 2.10 let $u \in C_0^\infty(R^n)$ and $\lambda > 0$, and set

$$H^\lambda = \{x \in R^n : (M^* u)(x) < \lambda\}$$

Then for every $x, y \in H^\lambda$ we have

$$\frac{|u(y) - u(x)|}{|y - x|} \le C(n)\lambda$$

For the proof, see [1].

Lemma 2.11 Let X be a metric space, E a subspace of X, and k, a positive real number. Then any k-Lipschitz mapping from E into R can be extended by a k-Lipschitz mapping from X into R.

For the proof, see [8, page 298].

Lemma 2.12 (see Minty [12], Lions [10]) Let $(\cdot,\cdot)$ be the inner product in R^k, $k \geq 1$. $A : R^k \to R^k$ be a continuous map such that

$$(Ax, x) \geq 0$$

whenever $x \in \partial B_R(0)$. Then there exists $x^* \in \bar{B}_R(0)$, such that

$$Ax^* = 0$$

We conclude this preliminary section by exhibiting several examples.

Example 2.13 For $P \in M^{2\times 2}$, define

$$W_\nu(P) = (\det P)^2 + \sum_{i,\alpha=1}^{2} \nu(P^i_\alpha)^4$$

We will show the following two facts:

i) for any $\nu > 0$, $W_\nu(P)$ is not convex;

ii) for $\nu \geq 1/2$ at least, $W_\nu(P)$ is semiconvex.

Proof of i) Since

$$\frac{\partial^2 W(P)}{\partial P^i_\alpha \partial P^j_\beta j} Q^i_\alpha Q^j_\beta = 2C^2 + 4(\det P)(\det Q) + 12\nu \sum_{i,\alpha=1}^{2} (P^i_\alpha)^2 (Q^i_\alpha)^2$$

where

$$C = \begin{vmatrix} P^1_1 & P^1_2 \\ Q^2_1 & Q^2_2 \end{vmatrix} + \begin{vmatrix} Q^1_1 & Q^1_2 \\ P^2_1 & P^2_2 \end{vmatrix}$$

We set

$$P = \begin{pmatrix} a & 0 \\ 0 & a \end{pmatrix}, \quad Q = \begin{pmatrix} 0 & b \\ b & 0 \end{pmatrix} \quad a, b \neq 0$$

then

$$C = 0, \sum_{i,\alpha=1}^{2} (P^i_\alpha)^2 (Q^i_\alpha)^2 = 0, \det(P) = a^2, \det(Q) = -b^2$$

so that

$$\frac{\partial^2 W_\nu(P)}{\partial P^i_\alpha \partial P^j_\beta} Q^i_\alpha Q^j_\beta = -(ab)^2 < 0. \qquad q.e.d..$$

Proof of ii). Set

$$J(P,Q) = \left(\frac{\partial W(P+Q)}{\partial P^i_\alpha} - \frac{\partial W(P)}{\partial P^i_\alpha}\right) Q^i_\alpha$$

Then

$$J(P,Q) = 2(\det Q + C)(2\det Q + C) + 4(\det P)(\det Q) + \nu \sum_{i,\alpha=1}^{2} (Q^i_\alpha)^4 +$$

$$+ 12\nu \sum_{i,\alpha=1}^{2} (P^i_\alpha Q^i_\alpha + (1/2)(Q^i_\alpha)^2)^2$$

Write

$$I(P,Q) = (\det Q + C)(2\det Q + C) + (\nu/2)\sum_{i,\alpha=1}^{2}(Q_\alpha^i)^4$$

and we prove that $I(P,Q) \geq 0$ whenever $\nu \geq 1/2$ for every $P, Q \in M^{N\times n}$. Since $I(P,Q)$ can be written as

$$I(P,Q) = C^2 + 3C(\det Q) + 2(\det Q)^2 + (\nu/2)\sum_{i,\alpha=1}^{2}(Q_\alpha^i)^4$$

$I(P,Q)$ will be non-negative as a quadratic function of C if

$$4((2\det Q)^2 + (\nu/2)\sum_{i,\alpha=1}^{2}(Q_\alpha^i)^4) - 9(\det Q)^2 \geq 0$$

i.e.,

$$2\nu\sum_{i,\alpha=1}^{2}(Q_\alpha^i)^4 \geq (\det Q)^2$$

This is valid if $\nu \geq 1/2$.

Now let $z \in C_0^1(G;R^2)$, G being an open subset of R^2 and $Q = Dz(x)$, then

$$\int_G \det(Dz(x))dx = 0, \int_G C dx = 0, \int_G z_{,\alpha}^i(x)dx = 0$$

Hence

$$\begin{aligned}\int_G \frac{\partial W(P+Dz(x))}{\partial P_\alpha^i} z_{,\alpha}^i(x)dx &= \int_G J(P,Dz(x))dx \\ &\geq 2\int_G I(P,Dz(x))dx \geq 0\end{aligned}$$

ii) is proved.

<u>Remark 2.14</u> When $\nu > 1/2$, we have moreover

$$\int_G W_{\nu,_{P_\alpha^i}}(P+Dz(x))z_{,\alpha}^i(x)dx \geq (\nu - 1/2)\int_G \sum_{i,\alpha=1}^{2}(z_{,\alpha}^i(x))^4dx$$

and for all $\nu \geq 0$, we have

$$\frac{\partial W_\nu(P)}{\partial P_\alpha^i}P^i \geq 4\nu\sum_{i,\alpha=1}^{2}(P_\alpha^i)^4$$

<u>Example 2.15</u> Let $P \to (A_\alpha^i(P))$ be a C^1 map from $M^{2\times 2}$ into $M^{2\times 2}$, such that

$$\lambda|Q|^2 \leq \frac{\partial A_\alpha^i(P)}{\partial P_\beta^j}Q_\alpha^i Q_\beta^j \leq \Lambda|Q|^2$$

for all P, $Q \in M^{2\times 2}$, where $0 < \lambda < \Lambda$ and $W_\nu(P)$ be the function defined in Example 2.13. Then the map

$$P \to (\frac{\partial W_\nu(P)}{\partial P_\alpha^i} + A_\alpha^i(P))$$

is quasimonotone when $\nu \geq 1/2$ and not necessarily a potential map.

Remark 2.16 Based on Example 2.13, we can construct a more general example. Let $W : M^{3\times 3} \to R$ be defined as

$$W_\nu(P) = \sum_{i,\alpha=1}^{3} (C^i_\alpha((adjP)^i_\alpha)^2) + (\nu/4)(\sum_{\substack{j\neq i\\ \beta\neq\alpha}} (P^j_\beta)^4)$$

where $C^i_\alpha > 0$, $adjP$ is the cofactor of matrix P. From Example 2.13, we know that W_ν are not convex for all $\nu > 0$. W_ν, however, is semiconvex when $\nu \geq 2\max_{i,\alpha}(C^i_\alpha)$.

3. The Existence Theorem

Let $\Omega \subset R^n$ be a bounded Lipschitz domain with $n \geq 2$. In this section we study the following system of equations:

$$\frac{\partial A^i_\alpha}{\partial x^\alpha}(x, u(x), Du(x)) = B^i(x, u(x), Du(x)), x \in \Omega, i = 1, \cdots, N; \tag{3.1}$$

with boundary values

$$u^i(x) = 0 \qquad x \in \partial\Omega, i = 1, \cdots, N. \tag{3.2}$$

where $u : \Omega \to R^N$ is a vector valued function.

We say that u is a weak solution of Problem (3.1) and (3.2) if

$$u \in W^{1,p}_0(\Omega; R^N) \tag{3.3}$$

(for simplicity we assume $p \geq 2$) and

$$\int_\Omega (A^i_\alpha(x, u(x), Du(x))z^i_{,\alpha}(x) + B^i(x, u(x), Du(x))z^i(x))dx = 0 \tag{3.4}$$

for all $z \in W^{1,p}_0(\Omega; R^N)$.

We study Problem (3.3) and (3.4) under the following assumptions:

(H1) $A^i_\alpha : \Omega \times R^N \times M^{N\times n} \to R, B^i : \Omega \times R^N \times M^{N\times n} \to R$ are Caratheodory functions, $i = 1, \cdots, N; \alpha = 1, \cdots, n$.

(H2) $|A(x, s, P)| \leq C_1|P|^{p-1} + C_2|s|^{p-1} + g(x)$, where $g \in L^{p*}(\Omega)$, $C_1, C_2 \geq 0$, C_2 small.

(H3) $|B(x, s, P)| \leq C'_1|P|^{p-1} + C'_2|s|^{p-1} + \bar{g}(x)$ where $\bar{g} \in L^{p*}(\Omega)$, $1/p + 1/p^* = 1, C'_1, C'_2 \geq 0$ and small.

(H4) (coerciveness)

$$A^i_\alpha(x, s, P)P^i_\alpha \geq \lambda_0|P|^P - c|s|^p + h(x)$$

where $\lambda_0 > 0; c \geq 0$ small and $h \in L^1(\Omega)$.

(H5) (ellipticity condition).

For almost every $x_0 \in \Omega$, every $s_o \in R^N$, the map $P \to A(x_0, s_0, P)$ is quasimonotone with

$$\int_G A^i_\alpha(x_0, s_0, P_0 + Dz(x))z^i_{,\alpha}(x)dx \geq \nu \int |Dz(x)|^p dx \tag{3.5}$$

for every $P_0 \in M^{N\times n}$, every open set $G \subset R^n$ and $z \in C^1_0(G; R^N)$, where $\nu > 0$.

Now we are in the position to state our main theorem.

<u>Theorem 3.1</u> Under the assumptions (H1)-(H5), System (3.1)-(3.2) possesses at least one weak solution satisfying (3.3)-(3.4).

<u>Remark 3.2</u> We can treat general Dirichlet problem of (3.1) with $u = u_0$ on $\partial\Omega$ in the sense $u - u_0 \in W_0^{1,p}(\Omega; R^N), u_0 \in W_0^{1,p}(\Omega; R^N)$ using the same method.

<u>Remark 3.3</u> Assumption (H5) can not imply monotonicity of (A_α^i), so that the classical monotone operator methods can not be used directly in the proof of the theorem. We prove the theorem by means of a method suggested by Acerbi & Fusco's paper (see [1]).

<u>Proof of Theorem 3.1</u>

Set $V = W_0^{1,p}(\Omega; R^N)$ and for $u \in V$, define a map $T : V \to V^*$ as

$$(Tu, w) = \int_\Omega (A_\alpha^i(x, u(x), Du(x))w^i_{,\alpha}(x) + B^i(x, u(x), Du(x))w^i(x))dx \tag{3.6}$$

for every $w \in V$. By (H1)-(H3), we know that T is strong - weak continuous and is continuous when it is restricted to finite dimentional subspace of V. We are to prove that there exists $u \in V$, such that $(Tu, w) = 0$ for all $w \in V$.

We prove this in steps.

<u>Step 1</u> We show that T is coercive, i.e.,

$$\lim_{\|u\|_V \to \infty} (Tu, u)/\|u\|_V = +\infty \tag{3.7}$$

By (H1)-(H4),

$$\begin{aligned}(Tu, u) &\geq \int_\Omega (\lambda_0|Du|^p - c|u|^p + h(x) - C_1'|Du|^{p-1}|u| - C_2'|u|^p - \bar{g}(x)|u|)dx \\ &\geq \int_\Omega ((\lambda_0 - C_1')|Du|^p - (c + C_1' + C_2' + \mu)|u|^p + h - c(\mu)\bar{g}^{p^*})dx \\ &\geq \int_\Omega (\lambda_0 - C_1' - C^*(c + C_1' + C_2' + \mu)|Du|^p + h(x) - C(\mu)\bar{g}^{p^*}(x))dx\end{aligned}$$

where C^* is the imbedding coefficient in Rillich Theorem.

Therefore we have (3.7) whenever $c_0 = (C_1' - C^*(c + C_1' + C_2') > 0$

<u>Step 2</u> A Galerkin type approximation. Let $(w_k)_{k=1}$ be a basis of V such that finite generated subspaces of (w_k) are dense in V. Now let B_s be the subspace of V spanned by $w_1, \cdots, w_s$. Then by the coerciveness of T, lemma 2.12 and the standard method used in Monty [12], Lions [10], Morrey [14], Necas [15], there exists $u_s \in B_s$ such that for all $w \in B_s$,

$$(Tu_s, w) = 0$$

and $\|u_s\|_V \leq C$, (C is independent of s). Since V is reflexive, we can extract a subsequence (u_k), such that $u_k \to u_0$ weakly in V and $Tu_k \to \varsigma$ weakly in V^* and $(\varsigma, w) = 0$, w belongs to a dense set of V, then by the weak continuity of $(\varsigma, .)$ for the fixed $\varsigma, (\varsigma, w) = 0$ for all $w \in V$. Thus

$$(Tu_k, u_k - u_0 = (Tu_k, u_k) - (Tu_k, u_0) = -(Tu_k, u_0) \to 0, \text{as } k \to \infty$$

Write

$$z_k = u_k - u_0$$

then $z_k \to 0$ weakly in V as $k \to \infty$. Now recall (3.6), we have

$$\begin{aligned}(Tu, u_k - u_0) = \int_\Omega (&A_\alpha^i(x, u_0 + z_k, Du_0 + Dz_k)z^i_{k,\alpha} \\ &+ B^i(x, u_0 + z_k, Du_0 + Dz_k)z_k^i)dx \to 0, \text{as } k \to \infty\end{aligned}$$

By Sobolev's Lemma, $z_k \to 0$ in $L^p(\Omega; R^N)$ strongly and by (H3),

$$\int_\Omega B^i(x, u_0 + z_k, Du_0 + Dz_k) z_k^i dx \to 0, \text{as } k \to \infty$$

This implies

$$\int_\Omega A_\alpha^i(x, u_0 + z_k, Du_0 + Dz_k) z_{k,\alpha}^i dx \to 0, \text{as } k \to \infty \tag{3.8}$$

If we can show that there exists a subsequence of (z_k) which converges to zero strongly in V. Then by the demicontinuity property of T (see [15]), we have $Tu_k \to Tu_0 = \varsigma$ weakly in V^* as $k \to \infty$, and u_0 is what we need. Hence we will find out such a subsequence.

Step 3 Approximate (z_k) by a sequence (g_k) (possibly different from a subsequence) in $W^{1,\infty}(\Omega; R^N)$.

For every measurable set $S \subset \Omega$, define

$$F(v; S) = \int_S A_\alpha^i(x, u_0 + v, Du_0 + Dv) v_{,\alpha}^i dx$$

for $v \in W_0^{1,p}(\Omega; R^N)$.

Since $C_0^\infty(\Omega; R^N)$ is dense in $W_0^{1,p}(\Omega; R^N)$, (H1) and (H2) hold, $F(v; \Omega)$ is continuous in strong topology of $W_0^{1,p}(\Omega; R^N)$, there exists $(f_k) \subset C_0^\infty(\Omega; R^N)$ such that

$$\|f_k - z_k\|_v < 1/k, |F(f_k; \Omega) - F(z_k; \Omega)| < 1/k$$

hence we may assume the sequence (z_k) to be in $C_0^\infty(\Omega; R^N)$ and bounded in $W_0^{1,p}(\Omega; R^N)$.

Now we extend z_k to be defined on whole R^n by defining $z_k(x) = 0$ when $x \in R^n \setminus \Omega$. Thus $(z_k) \subset W_0^{1,p}(R^n; R^N)$ bounded, $\text{supp}(z_k) \subset \Omega$.

Let $\eta : R^+ \mapsto R^+$ be a continuous increasing function with $\eta(0) = 0$, such that for every measurable set $B \subset \Omega$,

$$\sup_k \int_B (g(x)^{p^*} + |h(x)| + 1 + C(|u_0|^p + |z_k|^p + |Du_0|^p)) dx \leq \eta(\text{meas}(B))$$

where $C = C_1 + C_2$ are given in (H2) (note that $z_k \to 0$ in L^p strongly).

Let (ε_j) be a positive decreasing sequence, $\varepsilon_j \to 0$ as $j \to \infty$. Fix ε_1 and apply Lemma 2.6 to each of the N sequences $((M^* z_k^i)^p)$, $1 \leq i \leq N$. This give s a subsequence (z_{k_1}), a set $A_{\varepsilon_1} \subset \Omega$ with $\text{meas}(A_{\varepsilon_1}) < \varepsilon_1$, and a real number $\delta_1 > 0$, such that

$$\int_B (M^* z_{k_1}^i)^p dx < \varepsilon_1$$

for all k_1, for $1 \leq i \leq N$ and for every $B \subset \Omega \setminus A_{\varepsilon_1}$, with $\text{meas}(B) < \delta_1$. By Lemma 2.8, we may take $\lambda > 0$ so large that for all i, k_1

$$\text{meas}(\{x \in R^n : (M^* z_{k_1}^i)(x) \geq \lambda\}) \leq \min(\varepsilon_1, \delta_1) \tag{3.9}$$

For all i, k_1, set

$$H_{i,k_1}^\lambda = \{x \in R^n : (M^* z_{k_1}^i)(x) < \lambda\}, \quad H_{k_1}^\lambda = \bigcap_{i=1}^N H_{i,k_1}$$

Lemma 2.10 ensures that for all $x, y \in H^{\lambda}_{k_1}$ and $1 \le i \le N$,

$$\frac{|z^i_{k_1}(y) - z^i_{k_1}(x)|}{|y-x|} \le C(n)\lambda$$

Let $g^i_{k_1}$ be a Lipschitz function extending $z^i_{k_1}$ outside $H^{\lambda}_{k_1}$ with Lipschitz constant not greater than $C(n)\lambda$ (Lemma 2.11). Since $H^{\lambda}_{k_1}$ is an open set, we have

$$g^i_{k_1}(x) = z^i_{k_1}(x), Dg^i_{k_1}(x) = Dz^i_{k_1}(x)$$

for all $x \in H^{\lambda}_{k_1}$, and

$$\|Dg^i_{k_1}\|_{L^\infty(R^n)} \le C(n)\lambda$$

We may also assume

$$\|g^i_{k_1}\|_{L^\infty(R^n)} \le \|z^i_{k_1}\|_{L^\infty(H^{\lambda}_{k_1})} \le \lambda, \|g_{k_1}\|_{W^{1,p}(\Omega;R^N)} \le C$$

(Lemma 2.8). We may suppose that at least for a subsequence (still denote it by $(g^i_{k_1})$)

$$g^i_{k_1} \to v^i(\text{weak}^*) \text{ in } W^{1,\infty}(\Omega), \text{as } k_1 \to \infty \tag{3.10}$$

for $1 \le i \le N$. Put

$$(g^1_{k_1}, \cdots, g^N_{k_1}) = g_{k_1}, (v^1, \cdots, v^N) = v$$

we have

$$\begin{aligned} F(z_{k_1};\Omega) &= F(g_{k_1};(\Omega \setminus A_{\varepsilon_1}) \bigcap H^{\lambda}_{k_1}) + F(z_{k_1}; A_{\varepsilon_1} \bigcup (\Omega \setminus H^{\lambda}_{k_1})) \\ &= F(g_{k_1};\Omega \setminus A_{\varepsilon_1}) - F(g_{k_1};(\Omega \setminus A_{\varepsilon_1}) \setminus H^{\lambda}_{k_1}) + F(z_{k_1}; A_{\varepsilon_1} \bigcup (\Omega \setminus H^{\lambda}_{k_1})) \end{aligned} \tag{3.11}$$

Since

$$\text{meas}((\Omega \setminus A_{\varepsilon_1}) \setminus H^{\lambda}_{k_1}) \le \sum_{i=1}^{N} \text{meas}((\Omega \setminus A_{\varepsilon_1}) \setminus H^{\lambda}_{i,k_1}) \le \text{N}\min(\varepsilon_1, \delta_1) \tag{3.12}$$

by (H2), (H4) and by our choice of $A\varepsilon_1$, we obtain

$$\begin{aligned} |F(g_{k_1},(\Omega \setminus A_{\varepsilon_1}) \setminus H^{\lambda}_{k_1})| &\le \int_{(\Omega\setminus A_{\varepsilon_1})\setminus H^{\lambda}_{k_1}} |A^i_\alpha(x, u_0 + g_{k_1}, Du_0 + Dg_{k_1})g^i_{k_1,\alpha}|dx \\ &\le 2^{p-1}C(N,n)(\eta(N\varepsilon_1)) + (C_1 + C_2)\int_{(\Omega\setminus A_{\varepsilon_1})\setminus H^{\lambda}_{k_1}} (|g_{k_1}|^p + |Dg_{k_1}|^p)dx \\ &\le 2^{p-1}C(N,n)(\eta(N\varepsilon_1)) + C(n,\Omega,C_1+C_2)\lambda^p \text{meas}((\Omega \setminus A_{\varepsilon_1}) \setminus H^{\lambda}_{k_1}) \\ &\le 2^{p-1}C(N,n)(\eta(N\varepsilon_1)) + C(n,\Omega,C_1+C_2)\sum_{i=1}^{N}\int_{\Omega\setminus A_{\varepsilon_1})\setminus H^{\lambda}_{i,k_1}} ((M^* z^i_{k_1})(x))^p dx \\ &\le 2^{p-1}C(N,n)(\eta(N\varepsilon_1)) + NC(n,\Omega,C_1+C_2)\varepsilon_1 = V_1(\varepsilon_1) \end{aligned} \tag{3.12'}$$

$$\begin{aligned} &F(z_{k_1};(A_{\varepsilon_1} \bigcup (\Omega \setminus H^{\lambda}_{k_1})) \\ &= \int_{A_{\varepsilon_1} \bigcup (\Omega\setminus H^{\lambda}_{k_1})} A^i_\alpha(x, u_0 + z_{k_1}, Du_0 + Dz_{k_1}) z^i_{k_1,\alpha} dx \\ &= \int_{A_{\varepsilon_1} \bigcup (\Omega\setminus H^{\lambda}_{k_1})} A^i_\alpha(x, u_0 + z_{k_1}, Du_0 + Dz_{k_1})(u^i_{0,\alpha} + z^i_{k_1,\alpha}) dx - \\ &- \int_{A_{\varepsilon_1} \bigcup (\Omega\setminus H^{\lambda}_{k_1})} A^i_\alpha(x, u_0 + z_{k_1}, Du_0 + Dz_{k_1}) u^i_{0,\alpha} dx \end{aligned}$$

$$(3.12") \qquad \begin{aligned} &\geq (\lambda_0/2^{p-1} - \mu)|||Dz_{k_1}|||^p_{L^p(A_{\varepsilon_1} \bigcup (\Omega \setminus H^\lambda_{k_1}))} - \\ &- C(\mu, p, C_1, C_2, c)\eta(\text{meas}(A_{\varepsilon_1} \bigcup (\Omega \setminus H^\lambda_{k_1}))) \end{aligned}$$

for every $\mu > 0$. Choose $0 < \mu < 1/2^{(p-1)2}$, we have

$$\begin{aligned} &F(z_{k_1}; A_{\varepsilon_1} \bigcup (\Omega \setminus H^\lambda_{k_1})) \\ \geq &\lambda_0/2^{(p-1)\cdot 2}|||Dz|||^p_{L^p(A_{\varepsilon_1} \bigcup (\Omega \setminus H^\lambda_{k_1})} - V_2(\varepsilon_1) \end{aligned}$$

where $V_1(\varepsilon)$, $V_2(\varepsilon) \to 0$, as $\varepsilon \to 0_+, \lambda_0, C_1, C_2, c$ are the contants given in (H2) and (H4).

Denote $A_{\varepsilon_1} \bigcup (\Omega \setminus H^\lambda_{k_1}) = U^1_{\varepsilon_1, k_1}$, $\alpha_0 = \lambda_0/2^{(p-1)\cdot 2}$, $V_3(\varepsilon) = V_1(\varepsilon) + V_2(\varepsilon)$, we have

$$(3.13) \qquad F(z_{k_1}; \Omega) \geq F(g_{k_1}; \Omega \setminus A_{\varepsilon_1}) + \alpha_0 |||Dz_{k_1}|||^p_{L^p(U^1_{\varepsilon_1, k_1})} - V_3(\varepsilon_1)$$

<u>Step 4</u> Further approximations.

Now set

$$h_{k_1}(x) = g_{k_1}(x) - v(x)$$

where v is defined by (3.10), then

$$h_{k_1}(x) \to 0(\text{weak*}) \text{ in} W^{1,\infty}(\Omega; R^N) \text{as } k_1 \to \infty$$

and

$$\|h_{k_1}\|_{L^\infty(\Omega; R^N)} \leq 2C(n)\lambda, |||Dh_{k_1}|||_{L^\infty(\Omega)} \leq 2C(n)\lambda$$

where $C(n)$ is given in Lemma 2.10.

Define

$$G = \{x \in \Omega : v(x) \neq 0\}$$

then by Acerbi & Fusco [1, page 139-140], we have

$$\text{meas}(G) \leq (N+1)\varepsilon_1$$

and

$$\begin{aligned} F(g_{k_1}; \Omega \setminus A_{\varepsilon_1}) &= F(h_{k_1}; (\Omega \setminus A_{\varepsilon_1}) \setminus G) + F(g_{k_1}; (\Omega \setminus A_{\varepsilon_1}) \bigcap H^\lambda_{k_1} \bigcap G) \\ &= F(h_{k_1}; (\Omega \setminus A_{\varepsilon_1}) \setminus G) + F(z_{k_1}; (\Omega \setminus A_{\varepsilon_1}) \bigcap H^\lambda_{k_1} \bigcap G) + F(g_{k_1}; (\Omega \setminus A_{\varepsilon_1}) \bigcap G \setminus H^\lambda_{k_1}) \end{aligned}$$

Denote

$$U^2_{\varepsilon_1} = (\Omega \setminus A_{\varepsilon_1}) \setminus G; U^3_{\varepsilon_1, k_1} = (\Omega \setminus A_{\varepsilon_1}) \bigcap H^\lambda_{k_1} \bigcap G; U^4_{\varepsilon_1, k_1} = (\Omega \setminus A_{\varepsilon_1}) \bigcap G \setminus H^\lambda_{k_1}$$

then a similar argument as in Step 3((3.12)") yields

$$F(z_{k_1}; U^3_{\varepsilon_1, k_1}) \geq \alpha_0 |||Dz_{k_1}|||^p_{L^p(U^3_{\varepsilon_1, k_1})} - V_4(\varepsilon_1)$$

and on $U^4_{\varepsilon_1, k_1}$ we have

$$(3.14) \qquad \int_{U^4_{\varepsilon_1, k_1}} (|g_{k_1}|^p + |Dg_{k_1}|^p) dx \leq NC(n, \Omega)\varepsilon_1$$

(by the above property of U^4_{k} and an estimate similar to (3.12)' on g_{k_1}) then

$$|F(g_{k_1}, U^4_{\varepsilon_1,k_1})| \leq C(C_1, C_2, p)(NC(n,\Omega)\varepsilon_1 + \eta((N+1)\varepsilon_1) = V_5(\varepsilon_1)$$

Thus we have

$$F(g_{k_1}, \Omega \setminus A_{\varepsilon_1}) \geq F(h_{k_1}; U^2_{\varepsilon_1}) + \alpha_0 |||Dz|||^p_{L^p(U^3_{\varepsilon_1,k_1})} - V_4(\varepsilon_1) - V_5(\varepsilon_1)$$

Denote

$$U^5_{\varepsilon_1,k_1} = U^3_{k_1} \bigcup (A_{\varepsilon_1} \bigcup (\Omega \setminus H^\lambda_{k_1}))$$

then (3.13) yields

$$F(z_{k_1}; \Omega) \geq F(h_{k_1}; U^2_{\varepsilon_1}) + \alpha_0 |||Dz_{k_1}|||^p_{L^p(U^5_{\varepsilon_1,k_1})} - V_6(\varepsilon_1) \tag{3.15}$$

where $V_6(\varepsilon) = V_3(\varepsilon) + V_4(\varepsilon) + V_5(\varepsilon)$.

Choose an open set $\Omega' \subset \Omega$ containing $U^2_{\varepsilon_1}$, such that

$$|F(h_{k_1}; \Omega') - F(h_{k_1}; U^2_{\varepsilon_1})| < \varepsilon_1$$

(this is possible since the functions h_{k_1} are uniformly bounded in $W^{1,\infty}(\Omega; R^N)$).

By (3.15) we now have

$$F(z_{k_1}; \Omega) \geq F(h_{k_1}; \Omega') + \alpha_0 |||Dz_{k_1}|||^p_{L^p(U^5_{\varepsilon_1,k_1})} - V_7(\varepsilon_1) \tag{3.16}$$

where $V_7(\varepsilon) = V_6(\varepsilon) + \varepsilon$.

Let Ω' be approximated by a sequence of hypercubes with edges parallel to the coordinate axes, i.e.,

$$\begin{cases} H_j = \bigcup_{s=1}^{I_j} D_{j,s} \\ meas(\Omega' \setminus H_j) \to 0 \text{ as } j \to \infty \\ meas(D_{j,s}) = 1/2^{jn} \quad 1 \leq s \leq I_j. \end{cases}$$

Let $j > 0$ be so large that

$$|F(h_{k_1}; \Omega') - F(h_{k_1}; H_j)| < \varepsilon_1, \|h_{k_1}\|^p_{W^{1,p}(\Omega' \setminus H_j; R^N)} < \varepsilon_1 \tag{3.17}$$

for all $k_1 > 0$ and

$$\text{meas}(\Omega' \setminus H_j) < \min(\varepsilon_1, \delta_1)$$

Then, from (3.17)

$$F(z_{k_1}; \Omega) \geq F(h_{k_1}; H_j) + \alpha_0 |||Dz_{k_1}|||^p_{L^p(U^5_{\varepsilon_1,k_1})} - V_8(\varepsilon_1) \tag{3.18}$$

where

$$V_8(\varepsilon) = V_7(\varepsilon) + \varepsilon$$

Put $M = 2C(n)\lambda \geq |||Dh_{k_1}|||_{L^\infty(\Omega)}$, and $\alpha > 1$ so large that if

$$E = \{x \in \Omega', a(x) \leq \alpha\}$$

then

$$\text{meas}(\Omega' \setminus E) \leq \varepsilon_1/M, \quad \int_{\Omega' \setminus E} a(x)dx \leq \varepsilon_1$$

where

$$a(x) = 2^{p-1}((1 + C_1 + C_2)|Du_0(x)|^p + C_2|u_0(x)|^p + g(x)^{p*})$$

By Lemma 2.4, there exists a compact set $K \subset 0$ such that the function $f(x, s, P)$ defined by

$$(x, s, P) \mapsto A^i_\alpha(x, u_0(x) + s, Du_0(x) + P)P^i_\alpha := f(x, s, P)$$

for $x \in \Omega$, $s \in R^N$, $P \in M^{N\times n}$, is continuous on $K \times R^N \times M^{N\times n}$ and

$$\text{meas}(H_j \setminus K) \le \varepsilon_1/(\alpha + M)$$

Dividing each $D_{j,s}$ into 2^{nm} hypercubes $Q^m_{h,s,j}$, $1 \le h \le 2^{nm}$ with edge length 2^{-jm}. To each j, s, m, h, fix $x^m_{h,s,j} \in Q^m_{h,s,j} \bigcap K \bigcap E$ if this set is non-empty, or $x^m_{h,s,j} \in Q^m_{h,s,j}$ such that

$$a(x^m_{h,s,j})\text{meas}(Q^m_{h,s,j}) \le \int_{Q^m_{h,s,j}} a(x)dx$$

if $Q^m_{h,s,j} \bigcap K \bigcap E = \phi$. Then

$$\begin{aligned} F(h_{k_1}; H_j) &\ge F(h_{k_1}; H_j \bigcap K \bigcap E) - \int_{H_j\setminus(K \bigcup E)} a(x)dx - \\ &- 2^{p-1}(1 + C_1 + C_2)\int_{H_j\setminus(K \bigcup E)} (|Dh_{k_1}|^p + |h_{k_1}|^p)dx \\ &= F(h_{k_1}; H_j \bigcap K \bigcap E) - V_9(\varepsilon_1) \\ &= a^j_{k_1} + b^{m,j}_{k_1} + c^{m,j}_{k_1} + d^{m,j}_{k_1} - V_9(\varepsilon_1) \end{aligned} \tag{3.19}$$

where

$$\begin{aligned} a^j_{k_1} &= \int_{H_j \bigcap K \bigcap E} (f(x, h_{k_1}(x), Dh_{k_1}(x)) - f(x, 0, Dh_{k_1}(x)))dx \\ b^{m,j}_{k_1} &= \sum_{h,s} \int_{Q^m_{h,s,j} \bigcap K \bigcap E} (f(x, 0, Dh_{k_1}(x)) - f(x^m_{h,s,j}, 0, Dh_{k_1}(x)))dx \\ c^{m,j}_{k_1} &= \sum_{h,s} \int_{Q^m_{h,s,j}} f(x^m_{h,s,j}, 0, Dh_{k_1}(x))dx \\ d^{m,j}_{k_1} &= -\sum_{h,s} \int_{Q^m_{h,s,j}\setminus(K \bigcup E)} f(x^m_{h,s,j}, 0, Dh_{k_1}(x))dx \end{aligned}$$

By the uniform continuity of f on the bounded set of $K \times R^N \times M^{N\times n}$ and (3.8), we have

$$\lim_{k_1\to\infty} a^j_{k_1} = 0, \quad \lim_{k_1\to\infty} F(z_{k_1}; \Omega) = 0$$

and the pointwise convergence of $u_0(x^m_{h,s,j})$, $Du_0(x^m_{h,s,j})$ imply $\lim_{m\to\infty} b^{m,j}_{k_1} = 0$, uniformly with respect to k_1, for fixed j, and

$$\begin{aligned} |d^{m,j}_{k_1}| &\le \sum_{h,s}(\int_{Q^m_{h,s,j}\setminus(K \bigcup E)} (a(x^m_{h,s,j}) + 2^p(1 + C_1 + C_2)M)dx \\ &\le C(\alpha + M)\text{meas}(H_j \setminus E) + C\int_{H_j\setminus E} \alpha dx \\ &\le C(C_1, C_2, p)\varepsilon_1 \end{aligned}$$

Therefore we may suppose that m is large enough to ensure that $|b_{k_1}^{m,j}| < \varepsilon_1$ for all $k_1 > 0$; $F(z_{k_1};\Omega) < \varepsilon_1$, and $|a_{k_1}^j| < \varepsilon_1$ as $k_1 > \bar{k}_1 > 0$ for some $\bar{k}_1$. Hence now we have by (3.8), (3.18) and (3.19)

$$\begin{aligned}\varepsilon_1 \geq F(z_{k_1};\Omega) \geq c_{k_1}^{m,j} + \alpha_0 \|\|Dz_{k_1}\|\|_{L^p(U^\delta_{s,k_1})}^p - V_8(\varepsilon_1) - \\ - V_9(\varepsilon_1) - 2\varepsilon_1 - C(C,C,p)\varepsilon_1 \\ = c_{k_1}^{m,j} + \alpha_0 \|\|Dz_{k_1}\|\|_{L^p(U^\delta_{k_1})}^p - V_{10}(\varepsilon_1)\end{aligned} \tag{3.20}$$

Since $h_{k_1} \to 0$ (weak*) in $W^{1,\infty}(\Omega;R^N)$ as $k_1 \to \infty$, we have

$$R_{h,s,j}^{k_1,m} = \|\|h_{k_1}\|\|_{L^\infty(Q^m_{h,s,j})} \to 0 \text{ as } k_1 \to \infty, \text{for fixed } m.$$

Define a hypercube $E_{h,s,j}^{k_1,m}$ contained in $Q_{h,s,j}^m$ with edge lengh $1/2^{jm} - 2R_{h,s,j}^{k_1,m}$, such that

$$\text{dist}(\partial Q_{h,s,j}^m; E_{h,s,j}^{k_1,m}) = R_{h,s,j}^{k_1,m}$$

Next define

$$f_{k_1}(x) = \begin{cases} 0 & x \in \partial Q_{h,s,j}^m \\ h_{k_1}(x) & x \in E_{h,s,j}^{k_1,m}. \end{cases}$$

Note that f_{k_1} is a Lipschitz mapping on the set where it is defined and its Lipschitz constant is not greater than $2C(n)\lambda$. By lemma 2.11, f_{k_1} can be extended to whole $Q_{h,s,j}^m$ as a Lipschitz mapping with the same Lipschitz constant. We still denote it by f_{k_1} and consider of it as defined on whole H_j. Then it is easy to verify (see [6]) that

$$Df_{k_1}(x) - Dh_{k_1}(x) \to 0, a.e. \text{in} H_j$$

therefore, there is some $\bar{\bar{k}}_1 > \bar{k}_1$, such that

$$\begin{aligned}&\|\|Df_{k_1} - Dh_{k_1}\|\|_{L^p_{(H_j)}}^p \leq \varepsilon_1/2, \\ &|\sum_{h,s} \int_{Q^m_{h,i,j}} f(x_{h,s,j}^m, 0, Dh_{k_1}) - f(x_{h,s,j}^m, 0, Df_{k_1}))dx| \leq \varepsilon_1/2\end{aligned} \tag{3.21}$$

whenever $k_1 > \bar{\bar{k}}_1$. Now by (H5),

$$\begin{aligned} c_{k_1}^{m,j} &= \sum_{h,s} \int_{Q^m_{h,s,j}} f(x_{h,s,j}^m, 0, Dh_{k_1})dx \\ &\geq \sum_{h,s} \int_{Q^m_{h,s,j}} f(x_{h,s,j}^m, 0, Df_{k_1})dx - \varepsilon_1/2 \\ &\geq \sum_{h,s} \nu \int_{Q^m_{h,s,j}} |Df_{k_1}|^p dx - \varepsilon_1/2 \\ &\geq \nu/2^{p-1} \|\|Dh_{k_1}\|\|_{L^p(H_j)}^p - \nu\varepsilon_1/2 - \varepsilon_1/2 \end{aligned} \tag{3.22}$$

(recall the definition of f and (H5)). Thus in (3.20),

$$\varepsilon_1 \geq \alpha_0 \|\|Dz_{k_1}\|\|_{L^p(U^\delta_{\varepsilon_1,k_1})}^p + \nu/2^{p-1} \|\|Dh_{k_1}\|\|_{L^p(H_j)}^p - (1+\nu)\varepsilon_1/2 - V_{10}(\varepsilon_1) \tag{3.23}$$

whenever $k_1 > \overline{\overline{k_1}}$.

Put $K(\varepsilon) = V_{10}(\varepsilon) + (3+\nu)\varepsilon/2)/\min(\alpha_0, \nu/2^{p-1})$, we have

$$\|\|Dh_{k_1}\|\|^p_{L^p(H_j)} + \|\|Dz_{k_1}\|\|^p_{L^p(U^5_{\varepsilon_1,k_1})} \le K(\varepsilon_1), \text{ as } k_1 > \overline{\overline{k_1}} \tag{3.24}$$

Recall (3.17) and combine it with (3.24) yields

$$\|\|Dh_{k_1}\|\|^p_{L^p(\Omega')} \le K(\varepsilon_1) + \varepsilon_1, \|\|Dz_{k_1}\|\|^p_{L^p(U^5_{\varepsilon_1,k_1})} \le K(\varepsilon_1)$$

By the definition of Ω', we have

$$\|\|Dg_{k_1}\|\|^p_{L^p(U^2_{\varepsilon_1})} \le K(\varepsilon_1) + \varepsilon_1$$

Since $Dg_{k_1}(x) = Dz_{k_1}(x)$, for all $x \in H^\lambda_{k_1}$, we have

$$\|\|Dz_{k_1}\|\|^p_{L^p(U^2_{\varepsilon_1} \bigcap H^\lambda_{k_1})} \le K(\varepsilon_1) + \varepsilon_1$$

By the definition of $U^2_{\varepsilon_1}$ and $U^5_{\varepsilon_1,k_1}$, we know that $(U^5_{\varepsilon_1} \bigcap H^\lambda_{k_1}) \bigcup U^5_{\varepsilon_1,k_1} = \Omega$; which implies that

$$\|\|Dz_{k_1}\|\|^p_{L^p(\Omega)} \le 2K(\varepsilon_1) + \varepsilon_1 = W(\varepsilon_1) \tag{3.25}$$

where $W(\varepsilon) \to 0$, as $\varepsilon \to 0_+$.

For $\varepsilon_2 > 0$, and sequence (z_{k_1}), repeating the above arguments we can extract another subsequence, say, (z_{k_2}), such that

$$\|\|Dz_{k_2}\|\|^p_{L^p(\Omega)} \le W(\varepsilon_2)$$

whenever $k_2 >$ some $\overline{\overline{k_2}} > \overline{\overline{k_1}}$. Then a Cantor-Hilbert diagonal argument gives a subsequence $(z_{k_i})_{i=1}^\infty$ of (z_k) which converges strongly in $W^{1,p}_0(\Omega; R^N)$ to zero as $i \to \infty$.

q.e.d.

Remark 3.4 If $\nu = 0$ in (H5), we can still show that a subsequence of (z_k), which is integral equicontinuous, can be extracted, i.e., for any $\varepsilon > 0$, there exists $\delta > 0$, such that

$$\int_B |Dz_{k_1}|^p dx < \varepsilon$$

for all k_i, whenever meas$(B) < \delta$, since now in (3.23) we only have

$$\|\|Dz_{k_1}\|\|^q_{L^p(U^5_{\varepsilon_1,k_1})} \le (V_{10}(\varepsilon_1) + 3\varepsilon_1/2)/\alpha_0$$

this implies

$$\|\|Dz_{k_1}\|\|^p_{L^p(A_*)} \le (V_{10}(\varepsilon_1) + 3\varepsilon_1/2)/\alpha_0$$

Remark 3.5 If the right-hand side of (H5) is $\nu \int_G |Dz(x)|^r dx$ with $1 \le r \le p$, a similar argument as we have done gives a subsequence $(z_{k_i}$ such that $a_{k_i} \to 0$ strongly in $W^{1,p}_0(\Omega; R^N)$. This combining with Remark 3.4 gives $z_{k_i} \to 0$ strongly in $W^{1,p}_0(\Omega; R^N)$, (possibly different from a subsequence).

References

1. E. Acerbi; N. Fusco: Semicontinuity problems in the calculus of variations, Arch. Rational Mech. Anal., 86, (1984), 125-145.

2. H. Brezis: Opérateurs maximaum monotones et semi-groupes de contractions dans les espaces de Hilbert. North Holland, (1973).

3. J. M. Ball: Convexity conditions and existence theorems in nonlinear elasticity, Arch. Rational Mech. Anal., 63 (1977), 337-403.

4. J. M. Ball: Constitutive inequalities and existence theorems in nonlinear elastostatics, Heriot-Watt Symposium (Edinburgh, 1976), Vol.I, 187-241, Res. Notes in Math., No.17, Pitman, London, 1977.

5. F. E. Browder: Existence theorems for nonlinear PDE, Proc. Symp. Pure Math., 16, Global Analysis, (ed. by S. S. Chern) AMS (1970), 1-60.

6. B. Dacorogna: Weak continuity and weak lower semicontinuity of non-linearfunctionals, Lecture Notes in Mathematics, Vol.922, Springer-Veriag, (1982).

7. G. Eisen: A selection lemma for sequences of measurable sets, and lower semicontinuity of multiple integrals, Manuscripta Math., 27 (1979), 73-79.

8. I. Ekeland; R. Teman: Convex analysis and variational problems, Nortt Holland, Amsterdam, 1976.

9. M. Giaquinta: Multiple integrals in the calculus of variations, and nonlinear elliptic systems, Princeton Univ. Press, Princeton, 1983.

10. L. Lions: Quelques methodes de resolution des problemes aux limites nonlineaires, Paris, Dunod-Gauthier Villars, 1969.

11. F. C. Liu: A Luzin type property of Sobolev functions, Indiana Univ. Math. J., 26 (1977), (645-651).

12. J.Minty: Monotone operator in Hilbert spaces. Duke Math. J., 29 (1962), 341-346.

13. C. B. Jr. Morrey: Quasiconvexity and the lower semicontinuity of multiple integrals, Pac. J. Math., 2 (1952), 25-53.

14. C. B. Jr. Morrey: Multiple integrals in the calculus of variations. Springer, New York, 1966.

15. J. Necas: Introduction to the theory of lnonlinear elliptic equations, Teubner-Texte zur Math. Band 52, (1983), Leipzig, BSB. B. G., Teubner Verlagsgesellschaft.

16. E. M. Stein: Singular integrals and differentiability properties of functions, Princeton Univ. Press, Princeton, 1970.

17. G. Truesdell: Some challenges offered to analysis by rational thermodynamics; in Contemporay Developments in Contiuum Mechanics and PDE, (1978), 495-603, North Holland Publishing Company.

18. C. Truesdell; W. Noll: The nonlinear field theories of mechanics, in Handbuch der Physik, Vol. III.3. ed. Flugge; Springer, Berlin, (1965).

19. I. M. Višik: Quasilinear strongly ellptic systems of partial differential equations in divergence form. Trudy. Mosk. Mat. Obšč. 12, (1963).

INITIAL VALUE PROBLEMS FOR A NONLINEAR SINGULAR INTEGRAL-DIFFERENTIAL EQUATION OF DEEP WATER

Zhou Yulin & Guo Boling
Institute of Applied Physics &
Computational Mathematics
Beijing, China

§1. Introduction

The equation, which discribes the propagation of internal waves in the stratified fluid of finite depth is first derived by R.I. Joseph[1,2] and can be expressed in the form[3,4]

$$u_t+2uu_x+Gu_{xx} = 0, \tag{1}$$

where $G(\cdot)$ is a singular integral operator defined by

$$Gu(x,t) = \frac{\lambda}{2}P\int_{-\infty}^{\infty}[\coth \frac{\lambda\pi}{2}(y-x)-\mathrm{sgn}(y-x)]u(y,t)dy, \tag{2}$$

$\frac{1}{\lambda}$ is the parameter characterizing the depth of fluid and P denotes the principle value of integral. For the shallow water limit as $\lambda\to\infty$, this equation reduces to the well-known Korteweg-de Vries equation

$$u_t+2uu_x+\frac{1}{3\lambda}u_{xxx} = 0, \tag{3}$$

For the deep water limit, the equation (1) reduces to the following form

$$u_t+2uu_x+Hu_{xx} = 0, \tag{4}$$

where H is the Hilbert transform

$$Hu(x,t) = \frac{1}{\pi}P\int_{-\infty}^{\infty}\frac{u(y,t)}{y-x}dy. \tag{5}$$

The equations (1) and (4) are the nonlinear partial differential equation with the singular integral operator. The equation (4) is the equation of deep water and is usually called the Benjamin-One equation.

The study of these equations is of great interest in the physical and mathematical point of view. For example, there are a great deal of works contributed to the soliton solutions and the behaviors of the solutions of the problems for the Korteweg-de Vries equations and their various generalizations[8-14].

Very recently there have been many investigations of the physical purpose for the nonlinear partial differential equation (4) of deep water. The Bäcklund transformations, the conservation laws, various soliton solutions and their interactions for the Benjamin-Ono equation (4) are studied in [15-19].

If the effect of the amplitude of the internal wave is taken into account in the deep fluid, the equation (4) has an additional linear term as follows[6]:

$$u_t+C_o u_x+2uu_x+Hu_{xx} = 0. \tag{6}$$

The purpose of the present work is to establish the mathematical theorems for the initial value problems of the nonlinear singular integral-differential equation of deep water

$$u_t+2uu_x+Hu_{xx}+b(x,t)u_x+C(x,t)u = f(x,t). \tag{7}$$

The existence and uniqueness theorems of the generalized and classical global solutions for the initial value problems of the Benjamin-Ono equation are proved. The solutions of the mentioned problems are approximated by the solutions of the initial value problems for the equation

$$u_t+2uu_x+Hu_{xx}-\varepsilon u_{xx}+b(x,t)u_x+C(x,t)u = f(x,t) \tag{8}$$

obtained by increasing a diffusion term εu_{xx} with small coefficient to the original equation (7). This equation is a nonlinear parabolic equation with the Hilbert integral transform term. The solution of the initial value problem for the nonlinear singular integral-differential equation (7) is built up by the limiting process of the vanishing of the diffusion coefficient $\varepsilon\to 0$. The estimations of the convergence speed are made in the order of the diffusion coefficient ε at the end of this work.

§2. Equations with Diffusion Term

In this section we are going to consider the solution of the problem for the nonlinear parabolic equation (8) in the domain $Q_T^* = \{x\in R,\ 0\leqq t\leqq T\}$ with the initial value condition

$$u(x,0) = \phi(x), \tag{9}$$

where $\phi(x)$ is a given initial function for $x\in R$ and $0<T<\infty$ a given constant.

We are looking for the solutions of mentioned problems in the space of functions with the derivatives of any order tending to zero as $|x|\to\infty$.

Let us begin with the initial value problems for linear parabolic equations.

Denote by $W_2^{(k,[\frac{k}{2}])}(Q_T^*)$ the functional spaces of functions $f(x,t)$ with derivatives $f_{x^r t^s}(x,t)\in L_2(Q_T^*)$ for $2s+r\leqq k$, where $k=0,1,\ldots$. Also denote by $W_\infty^{(k,[\frac{k}{2}])}(Q_T^*)$ the functional space of functions $f(x,t)$ with derivatives $f_{x^r t^s}(x,t)\in L_\infty(Q_T^*)$ for $2s+r\leqq k$, where $k=0,1,\ldots$. For $k=0$, $W_2^{(o,o)}(Q_T^*)\equiv L_2(Q_T^*)$ and $W_\infty^{(o,o)}(Q_T^*)\equiv L_\infty(Q_T^*)$.

We state the existence theorem of the initial value problem for the linear parabolic equations as follows:

THEOREM 1. Suppose that $b(x,t)$ and $C(x,t)\in W_\infty^{(k,[\frac{k}{2}])}(Q_T^*)$ and $f(x,t)\in W_2^{(k,[\frac{k}{2}])}(Q_T^*)$ and suppose that $\phi(x)\in H^{k+1}(R)$. The initial value problem (9) in the domain Q_T^* for the linear parabolic equation

$$Lu \equiv u_t-\varepsilon u_{xx}+b(x,t)u_x+C(x,t)u = f(x,t), \tag{10}$$

has a unique global solution $u(x,t)\in W_2^{(k+2,[\frac{k}{2}]+1)}(Q_T^*)$.

Let us now investigate the initial value problem (9) for the linear parabolic equation with singular integral operator

$$L_\lambda u \equiv Lu+\lambda Hu_{xx} \equiv u_t-\varepsilon u_{xx}+\lambda Hu_{xx}+b(x,t)u_x+C(x,t)u = f(x,t) \tag{11}$$

where $0\leqq\lambda\leqq 1$ is a parameter.

It is evident that for $\lambda=0$, the initial value problem (9) of linear parabolic equation $L_\circ u=f(x,t)$ has a unique solution $u(x,t)\in W_2^{(2,1)}(Q_T^*)$, under the assumptions that $b(x,t)$, $C(x,t)\in L_\infty(Q_T^*)$; $f(x,t)\in L_2(Q_T^*)$ and $\phi(x)\in H^1(R)$.

Let E be the set of value of $\lambda\in[0,1]$, for which the initial value problem (9) of the linear parabolic equation $L_\lambda u=f(x,t)$ containing singular integral operator λHu_{xx} has a unique global solution $u_\lambda(x,t)\in W_2^{(2,1)}(Q_T^*)$. Then the set E is nonempty in the segment $[0,1]$.

For the solutions $u_\lambda(x,t)$ of the initial value problem (9) of the linear parabolic equation (11), we can make some a priori estimations by usual energy method.

Under given conditions, the solutions $u_\lambda(x,t)$ for the initial value problem (9) of the linear parabolic equation (11) with singular integral operator have the estimations

$$\sup_{0\leqq t\leqq T}\| u(\cdot,t)\|_{H^1(R)}+\| u_{xx}\|_{L_2(Q_T^*)}+\|u_t\|_{L_2(Q_T^*)}\leqq C_1\{\| \phi \|_{H^1(R)}+\|f\|_{L_2(Q_T^*)}\}, \tag{12}$$

in the other words,

$$\|u\|_{W_2^{(2,1)}(Q_T^*)}\leqq C_1\{\|\phi\|_{H^1(R)}+\|f\|_{L_2(Q_T^*)}\}, \tag{13}$$

where C_1 depends on the norms $\|b\|_{L_\infty(Q_T^*)}$, $\|C\|_{L_\infty(Q_T^*)}$ and the diffusion coefficient $\varepsilon>0$ and is independent of $0\leqq\lambda\leqq 1$.

By mean of these estimations, we can prove $E\equiv[0,1]$, i.e., for any $\lambda\in[0,1]$, then for $\lambda=1$, the problem (9) and (11) has a unique generalized global solution.

THEOREM 2. Suppose that $b(x,t)$, $C(x,t)\in L_\infty(Q_T^*)$; $f(x,t)\in L_2(Q_T^*)$ and $\phi(x)\in H^1(R)$. The initial value problem (9) of the linear parabolic equation (11) has a unique generalized global golution $u(x,t)\in W_2^{(2,1)}(Q_T^*)$.

COROLLARY. Under the conditions of Theorem 2, the generalized global solution $u(x,t)$ of the initial value problem (9) and (11) has the estimation

$$\| u \|_{W_2^{(2,1)}(Q_T^*)} \leq K_1\{\| \phi \|_{H^1(R)} + \| f \|_{L_2(Q_T^*)}\},$$

where K_1 is a constant dependent on the norms $\| b \|_{L_\infty(Q_T^*)}$, $\| C \|_{L_\infty(Q_T^*)}$ and the diffusion coefficient $\varepsilon>0$.

COROLLARY. Suppose that $b(x,t), C(x,t) \in W_\infty^{(k,[\frac{k}{2}])}(Q_T^*)$, $f(x,t) \in W_2^{(k,[\frac{k}{2}])}(Q_T^*)$ and $\phi(x) \in H^{k+1}(R)$ for $k \geq 1$ integer. Then the unique global solution $u(x,t)$ of the initial value problem (9) and (11) belongs to the space $W_2^{(k+2,[\frac{k}{2}]+1)}(Q_T^*)$.

Now we turn to prove the existence of the generalized global solution for the initial value problem (9) of the nonlinear parabolic equation (8) with singular integral operator.

THEOREM 3. Suppose that $b(x,t), C(x,t) \in L_\infty(Q_T^*)$, $f(x,t) \in L_2(Q_T^*)$ and $\phi(x) \in H^1(R)$. The initial value problem (9) of the nonlinear parabolic equation (8) with the Hilbert operator, has a generalized global solution $u(x,t) \in W_2^{(2,1)}(Q_T^*)$, which satisfies the equation (8) in generalized sense and the initial condition (9) in classical sense. Furthermore there is the estimation

$$\sup_{0 \leq t \leq T} \|u(\cdot,t)\|_{H^1(R)} + \|u_{xx}\|_{L_2(Q_T^*)} + \| u_t \|_{L_2(Q_T^*)} \leq K_2\{\| \phi \|_{H^1(R)} + \|f\|_{L_2(Q_T^*)}\}, \quad (14)$$

where K_2 is a constant dependent on the norms of the coefficients $b(x,t)$ and $C(x,t)$, the diffusion coefficient $\varepsilon>0$ and $T>0$.

Proof. We want to prove the existence of the generalized global solution for the present problem by the fixedpoint technique.

We define a mapping T_λ: $B \to B$ of the functional space $B=L_\infty(Q_T^*)$ into itself with a parameter $0 \leq \lambda \leq 1$ as follows: For any $v(x,t) \in B$, let $u(x,t)$ be the unique generalized global solution of the linear parabolic equation

$$u_t - \varepsilon u_{xx} + Hu_{xx} + 2vu_x + bu_x + Cu = \lambda f \quad (15)$$

with the initial condition

$$u(x,0) = \lambda\phi(x). \quad (16)$$

Then $v(x,t) \in B$, thus obtained function $u(x,t) \in W_2^{(2,1)}(Q_T^*)$. Since the injecting operator $W_2^{(2,1)}(Q_T^*) \hookrightarrow B$ is compact, then the mapping T_λ: $B \to B$ defined by $u=T_\lambda v$ for $v \in B$ is completely continuous for any $0 \leq \lambda \leq 1$.

As $\lambda=0$, $T_0(B)=0$.

In order to justify the existence of the generalized global solution of the original problem (8) and (9), it is sufficient to prove the uniform boundedness in the base space B of all possible fixed points of the mapping T_λ: B→B with respect to the parameter $0 \leq \lambda \leq 1$, i.e., it needs to give a priori estimations of the solutions $u_\lambda(x,t)$ for the initial problem (16) for the nonlinear parabolic equation

$$u_t - \varepsilon u_{xx} + Hu_{xx} + 2uu_x + bu_x + Cu = \lambda f \tag{17}$$

with respect to the parameter $0 \leq \lambda \leq 1$.

Taking the scalar product of the function $u(x,t)$ and the equation (17) in Hilbert space, we get

$$\int_{-\infty}^{\infty} u(u_t - \varepsilon u_{xx} + Hu_{xx} + 2uu_x + bu_x + Cu - \lambda f)dx = 0.$$

By simple calculations as before, this can be replaced by the inequality

$$\frac{d}{dt}\| u(\cdot,t) \|^2_{L_2(R)} + \| u_x(\cdot,t) \|^2_{L_2(R)} \leq C_2\{\| u(\cdot,t) \|^2_{L_2(R)} + \| f(\cdot,t) \|^2_{L_2(R)}\}.$$

Hence we have the estimation

$$\sup_{0\leq t\leq T} \| u(\cdot,t) \|_{L_2(R)} + \|u_x \|_{L_2(Q_T^*)} \leq C_3\{\| \phi \|_{L_2(R)} + \|f \|_{L_2(Q_T^*)}\},$$

where C_2 and C_3 depend on $\| b \|_{L_\infty(Q_T^*)}$, $\| C \|_{L_\infty(Q_T^*)}$ and $\varepsilon>0$, but are independent of $0 \leq \lambda \leq 1$.

Again multiplying the equation (17) by u_{xx} and then integrating the resulting product with respect to $x \in R$, we obtain

$$\int_{-\infty}^{\infty} u_{xx}(u_t - \varepsilon u_{xx} + Hu_{xx} + 2uu_x + bu_x + Cu - \lambda f)dx = 0.$$

Similarly, we get the estimation

$$\sup_{0\leq t\leq T} \|u_x(\cdot,t) \|_{L_2(R)} + \| u_{xx} \|_{L_2(Q_T^*)} \leq C_4\{\| \phi \|_{H^1(R)} + \| f \|_{L_2(Q_T^*)}\},$$

where C_4 is independent of $0 \leq \lambda \leq 1$.

This shows that all possible solutions of initial value problem (17) and (16) are uniformly bounded in space $L_\infty(0,T;\, H^1(R))$ hence in space B with respect to $0 \leq \lambda \leq 1$.

Therefore the existence of the generalized global solution $u(x,t) \in W_2^{(2,1)}(Q_T^*)$ for the initial value problem (9) for the nonlinear parabolic equation (8) is proved.

Suppose that there are two generalized global solutions $u(x,t)$ and $v(x,t)$ in $L_\infty(0,T;\, H^2(R))$ for the initial value problem (9) for the nonlinear parabolic equation (8). The difference function $W(x,t)=u(x,t)-v(x,t)$ satifies the homogeneous linear equation

$$W_t - \varepsilon W_{xx} + HW_{xx} + (b+u+v)W_x + (C+u_x+v_x)W = 0$$

in generalized sense and the homogeneous initial condition

$W(x,0)=0.$

Forming the estimation formulars of the generalized global solution, gives $W(x,t)=0$, where the coefficients of W_x and W terms are bounded, since $u,v\in L_\infty(0,T;H^2(R))$.

Hence the solution of the problem (8) and (9) is unique.

The theorem is proved.

§3. A Priori Estimations

In order to obtain the global solution to the initial value problem (9) for the nonlinear singular integral-differential equation (7) by the limiting process as the coefficient $\varepsilon>0$ of the additional diffusion term tends to zero, we must derive a series of a priori uniform estimations for the solutions to the initial value problem (9) of the nonlinear parabolic equation (8) containing the Hilbert operator with respect to the coefficient $\varepsilon>0$.

LEMMA 3. Suppose that $b(x,t)$, $b_x(x,t)$, $C(x,t)\in L_\infty(Q_T^*)$ and $f(x,t)\in L_2(Q_T^*)$ and suppose that $\phi(x)\in L_2(R)$. The generalized global solutions $u_\varepsilon(x,t)\in W_2^{(2,1)}(Q_T^*)$ to the initial value problem (9) for the nonlinear parabolic equation (8) with Hilbert operator have the estimation

$$\sup_{0\le t\le T}\|u(\cdot,t)\|_{L_2(R)}\le K_3\{\|\phi\|_{L_2(R)}+\|f\|_{L_2(Q_T^*)}\}, \tag{18}$$

where K_3 is a constant independent of $\varepsilon>0$, but dependent on the norms $\|b_x\|_{L_\infty(Q_T^*)}$ and $\|C\|_{L_\infty(Q_T^*)}$.

By direct calculations and by use of the behaviors of the Hilbert operator, we can obtain the final equality:

$$\begin{aligned}
&\frac{d}{dt}\int_{-\infty}^{\infty}(u^4+2bu^3+3buHu_x+3u^2Hu_x+2u_x^2)dx\\
&=-4\varepsilon\|u_{xx}(\cdot,t)\|^2_{L_2(R)}+\varepsilon\int_{-\infty}^{\infty}\{6(uHu_x+H(uu_x))+3(bHu_x+H(bu)_x+4u^3+6bu^2\}u_{xx}dx+\\
&+\int_{-\infty}^{\infty}[(\tfrac{5}{2}b_x-4c)u_x^2+\tfrac{3}{2}b_x(Hu_x)^2-3b^2u_xHu_x]dx-16\int_{-\infty}^{\infty}bu^3u_xdx+\\
&+6\int_{-\infty}^{\infty}[(b_x-c)(u^2Hu_x+uH(uu_x))-b^2u^2u_x]dx+\\
&+3\int_{-\infty}^{\infty}[(-bc+b_t)uHu_x+(b_{xx}-4c_x)uu_x-bu_xH(b_xu)-cuH(bu_x)]dx+\\
&+6\int_{-\infty}^{\infty}f(uHu_x+H(uu_x))dx+\int_{-\infty}^{\infty}[3bfHu_x+3fH(bu_x)+4u_xf_x]dx-\\
&-\int_{-\infty}^{\infty}cu^4dx+\int_{-\infty}^{\infty}(2b_t-6bc)u^3dx+4\int_{-\infty}^{\infty}fu^3dx+6\int_{-\infty}^{\infty}bfu^2dx-3\int_{-\infty}^{\infty}cuH(b_xu)dx+\\
&+3\int_{-\infty}^{\infty}fH(b_xu)dx\,.
\end{aligned}\tag{19}$$

Let us now suppose that the coefficients $b(x,t)$ and $c(x,t)$, the free term $f(x,t)$ and the initial function $\phi(x)$ satisfy the following assumptions:

<1> $f(x,t) \in W_2^{(1,0)}(Q_T^*)$,

<2> $c(x,t) \in W_\infty^{(1,0)}(Q_T^*)$,

<3> $b(x,t) \in W_\infty^{(2,1)}(Q_T^*)$,

<4> $\phi(x) \in H^1(R)$.

Now we turn to simplify the equality (19). At first let us consider the first integral of the right-hand side of (19) with coefficient $\varepsilon>0$. Denote by $J(x,t)$ the expression in the curved parenthesis of this integral. Then

$$\left|\int_{-\infty}^{\infty} Ju_{xx}dx\right| \leqq \frac{1}{2}\|u_{xx}(\cdot,t)\|^2_{L_2(R)}+\frac{1}{2}\|J(\cdot,t)\|^2_{L_2(R)}, \tag{20}$$

where

$$\|J(\cdot,t)\|^2_{L_2(R)} \leqq C_7\int_{-\infty}^{\infty}\{u^2(Hu_x)^2+(H(uu_x))^2+b^2(Hu_x)^2+(H(bu)_x)^2+u^6+b^2u^4\}dx. \tag{21}$$

Hence we have

$$\begin{aligned}\int_{-\infty}^{\infty} u^2(Hu_x)^2dx &\leqq \|u(\cdot,t)\|^2_{L_\infty(R)}\|Hu_x(\cdot,t)\|^2_{L_2(R)}\\ &= \|u(\cdot,t)\|^2_{L_\infty(R)}\|u_x(\cdot,t)\|^2_{L_2(R)}\\ &\leqq \eta\|u(\cdot,t)\|^2_{L_2(R)}+C_5(\eta)\|u(\cdot,t)\|^{10}_{L_2(R)},\end{aligned}$$

$$\int_{-\infty}^{\infty}(H(uu_x))^2dx = \int_{-\infty}^{\infty} u^2u_x^2dx \leqq \eta\|u_{xx}(\cdot,t)\|^2_{L_2(R)}+C_5(\eta)\|u(\cdot,t)\|^{10}_{L_2(R)},$$

and so on. Substituting these estimation into (21), we get

$$\|J(\cdot,t)\|^2_{L_2(R)} \leqq 6C_6\eta\|u_{xx}(\cdot,t)\|^2_{L_2(R)}+C_7(\eta)\|u(\cdot,t)\|^2_{L_2(R)},$$

where $C_7(\eta)$ depends on the norms $\|b\|_{W_\infty^{(1,0)}(Q_T^*)}$ and $\sup\limits_{0\leqq t\leqq T}\|u(\cdot,t)\|_{L_2(R)}$. Let us take η to be so small that $6C_6\eta=1$. Hence

$$\|J(\cdot,t)\|^2_{L_2(R)} \leqq \|u_{xx}(\cdot,t)\|^2_{L_2(R)}+C_8\|u(\cdot,t)\|^2_{L_2(R)}$$

and then

$$\left|\int_{-\infty}^{\infty} Ju_{xx}dx\right| \leqq \|u_{xx}(\cdot,t)\|^2_{L_2(R)}+\frac{1}{2}C_8\|u(\cdot,t)\|^2_{L_2(R)}, \tag{22}$$

where C_8 is a constant dependent on the norms $\|b\|_{W_\infty^{(1,0)}(Q_T^*)}$ and $\sup\limits_{0\leqq t\leqq T}\|u(\cdot,t)\|_{L_2(R)}$.

Using the interpolation formulars, we can estimate some of the remaining terms of (19) as follows:

$$\begin{aligned}&\left|\int_{-\infty}^{\infty}[(\tfrac{5}{2}b_x-4C)u_x^2+\tfrac{3}{2}b_x(Hu_x)^2-3b^2u_xHu_x]dx\right|\\ &\qquad\leqq (4\|b_x\|_{L_\infty(Q_T^*)}+4\|c\|_{L_\infty(Q_T^*)}+3\|b\|^2_{L_\infty(Q_T^*)})\|u_x(\cdot,t)\|^2_{L_2(R)},\end{aligned}$$

$$|\int_{-\infty}^{\infty} bu^3 u_x dx| \leq C_9 \| b \|_{L_\infty(Q_T^*)} \| u(\cdot,t) \|^2_{L_2(R)} \| u_x(\cdot,t) \|^2_{L_2(R)} ,$$

and so on. Then the equality (19) is simplified and can be replaced by the inequality

$$\frac{d}{dt}\int_{-\infty}^{\infty}(u^4+2bu^3+3buHu_x+3u^2Hu_x+2u_x^2)dx+3\varepsilon \| u_{xx}(\cdot,t) \|^2_{L_2(R)}$$

$$\leq C_{10} \| u(\cdot,t) \|^2_{H^1(R)} + C_{11} \| f(\cdot,t) \|^2_{H^1(R)} , \tag{23}$$

where C_{10} and C_{11} are constants dependent on the norms $\| b \|_{W_\infty^{(2,1)}(Q_T^*)}$, $\| C \|_{W_\infty^{(1,0)}(Q_T^*)}$ and $\sup_{0\leq t\leq T} \| u(\cdot,t) \|_{L_2(R)}$, but independent of the diffusion coefficient.

Integrating both sides of the inequality (23) with respect to the time variable in the interval [0,t], we have

$$\| u_x(\cdot,t) \|^2_{L_2(R)} \leq C_{12}\{ \| \phi \|^2_{H^1(R)} + \| f \|^2_{W_2^{(1,0)}(Q_T^*)}\}$$

for any $0\leq t\leq T$, where C_{12} is independent of $\varepsilon>0$.

<u>LEMMA 4</u>. Suppose that $b(x,t)\in W_\infty^{(2,1)}(Q_T^*)$, $C(x,t)\in W_\infty^{(1,0)}(Q_T^*)$ and $f(x,t)\in W_2^{(1,0)}(Q_T^*)$ and suppose that $\phi(x)\in H^1(R)$. The generalized global solutions $u_\varepsilon(x,t)\in W_2^{(2,1)}(Q_T^*)$ to the initial value problem (9) for the nonlinear parabolic equation (8) with Hilbert operator have the estimation

$$\sup_{0\leq t\leq T} \| u_{\varepsilon x}(\cdot,t) \|_{L_2(R)} \leq K_4\{ \| \phi \|_{H^1(R)} + \| f \|_{W_2^{(1,0)}(Q_T^*)}\}, \tag{24}$$

where K_4 is a constant, independent of $\varepsilon>0$.

<u>LEMMA 5</u>. Under the conditions of Lemma 4, the solution $u_\varepsilon(x,t)\in W_2^{(2,1)}(Q_T^*)$ to the initial value problems (8) and (9) have the estimation

$$\| u_\varepsilon \|_{L_\infty(Q_T^*)} \leq K_5\{ \| \phi \|_{H^1(R)} + \| f \|_{W_2^{(1,0)}(Q_T^*)}\}, \tag{25}$$

where K_5 is a constant independent of $\varepsilon>0$.

In order to estimate $\| u_{\varepsilon xx}(\cdot,t) \|_{L_2(R)}$, by direct calculations we can obtain the following identity

$$\frac{d}{dt}\int_{-\infty}^{\infty}(2u_{xx}^2+5u_x^2Hu_x+10uu_xHu_{xx})dx+4\varepsilon \| u_{xxx}(\cdot,t) \|^2_{L_2(R)}$$

$$=-10\varepsilon\int_{-\infty}^{\infty}[uHu_{xxx}+H(uu_x)_{xx}+(u_xHu_x)_x-H(u_xu_{xxx})]u_{xx}dx -$$

$$-\int_{-\infty}^{\infty}[(6b_x+4C)u_{xx}^2 +10(bu+2u^2)u_{xx}Hu_{xx}+10bu_{xx}H(uu_{xx})]dx +$$

$$+10\int_{-\infty}^{\infty}[(2u+b)u_x Hu_x - bH(u_x)^2 + 2u_x H(uu_x)]u_{xx}dx -$$

$$-20\int_{-\infty}^{\infty}uu_x^2 Hu_{xx}dx - 10\int_{-\infty}^{\infty}bu_x H(u_x u_{xx})dx +$$

$$+\int_{-\infty}^{\infty}[10CuHu_x - 4(2b_{xx}+C_x)u_x]u_{xx}dx - \qquad (26)$$

$$-10\int_{-\infty}^{\infty}(b_x - c)uu_x Hu_{xx}dx - 10\int_{-\infty}^{\infty}(b_x+c)u_x H(uu_{xx})dx -$$

$$-10\int_{-\infty}^{\infty}cuH(u_x u_{xx})dx + 10\int_{-\infty}^{\infty}[H(u_x u_{xx}) - u_{xx}Hu_x]fdx -$$

$$-4\int_{-\infty}^{\infty}(c_{xx}u - f_{xx})u_{xx}dx - 10\int_{-\infty}^{\infty}(c_x u^2 - f_x u)Hu_{xx}dx -$$

$$-10\int_{-\infty}^{\infty}(c_x u + f_x)H(uu_{xx})dx -$$

$$-10\int_{-\infty}^{\infty}(b_x+c)u_x H(u_x^2)dx - 10\int_{-\infty}^{\infty}(c_x u - f_x)H(u_x^2)dx$$

Dinote J_k (k=1,2,...,15) the k-th integral term on the right-hand side of the identity (26).

For J_1, we have

$$J_1 = 20\varepsilon\int_{-\infty}^{\infty}(u_{xx}u_{xxx}Hu - uu_{xx}Hu_{xxx})dx.$$

Then

$$|J_1| \leqq 20\varepsilon\{\| Hu\|_{L_\infty(Q_T^*)}\| u_{xx}(\cdot,t)\|_{L_2(R)}\| u_{xxx}(\cdot,t)\|_{L_2(R)}$$

$$+\| u\|_{L_\infty(Q_T^*)}\| u_{xx}(\cdot,t)\|_{L_2(R)}\| Hu_{xxx}(\cdot,t)\|_{L_2(R)}\}$$

$$\leqq 4\varepsilon\| u_{xxx}(\cdot,t)\|^2_{L_2(R)} + 50\varepsilon(\| u\|_{L_\infty(Q_T^*)} + \| Hu\|_{L_\infty(Q_T^*)})\| u_{xx}(\cdot,t)\|^2_{L_2(R)},$$

where

$$\| Hu\|^2_{L_\infty(Q_T^*)} = \sup_{0\leqq t\leqq T}\| Hu(\cdot,t)\|^2_{L_\infty(R)}$$

$$\leqq C_{13}\sup_{0\leqq t\leqq T}\| Hu(\cdot,t)\|_{L_2(R)}\| Hu_x(\cdot,t)\|_{L_2(R)}$$

$$\leqq C_{13}\sup_{0\leqq t\leqq T}\| u(\cdot,t)\|_{L_2(R)}\sup_{0\leqq t\leqq T}\| u_x(\cdot,t)\|_{L_2(R)}$$

is bounded with respect to $\varepsilon>0$.

For J_2, we have

$$|J_2| \leqq \| 6b_x+4c\|_{L_\infty(Q_T^*)}\| u_{xx}(\cdot,t)\|^2_{L_2(R)} +$$

$$+10\| bu+2u^2\|_{L_\infty(Q_T^*)}\| u_{xx}(\cdot,t)\|_{L_2(R)}\| Hu_{xx}(\cdot,t)\|_{L_2(R)} +$$

$$+10\| b\|_{L_\infty(Q_T^*)}\| u_{xx}(\cdot,t)\|_{L_2(R)}\| H(u(\cdot,t)u_{xx}(\cdot,t)\|_{L_2(R)}$$

$$\leqq(6\| b_x\|_{L_\infty(Q_T^*)} + 4\| C\|_{L_\infty(Q_T^*)} + 20\| b\|_{L_\infty(Q_T^*)}\| u\|_{L_\infty(Q_T^*)} +$$

$$+20\| u \|^2_{L_\infty(Q_T^*)}) \| u_{xx}(\cdot,t) \|^2_{L_2(R)} .$$

Similarly for the remaining J's, we can obtain the following inequality in stead of the equality (26)

$$\frac{d}{dt}\int_{-\infty}^{\infty}(2u_{xx}^2+5u^2Hu_x+10uu_xHu_{xx})dx$$

$$\leq C_{14} \| u_{xx}(\cdot,t) \|^2_{L_2(R)}+C_{15} \| u_x(\cdot,t) \|^2_{L_2(R)}+C_{16} \| u(\cdot,t) \|^2_{L_2(R)}$$

$$+C_{17} \|f(\cdot,t)\|^2_{H^1(R)}+ \|f_{xx}(\cdot,t)\|^2_{L_2(R)},$$

where C's are constants dependent on the norms $\| b \|_{W_\infty^{(2,0)}(Q_T^*)}$, $\| C \|_{W_\infty^{(2,0)}(Q_T^*)}$ and

$\sup\limits_{0\leq t\leq T} \| u(\cdot,t) \|_{H^1(R)}$ and independent of $\varepsilon>0$. Integrating both sides of this inequality with respect to the variable t in the interval $[0,t)$ and regarding the estimations given in Lemmas 3 and 4, we have

LEMMA 6. Suppose that $b(x,t)\in W_\infty^{(2,1)}(Q_T^*)$, $C(x,t)\in W_\infty^{(2,0)}(Q_T^*)$ and $f(x,t)\in W_2^{(2,0)}(Q_T^*)$ and suppose that $\phi(x)\in H^2(R)$. Then the generalized global solutions $u_\varepsilon(x,t)\in W_2^{(2,1)}(Q_T^*)$ to the initial value problems (9) for the nonlinear parabolic equation (8) with Hilbert operator have the estimation

$$\sup_{0\leq t\leq T} \| u_{\varepsilon xx}(\cdot,t)\|_{L_2(R)} \leq K_6\{\| \phi \|_{H^2(R)}+\| f \|_{W_2^{(2,0)}(Q_T^*)}\}, \tag{27}$$

where the constant K_6 depends on the norms $\| b \|_{W_\infty^{(2,1)}(Q_T^*)}$, $\| C \|_{W_\infty^{(2,0)}(Q_T^*)}$, $\| \phi \|_{H^1(R)}$ and is independent of $\varepsilon>0$.

By means of equation (8) and the interpolation relations, we have the following lemmas as the immediate consequences of the previous lemmas.

LEMMA 7. Under the conditions of Lemma 6, the generalized global solutions $u_\varepsilon(x,t)\in W_2^{(2,1)}(Q_T^*)$ to the initial value problem (9) for the nonlinear parabolic equation (8) with Hilbert operator have the estimation

$$\sup_{0\leq t\leq T} \|u_{\varepsilon t}(\cdot,t) \|_{L_2(R)} \leq K_7\{\| \phi \|_{H^2(R)}+\| f \|_{W_2^{(2,0)}(Q_T^*)}\}, \tag{28}$$

where K_7 is a constant, dependent on the norms $\| b \|_{W_\infty^{(2,0)}(Q_T^*)}$, $\| C \|_{W_\infty^{(2,0)}(Q_T^*)}$ and $\| \phi \|_{H^1(R)}$, but independent of $\varepsilon>0$.

LEMMA 8. Under the conditions of Lemma 6, the generalized global solutions $u_\varepsilon(x,t)\in W_2^{(2,1)}(Q_T^*)$ to the initial value problem (8) and (9) have the estimation

$$\| u_{\varepsilon x} \|_{L_\infty(Q_T^*)} \leq K_8\{\| \phi \|_{H^2(R)}+\| f \|_{W_2^{(2,0)}(Q_T^*)}\} , \tag{29}$$

where K_8 is independent of $\varepsilon>0$.

§4. Generalized Solutions

From Lemmas 3-8, the set of functions $\{u_\varepsilon(x,t)\}$ is uniformly bounded in the functional space $Z=L_\infty(0,T;\ H^2(R))\cap W_\infty^{(1)}(0,T);L_2(R))$ with respect to the diffusion coefficient $\varepsilon>0$ of the nonlinear parabolic equation (8) with Hilbert operator. By means of interpolation formulars for the functional spaces, we have the following lemma of the uniform estimations with respect to $\varepsilon>0$.

LEMMA 9. Under the conditions of Lemma 6, the set $\{u_\varepsilon(x,t)\}$ of the generalized global solutions of initial value problems (8) and (9) has the following estimations:

$$|u_\varepsilon(\bar{x},t)-u_\varepsilon(x,t)| \leq K_9|\bar{x}-x|, \tag{30}$$

$$|u_\varepsilon(x,\bar{t})-u_\varepsilon(x,t)| \leq K_{10}|\bar{t}-t|^{3/4} \tag{31}$$

$$|u_{\varepsilon x}(\bar{x},t)-u_{\varepsilon x}(x,t)| \leq K_{11}|\bar{x}-x|^{\frac{1}{2}} \tag{32}$$

and

$$|u_{\varepsilon x}(x,\bar{t})-u_{\varepsilon x}(x,t)| \leq K_{12}|\bar{t}-t|^{\frac{1}{4}}, \tag{33}$$

where $\bar{x},x\in R$; $\bar{t},t\in[0,T]$, and the constants K's are independent of $\varepsilon>0$.

From the uniform estimations (30)-(33), the sets $\{u_\varepsilon(x,t)\}$ and $\{u_{\varepsilon x}(x,t)\}$ are uniformly bounded in the spaces $C_{x,t}^{(1,3/4)}(Q_T^*)$ and $C_{x,t}^{(\frac{1}{2},\frac{1}{4})}(Q_T^*)$ of Holder continuous functions respectively. By using analogous method in [11-14], the following theorems can be proved (Because the proofs are tedious, and the article have limited space, here one only state the results).

THEOREM 4. Suppose that $b(x,t)\in W_\infty^{(2,1)}(Q_T^*)$, $C(x,t)\in W^{(2,0)}(Q_T^*)$ and $f(x,t)\in W_2^{(2,0)}(Q_T^*)$ and suppose also that $\phi(x)\in H^2(R)$. For the initial value problem with initial condition (9) for the nonlinear singular integral-differential equation (7) with Hilbert operator, there exists at least one genrealized global solution $u(x,t)\in Z$, which satisfies the equation (7) in generalized sense and satisfies the initial condition (9) in classical sense.

THEOREM 5. Suppose that $b(x,t)\in W_\infty^{(1,0)}(Q_T^*)$ and $C(x,t)\in L_\infty(Q_T^*)$. The generalized global solution $u(x,t)\in Z$ for the initial problem (9) of the nonlinear singular integral-differential equation (7) is unique.

THEOREM 6. Under the conditions of Theorem 4, as $\varepsilon\to0$, the generalized global solution $u_\varepsilon(x,t)\in W_2^{(2,1)}(Q_T^*)$ to the initial problem (9) for the nonlinear parabolic equation (8) with Hilbert operator converges to the unique generalized global solution $u(x,t)\in Z$ to the initial problem (9) for the nonlinear singular integral-differential equation (7) in the sense that $\{u_\varepsilon(x,t)\}$ and $\{u_{\varepsilon x}(x,t)\}$ are uniformly convergent to $u(x,t)$ and $u_x(x,t)$ respectively in any compact set of Q_T^*, and

$\{u_{\varepsilon xx}(x,t)\}$ and $\{u_{\varepsilon t}(x,t)\}$ are weakly convergent to $u_{xx}(x,t)$ and $u_t(x,t)$ respectively in $L_p(0,T; L_2(R))$ for $2\leqq p>\infty$.

The following theorem is concerned with the estimation of the rate of convetence.

THEOREM 7. Under the conditions of Theorem 4, for the generalized global solutions $u_\varepsilon(x,t)\in W_2^{(2,1)}(Q_T^*)$ and $u(x,t)\in Z$ to the initial problems (9) for the nonlinear parabolic equation (8) and the nonlinear integral-differential equation (7) respectively, there are the estimations for the rate of convergence in terms of the power of the diffusion coefficient $\varepsilon>0$ as follows:

$$\sup_{o\leqq t\leqq T} \| u_\varepsilon(\cdot,t)-u(\cdot,t) \|_{L_2(R)} \leqq K_{13}\varepsilon, \tag{34}$$

$$\| u_\varepsilon - u \|_{L_\infty(Q_T^*)} \leqq K_{14}\varepsilon^{3/4}, \tag{35}$$

$$\sup_{o\leqq t\leqq T} \|u_{\varepsilon x}(\cdot,t)-u_x(\cdot,t)\|_{L_2(R)} \leqq K_{15}\varepsilon^{\frac{1}{2}}, \tag{36}$$

and

$$\| u_{\varepsilon x}-u_x \|_{L_\infty(Q_T^*)} \leqq K_{16}\varepsilon^{\frac{1}{4}}, \tag{37}$$

where K's are the constants independent of $\varepsilon>0$.

THEOREM 8. Suppose that the conditions of Theorem 4 are satisfied for any value of $T>0$. The initial value problem (9) for the nonlinear integral-differential equation (7) has a unique generalized global solution $u(x,t)\in L_{\infty loc}(R_+; H^2(R))\cap W^{(1)}_{\infty loc}\cdot(R_+; L_2(R)$ in the infinite demain $Q_\infty^*=\{x\in R;\ t\in R_+\}$.

THEOREM 9. Suppose that $\phi(x)\in H^M(R)$ for $M\geqq 2$. The initial value problem (9) for the Benjamin-Ono equation (4) has a unique global solution $u(x,t)\ \bigcap_{k=0}^{[\frac{M}{2}]} W^{(k)}_{\infty,loc}(R_+;H^{M-2k}\cdot(R))$, which has the derivatives $u_{x^r t^s}(x,t)\in L_{\infty,loc}(R_+;L_2(R))$ for $0\leqq 2s+r\leqq M$.

REFERENCE

[1] R. I. Joseph, Solitary Waves in a Finite Depth Fluid, J. Phys. A: Math. Gen., 10 (1977), L225-L227.

[2] R. I. Joseph & R. Egri, Multi-Soliton Solutions in a Finite Depth Fluid. J. Phys. A : Math. Gen., 11 (1978), L97-L102.

[3] Y. Matsuno, Exact Multi-Soliton Solution for Nonlinear Waves in a Stratified Fluid of Finite Depth, Phys. Lett., 74 A (1979), 233-235.

[4] J. Satsumy, M. J. Ablowitz & Y. Kodama, On a Interval Wave Equation Describing a Stratified Fluid with Finite Depth, Phys. Lett., 73 A (1979), 283-284.

[5] D. J. Korteweg & G. de Vries, On the Change of Form of Long Waves Advancing in a Rectangular Canal and a New Type of Long Stationary Waves, Phil. Mag., (5), 39 (1895), 422-433.

[6] T. B. Benjamin, Internal Waves of Permanent Form in Fluids of Great Depth, J. Fluid Meth., 29(1967), 559-592.

[7] H. Ono, Algebraic Solitary Waves in Stratified Fluids, J. Phys. Soc. Japan, 39 (1975), 1082-1091.

[8] J. Satsuma & D. J. Kaup, A Backlund Transformation for a Higher Order Korteweg-De Vries Equation. J. Phys. Soc. Japan, 43 (1977), 692-697.

[9] J. Satsuma & R. Hirota, A Coupled KdV Equation in One Case of Four-Reduction of the KP Hierarchy, J. Phys. Soc. Japan, 51 (1982), 3390-3397.

[10] R. Hirota & M. Ito, Resonance of Solitons in One Dimension, J. Phys. Soc. Japan, 52 (1983), 744-748.

[11] Zhou Yu-lin & Guo Bo-ling, Periodic Boundary Problem and Initial Value Problem for the Generalized Korteweg-de Vries Systems of Higher Order, Acta Mathematica Sinica, 27 (1984), 154-176. (in Chinese).

[12] Zhou Yu-lin & Guo Bo-ling, On the System of the Generalized Korteweg-de Vries Equations, Proc. of the 1982 Symposium of Diff. Geom. & Diff. Eq.

[13] Zhou Yu-lin & Guo Bo-ling, A Class of General Systems of KdV Type, (I) Weak Solutions with Derivative $u_x p$, Acta Math. Appl. Sinica, 1 (1984), 153-162.

[14] Zhou Yu-lin & Guo Boling, Existence of Global Weak Solutions for Generalized Korteweg-de Vries Systems with Several Variables, Scientia Sinica (ser. A), 29 (1986), 375-390.

[15] A. Nakamuro, Bäcklund Transform and Conservation Laws of the Benjamin-Ono Equation, J. Phys. Soc. Japan, 47 (1979), 1335-1340.

[16] J. Satsuma & Y. Ishimori, Periodic Wave and Rational Soliton Solutions of the Benjamin-Ono Equation, J. Phys. Soc, Japan, 46 (1979), 681-687.

[17] Y. Matsuno, Interaction of the Benjamin-Ono Solutions J. Phys. A : Math. Gen., 13 (1980), 1519-1536.

[18] Y. Matsuno, Soliton and Algebraic Equation, J. Phys. Soc. Japan, 51 (1982), 3375-3380.

[19] Y. Matsuno, Recurrence Formula and Conserved Quanlity of the Benjamin-Ono Equation, J. Phys. Soc. Japan, 52 (1983), 2955-2958.

A LIST OF THE PAPERS PRESENTED AT THE DD7 SYMPOSIUM

S. Agmon, On Green's Functions and Generalized Eigenfunctions of Schrödinger Operators.

F. Almgren, W. Browder & E.H. Lieb, Co-area, Liquid Crystals and Minimal Surface.

R. Beals, Non-elliptic Problem and Complex Analysis.

Chang Kung-ching, Remarks on Saddle Points in the Calculus of Variations.

Chen Hua, Well-Posedness in Gevrey Classes of the Cauchy Problem for a Class of Totally Characteristic Hyperbolic Operators.

Chen Shuxing, Smoothness of Shock Front Solutions for System of Conservation Laws.

Chen Yazhe, On Degenerate Monge-Ampere Equations.

Chen Yunmei & Gao Jianmin, Global Existence of Solutions for Quasilinear Hyperbolic Equations in an Exterior Domain.

Chen Zhimin & Shi Shuzhong, Nagumo Type Conditions for Semilinear Parabolic Systems.

Chen Zu-chi, Inequalities for Eigenvalues of a Class of Polyharmonic Operators.

Chen Zu-chi & Qian Chun-lin, The Mixed Boundary Value Problem for Degenerate Quasilinear Parabolic Equations.

Chi Min-you, Morse Type Non-Nilsson Solutions of Fuchsian Type Partial Differential Equations.

Ding Weiyue, On a Conformally Invariant Elliptic Equation on $\mathbb{R}^n$.

Ding Xiaxi, Chen Guiqiang & Luo Peizhu, Convergence of the Lax-Friedrichs Scheme for Isentropic Gas Dynamics.

Guangchang Dong, On the Estimation of Derivatives for Solutions of Fully Nonlinear Parabolic Equation.

Fu Hong-Yuan, Initial and Boundary Problems for the Degenerate or Singular Systems of the Filtration Type.

Gu Chaohao, Explicit and Unified Form of Bäcklund Transformations.

Gu Liankun, Yang Shixin, Cao Zhenzong & Chen Longsheng, Boundary Value Problems of Degenerate Quasilinear Elliptic Equations and the Multiplicity of the Generalized Solutions.

Guo Boling, The Initial Value Problem and Initial-Boundary Value Problem for the System of Generalized Pekar-Choquard Nonlinear Schrödinger Equations in Three Dimensions.

Guo Boling & Shen Longjun, The Global Solution for the Initial Value Problem of the System of Schrödinger-Boussinesq Type Equations in Three Diminsions.

Guo Bo-ling & Wang Li-ring, The Periodic Solutions for the Nonlinear Systems of Schrödinger Equations and the KdV Equation.

He Ping-fan, Global Solutions for a Class of Coupled KdV System.

Hong Chongwei, Equation $\Delta u+K(x)e^{2u}=f(x)$ on R^2 Via Stereographic Projection.

Hong Min-chun, Existence and Partial Regularity in the Calculus of Variations.

Huang Qingbo, The Boundary Value Problem for a Class of Nonlinear Degenerate and Singular Diffusion Equations.

Huang Sixun, Some Operators, Their Properties and Application of PDE in a Douglis Algebra.

Huang Yumin, On Interior Regularity of the Solutions of a Class of Hypoelliptic Differential Equations.

Jiang Lumin, The Existence of Global Solutions to the Cauchy Problem for $U_t-\Delta U=f(U,U_t,D_xU,D_x^2U)$.

Jiang Meiyue, An Existence Theorem for Periodic Solutions of Second Order Hamiltonian System.

Jin Zhiren, Conformal Deformation of Riemannian Metrics on a Class of Noncompact Complete Reimannian Manifolds and the Yamabe Problem.

H. Lewy & Tang Zhiyuan, On Free Boundary Problems in Two Dimensions.

Li Chengzhang, Sharp Garding Inequality for Paradifferential Operators.

Li Hui-lai, Free Boundary Problems for Degenerate Parabolic Equations.

Li Ta-tsien & Zhao Yanchun, Global Perturbation of the Riemann Problem for the System of One-Dimensional Isentropic Flow.

Li Zhengyuan, Existence and Stability of Nonnegative Steady-State Solutions for a Class of Systems of Reaction-Diffusion Equations.

Lin Zhenguo, The Method of Partial Viscous Vanishing for Quasilinear Hyperbolic-Parabolic Coupled System.

Liu Baoping, Almost Periodic Forced Vibrations for a Semilinear Wave Equation.

Liu Linqi, Optimal Propagation of Singularities for Semilinear Hyperbolic Differential Equations.

Liu Xiyuan, Remarks on Differentiability of the Free-Boundary Operator for One-Phase Stefan Problems.

Liu Yacheng, The Global Solvability for a Class of Evolution Equations (Systems) with Nonlinear Strong Dissipation.

Luo Xue-bo, Solvability in $\mathcal{S}'$ of Some LPDEs

Ma Li, A Remark of the Positive Eigen-Solution of Quasi-Linear Elliptic Equation Involving Sobolev Critical Exponential.

Ma Runian & Zhu Xiping, On the Muliple Solutions of Dirichlet Problems for Nonlinear Elliptic Equations (Critical Increasing Case).

Mu Mu, On teh Boundary Value Problem for a Degenerate Elliptic Equation.

Pan Xiu-de & Xu Bao-zhi, The KDV Equation with Nonuniformity Terms.

Pan Zu-liang & Zheng Ke-jie, Lie's Transformation Group and Similarity Solutions of the S3 Equation.

Qin Tiehu, The Boundary Value Problem for Nonlinear Hyperbolic Equations with Dissipative Boundary Conditions.

Qing Jie, A Priori Estimates for Positive Solutions of Semilinear Elliptic Systems.

Qiu Qing-jiu & Qian Si-xin, Analysis of C^∞ Singularities for a Class of Operators with Varying Multiple Characteristics.

Shen Yaotian & Guo Xinkang, On the Existence of Infinitely Many Critical Points of the Even Functional $\int_\Omega F(x,u,Du)dx$ in $W_0^{1,p}$.

Sun Hesheng, Global Solutions of Various Nonlinear Evolutional Systems of PDEs.

Tan Yongji, An Inverse Probelm for Nonlocal Elliptic BVP and Resistivity Identificating.

Tang Xianjiang, Mixed Problems for Second Order Hyperbolic Equation with a Singular Oblique Derivative.

J.E. Taylor, Local Ellipticity of F and Regularity of F Minimizing Currents.

F. Tomi & A.J. Tromba, A Geometric Proof of the Mumford Compactness Theorem.

Wang Guanglie, Harnack Inequalities for Functions in de Giorgi Parabolic Class.

Wang Jianhua & Li Caizhong, Globally Smooth Resolvability and Formation of Singularities of Solutions for Certain Quasilinear Hyperbolic Systems with a Dissipative Term.

Wang Rouhwai, Some Supplements to ADN Theory about Elliptic and Parabolic BVPs.

Wang Zhiqiang, Equivariant Three Solution Theorem and its Applications.

Wu Fantong, A Class Diffractive Boundary Value Problem with Multiple Characteristic.

Wu Lancheng, Existence, Uniqueness and Regularity of the Minimizer of a Certain Functional.

Xiang Xiao-ling, Some Theorems to Judge the Globosity of the Semiflows.

Xie Hong-zheng, One Free Boundary Problem for a Degenerate Parabolic Equation.

Xu Chao-jiang, Subelliptic Operators and Regularity of the Solutions of Partial Differential Equations of the Second Order in R^2.

Yan Ziqian, Everywhere Regularity for Solutions to Quasilinear Elliptic Systems of Triangular Form.

Ye Qixiao & Wang Mingxin, Travelling Wave Front Solutions of a Model of Belousov-Zhabotinskii Chemical Reaction.

Zhang Kewei, On the Dirichlet Problem of a Class of Elliptic Systems.

Zheng Sining, A Reaction-Diffusion System of a Competitor-Competitor-Mutualist Model.

Zhou Yulin & Guo Boling, Initial Value Problems for a Nonlinear Singular Integral-Differential Equation of Deep Water.

LECTURE NOTES IN MATHEMATICS
Edited by A. Dold and B. Eckmann

Some general remarks on the publication of proceedings of congresses and symposia

Lecture Notes aim to report new developments - quickly, informally and at a high level. The following describes criteria and procedures which apply to proceedings volumes.

1. One (or more) expert participant(s) of the meeting should act as the responsible editor(s) of the proceedings. They select the papers which are suitable (cf. points 2, 3) for inclusion in the proceedings, and have them individually refereed (as for a journal). It should not be assumed that the published proceedings must reflect conference events faithfully and in their entirety. Contributions to the meeting which are not included in the proceedings can be listed by title. The series editors will normally not interfere with the editing of a particular proceedings volume - except in fairly obvious cases, or on technical matters, such as described in points 2, 3. The names of the responsible editors appear on the title page of the volume.

2. The proceedings should be reasonably homogeneous (concerned with a limited area). For instance, the proceedings of a congress on "Analysis" or "Mathematics in Wonderland" would normally not be sufficiently homogeneous.

   One or two longer survey articles on recent developments in the field are often very useful additions to such proceedings - even if they do not correspond to actual lectures at the congress. An extensive introduction on the subject of the congress would be desirable.

3. The contributions should be of a high mathematical standard and of current interest. Research articles should present new material and not duplicate other papers already published or due to be published. They should contain sufficient information and motivation and they should present proofs, or at least outlines of such, in sufficient detail to enable an expert to complete them. Thus resumes and mere announcements of papers appearing elsewhere cannot be included, although more detailed versions of a contribution may well be published in other places later.

   Surveys, if included, should cover a sufficiently broad topic, and should in general not simply review the author's own recent research. In the case of surveys, exceptionally, proofs of results may not be necessary.

   The editors of a volume are strongly advised to inform contributors about these points at an early stage.

.../...

4. Proceedings should appear soon after the meeeting. The publisher should, therefore, receive the complete manuscript within nine months of the date of the meeting at the latest.

5. Plans or proposals for proceedings volumes should be sent to one of the editors of the series or to Springer-Verlag Heidelberg. They should give sufficient information on the conference or symposium, and on the proposed proceedings. In particular, they should contain a list of the expected contributions with their prospective length. Abstracts or early versions (drafts) of some of the contributions are very helpful.

6. Lecture Notes are printed by photo-offset from camera-ready typed copy provided by the editors. For this purpose Springer-Verlag provides editors with technical instructions for the preparation of manuscripts and these should be distributed to all contributing authors. Springer-Verlag can also, on request, supply stationery on which the prescribed typing area is outlined. Some homogeneity in the presentation of the contributions is desirable.

   Careful preparation of manuscripts will help keep production time short and ensure a satisfactory appearance of the finished book. The actual production of a Lecture Notes volume normally takes 6 -8 weeks.

   Manuscripts should be at least 100 pages long. The final version should include a table of contents.

7. Editors receive a total of 50 free copies of their volume for distribution to the contributing authors, but no royalties. (Unfortunately, no reprints of individual contributions can be supplied.) They are entitled to purchase further copies of their book for their personal use at a discount of 33 1/3%, other Springer mathematics books at a discount of 20% directly from Springer-Verlag.

   Commitment to publish is made by letter of intent rather than by signing a formal contract. Springer-Verlag secures the copyright for each volume.

Vol. 1145: G. Winkler, Choquet Order and Simplices. VI, 143 pages. 1985.

Vol. 1146: Séminaire d'Algèbre Paul Dubreil et Marie-Paule Malliavin. Proceedings, 1983–1984. Edité par M.-P. Malliavin. IV, 420 pages. 1985.

Vol. 1147: M. Wschebor, Surfaces Aléatoires. VII, 111 pages. 1985.

Vol. 1148: Mark A. Kon, Probability Distributions in Quantum Statistical Mechanics. V, 121 pages. 1985.

Vol. 1149: Universal Algebra and Lattice Theory. Proceedings, 1984. Edited by S. D. Comer. VI, 282 pages. 1985.

Vol. 1150: B. Kawohl, Rearrangements and Convexity of Level Sets in PDE. V, 136 pages. 1985.

Vol 1151: Ordinary and Partial Differential Equations. Proceedings, 1984. Edited by B.D. Sleeman and R.J. Jarvis. XIV, 357 pages. 1985.

Vol. 1152: H. Widom, Asymptotic Expansions for Pseudodifferential Operators on Bounded Domains. V, 150 pages. 1985.

Vol. 1153: Probability in Banach Spaces V. Proceedings, 1984. Edited by A. Beck, R. Dudley, M. Hahn, J. Kuelbs and M. Marcus. VI, 457 pages. 1985.

Vol. 1154: D.S. Naidu, A.K. Rao, Singular Pertubation Analysis of Discrete Control Systems. IX, 195 pages. 1985.

Vol. 1155: Stability Problems for Stochastic Models. Proceedings, 1984. Edited by V.V. Kalashnikov and V.M. Zolotarev. VI, 447 pages. 1985.

Vol. 1156: Global Differential Geometry and Global Analysis 1984. Proceedings, 1984. Edited by D. Ferus, R.B. Gardner, S. Helgason and U. Simon. V, 339 pages. 1985.

Vol. 1157: H. Levine, Classifying Immersions into $\mathbb{R}^4$ over Stable Maps of 3-Manifolds into $\mathbb{R}^2$. V, 163 pages. 1985.

Vol. 1158: Stochastic Processes – Mathematics and Physics. Proceedings, 1984. Edited by S. Albeverio, Ph. Blanchard and L. Streit. VI, 230 pages. 1986.

Vol. 1159: Schrödinger Operators, Como 1984. Seminar. Edited by S. Graffi. VIII, 272 pages. 1986.

Vol. 1160: J.-C. van der Meer, The Hamiltonian Hopf Bifurcation. VI, 115 pages. 1985.

Vol. 1161: Harmonic Mappings and Minimal Immersions, Montecatini 1984. Seminar. Edited by E. Giusti. VII, 285 pages. 1985.

Vol. 1162: S.J.L. van Eijndhoven, J. de Graaf, Trajectory Spaces, Generalized Functions and Unbounded Operators. IV, 272 pages. 1985.

Vol. 1163: Iteration Theory and its Functional Equations. Proceedings, 1984. Edited by R. Liedl, L. Reich and Gy. Targonski. VIII, 231 pages. 1985.

Vol. 1164: M. Meschiari, J.H. Rawnsley, S. Salamon, Geometry Seminar "Luigi Bianchi" II – 1984. Edited by E. Vesentini. VI, 224 pages. 1985.

Vol. 1165: Seminar on Deformations. Proceedings, 1982/84. Edited by J. Ławrynowicz. IX, 331 pages. 1985.

Vol. 1166: Banach Spaces. Proceedings, 1984. Edited by N. Kalton and E. Saab. VI, 199 pages. 1985.

Vol. 1167: Geometry and Topology. Proceedings, 1983–84. Edited by J. Alexander and J. Harer. VI, 292 pages. 1985.

Vol. 1168: S.S. Agaian, Hadamard Matrices and their Applications. III, 227 pages. 1985.

Vol. 1169: W.A. Light, E.W. Cheney, Approximation Theory in Tensor Product Spaces. VII, 157 pages. 1985.

Vol. 1170: B.S. Thomson, Real Functions. VII, 229 pages. 1985.

Vol. 1171: Polynômes Orthogonaux et Applications. Proceedings, 1984. Edité par C. Brezinski, A. Draux, A.P. Magnus, P. Maroni et A. Ronveaux. XXXVII, 584 pages. 1985.

Vol. 1172: Algebraic Topology, Göttingen 1984. Proceedings. Edited by L. Smith. VI, 209 pages. 1985.

Vol. 1173: H. Delfs, M. Knebusch, Locally Semialgebraic Spaces. XVI, 329 pages. 1985.

Vol. 1174: Categories in Continuum Physics, Buffalo 1982. Seminar. Edited by F.W. Lawvere and S.H. Schanuel. V, 126 pages. 1986.

Vol. 1175: K. Mathiak, Valuations of Skew Fields and Projective Hjelmslev Spaces. VII, 116 pages. 1986.

Vol. 1176: R.R. Bruner, J.P. May, J.E. McClure, M. Steinberger, H_∞ Ring Spectra and their Applications. VII, 388 pages. 1986.

Vol. 1177: Representation Theory I. Finite Dimensional Algebras. Proceedings, 1984. Edited by V. Dlab, P. Gabriel and G. Michler. XV, 340 pages. 1986.

Vol. 1178: Representation Theory II. Groups and Orders. Proceedings, 1984. Edited by V. Dlab, P. Gabriel and G. Michler. XV, 370 pages. 1986.

Vol. 1179: Shi J.-Y. The Kazhdan-Lusztig Cells in Certain Affine Weyl Groups. X, 307 pages. 1986.

Vol. 1180: R. Carmona, H. Kesten, J.B. Walsh, École d'Été de Probabilités de Saint-Flour XIV – 1984. Édité par P.L. Hennequin. X, 438 pages. 1986.

Vol. 1181: Buildings and the Geometry of Diagrams, Como 1984. Seminar. Edited by L. Rosati. VII, 277 pages. 1986.

Vol. 1182: S. Shelah, Around Classification Theory of Models. VII, 279 pages. 1986.

Vol. 1183: Algebra, Algebraic Topology and their Interactions. Proceedings, 1983. Edited by J.-E. Roos. XI, 396 pages. 1986.

Vol. 1184: W. Arendt, A. Grabosch, G. Greiner, U. Groh, H.P. Lotz, U. Moustakas, R. Nagel, F. Neubrander, U. Schlotterbeck, One-parameter Semigroups of Positive Operators. Edited by R. Nagel. X, 460 pages. 1986.

Vol. 1185: Group Theory, Beijing 1984. Proceedings. Edited by Tuan H.F. V, 403 pages. 1986.

Vol. 1186: Lyapunov Exponents. Proceedings, 1984. Edited by L. Arnold and V. Wihstutz. VI, 374 pages. 1986.

Vol. 1187: Y. Diers, Categories of Boolean Sheaves of Simple Algebras. VI, 168 pages. 1986.

Vol. 1188: Fonctions de Plusieurs Variables Complexes V. Séminaire, 1979–85. Edité par François Norguet. VI, 306 pages. 1986.

Vol. 1189: J. Lukeš, J. Malý, L. Zajíček, Fine Topology Methods in Real Analysis and Potential Theory. X, 472 pages. 1986.

Vol. 1190: Optimization and Related Fields. Proceedings, 1984. Edited by R. Conti, E. De Giorgi and F. Giannessi. VIII, 419 pages. 1986.

Vol. 1191: A.R. Its, V.Yu. Novokshenov, The Isomonodromic Deformation Method in the Theory of Painlevé Equations. IV, 313 pages. 1986.

Vol. 1192: Equadiff 6. Proceedings, 1985. Edited by J. Vosmansky and M. Zlámal. XXIII, 404 pages. 1986.

Vol. 1193: Geometrical and Statistical Aspects of Probability in Banach Spaces. Proceedings, 1985. Edited by X. Femique, B. Heinkel, M.B. Marcus and P.A. Meyer. IV, 128 pages. 1986.

Vol. 1194: Complex Analysis and Algebraic Geometry. Proceedings, 1985. Edited by H. Grauert. VI, 235 pages. 1986.

Vol. 1195: J.M. Barbosa, A.G. Colares, Minimal Surfaces in $\mathbb{R}^3$. X, 124 pages. 1986.

Vol. 1196: E. Casas-Alvero, S. Xambó-Descamps, The Enumerative Theory of Conics after Halphen. IX, 130 pages. 1986.

Vol. 1197: Ring Theory. Proceedings, 1985. Edited by F.M.J. van Oystaeyen. V, 231 pages. 1986.

Vol. 1198: Séminaire d'Analyse, P. Lelong – P. Dolbeault – H. Skoda. Seminar 1983/84. X, 260 pages. 1986.

Vol. 1199: Analytic Theory of Continued Fractions II. Proceedings, 1985. Edited by W.J. Thron. VI, 299 pages. 1986.

Vol. 1200: V.D. Milman, G. Schechtman, Asymptotic Theory of Finite Dimensional Normed Spaces. With an Appendix by M. Gromov. VIII, 156 pages. 1986.

Vol. 1201: Curvature and Topology of Riemannian Manifolds. Proceedings, 1985. Edited by K. Shiohama, T. Sakai and T. Sunada. VII, 336 pages. 1986.

Vol. 1202: A. Dür, Möbius Functions, Incidence Algebras and Power Series Representations. XI, 134 pages. 1986.

Vol. 1203: Stochastic Processes and Their Applications. Proceedings, 1985. Edited by K. Itô and T. Hida. VI, 222 pages. 1986.

Vol. 1204: Séminaire de Probabilités XX, 1984/85. Proceedings. Edité par J. Azéma et M. Yor. V, 639 pages. 1986.

Vol. 1205: B.Z. Moroz, Analytic Arithmetic in Algebraic Number Fields. VII, 177 pages. 1986.

Vol. 1206: Probability and Analysis, Varenna (Como) 1985. Seminar. Edited by G. Letta and M. Pratelli. VIII, 280 pages. 1986.

Vol. 1207: P.H. Bérard, Spectral Geometry: Direct and Inverse Problems. With an Appendix by G. Besson. XIII, 272 pages. 1986.

Vol. 1208: S. Kaijser, J.W. Pelletier, Interpolation Functors and Duality. IV, 167 pages. 1986.

Vol. 1209: Differential Geometry, Peñíscola 1985. Proceedings. Edited by A.M. Naveira, A. Ferrández and F. Mascaró. VIII, 306 pages. 1986.

Vol. 1210: Probability Measures on Groups VIII. Proceedings, 1985. Edited by H. Heyer. X, 386 pages. 1986.

Vol. 1211: M.B. Sevryuk, Reversible Systems. V, 319 pages. 1986.

Vol. 1212: Stochastic Spatial Processes. Proceedings, 1984. Edited by P. Tautu. VIII, 311 pages. 1986.

Vol. 1213: L.G. Lewis, Jr., J.P. May, M. Steinberger, Equivariant Stable Homotopy Theory. IX, 538 pages. 1986.

Vol. 1214: Global Analysis – Studies and Applications II. Edited by Yu.G. Borisovich and Yu.E. Gliklikh. V, 275 pages. 1986.

Vol. 1215: Lectures in Probability and Statistics. Edited by G. del Pino and R. Rebolledo. V, 491 pages. 1986.

Vol. 1216: J. Kogan, Bifurcation of Extremals in Optimal Control. VIII, 106 pages. 1986.

Vol. 1217: Transformation Groups. Proceedings, 1985. Edited by S. Jackowski and K. Pawalowski. X, 396 pages. 1986.

Vol. 1218: Schrödinger Operators, Aarhus 1985. Seminar. Edited by E. Balslev. V, 222 pages. 1986.

Vol. 1219: R. Weissauer, Stabile Modulformen und Eisensteinreihen. III, 147 Seiten. 1986.

Vol. 1220: Séminaire d'Algèbre Paul Dubreil et Marie-Paule Malliavin. Proceedings, 1985. Edité par M.-P. Malliavin. IV, 200 pages. 1986.

Vol. 1221: Probability and Banach Spaces. Proceedings, 1985. Edited by J. Bastero and M. San Miguel. XI, 222 pages. 1986.

Vol. 1222: A. Katok, J.-M. Strelcyn, with the collaboration of F. Ledrappier and F. Przytycki, Invariant Manifolds, Entropy and Billiards; Smooth Maps with Singularities. VIII, 283 pages. 1986.

Vol. 1223: Differential Equations in Banach Spaces. Proceedings, 1985. Edited by A. Favini and E. Obrecht. VIII, 299 pages. 1986.

Vol. 1224: Nonlinear Diffusion Problems, Montecatini Terme 1985. Seminar. Edited by A. Fasano and M. Primicerio. VIII, 188 pages. 1986.

Vol. 1225: Inverse Problems, Montecatini Terme 1986. Seminar. Edited by G. Talenti. VIII, 204 pages. 1986.

Vol. 1226: A. Buium, Differential Function Fields and Moduli of Algebraic Varieties. IX, 146 pages. 1986.

Vol. 1227: H. Helson, The Spectral Theorem. VI, 104 pages. 1986.

Vol. 1228: Multigrid Methods II. Proceedings, 1985. Edited by W. Hackbusch and U. Trottenberg. VI, 336 pages. 1986.

Vol. 1229: O. Bratteli, Derivations, Dissipations and Group Actions on C*-algebras. IV, 277 pages. 1986.

Vol. 1230: Numerical Analysis. Proceedings, 1984. Edited by J.-P. Hennart. X, 234 pages. 1986.

Vol. 1231: E.-U. Gekeler, Drinfeld Modular Curves. XIV, 107 pages. 1986.

Vol. 1232: P.C. Schuur, Asymptotic Analysis of Soliton Problems. VIII, 180 pages. 1986.

Vol. 1233: Stability Problems for Stochastic Models. Proceedings, 1985. Edited by V.V. Kalashnikov, B. Penkov and V.M. Zolotarev. VI, 223 pages. 1986.

Vol. 1234: Combinatoire énumérative. Proceedings, 1985. Edité par G. Labelle et P. Leroux. XIV, 387 pages. 1986.

Vol. 1235: Séminaire de Théorie du Potentiel, Paris, No. 8. Directeurs: M. Brelot, G. Choquet et J. Deny. Rédacteurs: F. Hirsch et G. Mokobodzki. III, 209 pages. 1987.

Vol. 1236: Stochastic Partial Differential Equations and Applications. Proceedings, 1985. Edited by G. Da Prato and L. Tubaro. V, 257 pages. 1987.

Vol. 1237: Rational Approximation and its Applications in Mathematics and Physics. Proceedings, 1985. Edited by J. Gilewicz, M. Pindor and W. Siemaszko. XII, 350 pages. 1987.

Vol. 1238: M. Holz, K.-P. Podewski and K. Steffens, Injective Choice Functions. VI, 183 pages. 1987.

Vol. 1239: P. Vojta, Diophantine Approximations and Value Distribution Theory. X, 132 pages. 1987.

Vol. 1240: Number Theory, New York 1984–85. Seminar. Edited by D.V. Chudnovsky, G.V. Chudnovsky, H. Cohn and M.B. Nathanson. V, 324 pages. 1987.

Vol. 1241: L. Gårding, Singularities in Linear Wave Propagation. III, 125 pages. 1987.

Vol. 1242: Functional Analysis II, with Contributions by J. Hoffmann-Jørgensen et al. Edited by S. Kurepa, H. Kraljević and D. Butković. VII, 432 pages. 1987.

Vol. 1243: Non Commutative Harmonic Analysis and Lie Groups. Proceedings, 1985. Edited by J. Carmona, P. Delorme and M. Vergne. V, 309 pages. 1987.

Vol. 1244: W. Müller, Manifolds with Cusps of Rank One. XI, 158 pages. 1987.

Vol. 1245: S. Rallis, L-Functions and the Oscillator Representation. XVI, 239 pages. 1987.

Vol. 1246: Hodge Theory. Proceedings, 1985. Edited by E. Cattani, F. Guillén, A. Kaplan and F. Puerta. VII, 175 pages. 1987.

Vol. 1247: Séminaire de Probabilités XXI. Proceedings. Edité par J. Azéma, P.A. Meyer et M. Yor. IV, 579 pages. 1987.

Vol. 1248: Nonlinear Semigroups, Partial Differential Equations and Attractors. Proceedings, 1985. Edited by T.L. Gill and W.W. Zachary. IX, 185 pages. 1987.

Vol. 1249: I. van den Berg, Nonstandard Asymptotic Analysis. IX, 187 pages. 1987.

Vol. 1250: Stochastic Processes – Mathematics and Physics II. Proceedings 1985. Edited by S. Albeverio, Ph. Blanchard and L. Streit. VI, 359 pages. 1987.

Vol. 1251: Differential Geometric Methods in Mathematical Physics. Proceedings, 1985. Edited by P.L. García and A. Pérez-Rendón. VII, 300 pages. 1987.

Vol. 1252: T. Kaise, Représentations de Weil et GL_2 Algèbres de division et GL_n. VII, 203 pages. 1987.

Vol. 1253: J. Fischer, An Approach to the Selberg Trace Formula via the Selberg Zeta-Function. III, 184 pages. 1987.

Vol. 1254: S. Gelbart, I. Piatetski-Shapiro, S. Rallis. Explicit Constructions of Automorphic L-Functions. VI, 152 pages. 1987.

Vol. 1255: Differential Geometry and Differential Equations. Proceedings, 1985. Edited by C. Gu, M. Berger and R.L. Bryant. XII, 243 pages. 1987.

Vol. 1256: Pseudo-Differential Operators. Proceedings, 1986. Edited by H.O. Cordes, B. Gramsch and H. Widom. X, 479 pages. 1987.

Vol. 1257: X. Wang, On the C*-Algebras of Foliations in the Plane. V, 165 pages. 1987.

Vol. 1258: J. Weidmann, Spectral Theory of Ordinary Differential Operators. VI, 303 pages. 1987.